1709

Microwave Study of Chemical Structures
and Reactions

Microwave Study of Chemical Structures and Reactions

P. HEDVIG, D. Phil.
Senior research officer, Plastics Research Institute, Budapest

and

G. ZENTAI, B. Eng.
Lecturer in electric engineering, Technical University of Budapest

English translation edited by

E. D. MORGAN, B.Sc. (Dal.), M.A., D.Phil. (Oxon.), F.R.I.C.
Lecturer in Chemistry, University of Keele

LONDON   ILIFFE BOOKS LTD

QD
95
H3713
1969b

ILIFFE BOOKS LTD
42 RUSSEL SQUARE
LONDON, W.C. 1

© Akadémiai Kiadó, Budapest 1969

First published in Great Britain in 1969 by Iliffe Books Ltd., in co-edition
with Akadémiai Kiadó, Budapest
Publishing House of the Hungarian Academy of Sciences

592 01223 9

Printed in Hungary

Contents

1 **INTRODUCTION** 9

  1.1 Microwave rotation–inversion spectroscopy 10
  1.2 Electron spin resonance 15
  1.3 Nuclear magnetic resonance 22
  1.4 Nuclear quadrupole resonance 26
  1.5 Dielectric spectroscopy 28
  1.6 Electrical conduction in organic compounds 31
     References 33

2 **EXPERIMENTAL METHODS** 35

  2.1 Microwave molecular spectrometers 35
    2.1.1 Principle of operation 35
    2.1.2 Microwave components 38
    2.1.3 Sensitivity 40
    2.1.4 Resolution 42
    2.1.5 Stark modulation systems 43
    2.1.6 Lock-in detector 45
    2.1.7 Emission spectra 47
  2.2 Electron spin resonance technique 48
    2.2.1 Principle of operation 48
    2.2.2 Cavity resonators 52
    2.2.3 Automatic frequency control 54
    2.2.4 Sensitivity 55
    2.2.5 Resolution 61
    2.2.6 Operation at unusual conditions 71
    2.2.7 Errors in operation 78
  2.3 Nuclear resonance technique 80
    2.3.1 Principle of operation 80
    2.3.2 Sensitivity 96
    2.3.3 Typical NMR spectrometer 98
  2.4 Dynamic methods 99
    2.4.1 Adiabatic fast passage 100
    2.4.2 Saturation 103

2.4.3 Spin–echo technique 104
2.4.4 Double resonance methods 108
2.4.5 Nuclear–nuclear double resonances 113
2.4.6 Double resonance in microwave molecular spectroscopy 116
2.5 Dielectric spectroscopy technique 119
    2.5.1 General problems 119
    2.5.2 Molecular relaxation times 122
    2.5.3 Measurement at intermediate frequencies 125
    2.5.4 Measurement at microwaves 128
    2.5.5 Dielectric spectroscopy at very low frequencies 131
    2.5.6 Measurement of d.c. conductivities 134
    References 139

3 FREE RADICALS 141

3.1 Determination of radical concentrations 142
    3.1.1 Choice of standard materials 144
    3.1.2 Examples of standard materials 146
    3.1.3 Positioning of standard materials 152
3.2 Determination of radical structures 155
    3.2.1 Hyperfine interaction 156
    3.2.2 Hyperfine interaction with many nuclei 161
3.3 Radicals in the solid state 174
    3.3.1 Trapping 175
    3.3.2 Radicals in single crystals 177
    3.3.3 Radicals in polycrystalline and amorphous media 184
    3.3.4 Electron-nuclear double resonance 195
    3.3.5 Radicals formed by mechanical treatment 197
3.4 Radicals in the liquid state 199
    3.4.1 Solvent effects 200
    3.4.2 Typical radicals in solution 208
    3.4.3 Aliphatic radicals 214
3.5 Gaseous radicals 217
    3.5.1 ESR of atomic gases 219
    References 224

4 STRUCTURE DETERMINATION AND ANALYSIS 226

4.1 General ideas 226
    4.1.1 Diamagnetic shielding 226
    4.1.2 Spin–spin coupling 230
    4.1.3 Anisotropy of local fields 234
4.2 Determination of hydrocarbon groups 237

    4.2.1 Chemical shifts in hydrocarbons   237
    4.2.2 Spin–spin coupling in hydrocarbon groups   250
    4.2.3 Illustrative examples   254
    4.2.4 ^{13}C resonances   257
4.3 Determination of other groups   261
    4.3.1 Determination of groups X—H   261
    4.3.2 Groups containing fluorine   270
    4.3.3 Groups containing phosphorus   274
    4.3.4 Groups containing boron   276
    4.3.5 Groups containing other nuclei   279
4.4 Stereochemical analysis   281
    4.4.1 Rotational isomers   283
    4.4.2 Cyclohexane-type stereoisomers   295
    4.4.3 Steroid structures   301
    4.4.4 Methyl-group shifts in steroids   303
4.5 Polymers   306
    4.5.1 Stereochemical study of polymers   310
    4.5.2 Dielectric spectroscopy of polymers   312
    References   317

5 REACTIONS   319

5.1 Simple radical reactions. Oxidation   320
    5.1.1 Oxidation of hydroquinones   321
    5.1.2 Oxidation of radicals trapped in the solid phase   326
5.2 Chemical exchange   331
    5.2.1 Electron exchange reactions   331
    5.2.2 Electron transfer   332
    5.2.3 Proton exchange   334
    5.2.4 Intermolecular cation exchange investigated by ESR   338
5.3 Photolysis and radiolysis   340
    5.3.1 Detection of primary radicals   343
    5.3.2 ESR spectra of atoms abstracted by irradiation   344
    5.3.3 ESR spectra of other primary radicals   346
    5.3.4 Detection of radical intermediates   348
    5.3.5 Radicals formed by addition of hydrogen   350
    5.3.6 Kinetic measurements   353
    5.3.7 Non-steady state measurements   356
    5.3.8 Investigation of energy transfer by ESR   357
5.4 Polymerization   359
    5.4.1 Detection of initiating radicals   361
    5.4.2 Intermediate radicals trapped in viscous media   363
    5.4.3 Radical intermediates in the liquid state   369

	5.4.4 Kinetic studies	375
	5.4.5 Change of physical structure during solid state polymerization	383
	5.4.6 Destruction of polymers	386
5.5	Catalysis	396
	5.5.1 Adsorption	397
	5.5.2 Investigation of chromium oxide catalysts by ESR	401
	5.5.3 ESR study of Ziegler–Natta catalyst systems	404
	5.5.4 Anionic catalysts. Dimerization of 1,1-diphenylethylene	409
	5.5.5 Catalysis by organic semiconductors	410
5.6	Biochemical reactions	415
	5.6.1 Photosynthesis	416
	5.6.2 Enzyme reactions	419
	5.6.3 Charge transfer complexes	421
	5.6.4 Radiolysis in biological systems	425
	References	428
Appendix		431
Bibliography		435
Index		439

1
Introduction

In the last two decades a powerful new method has been developed in structural chemistry; that is, spectroscopy at radio and microwave frequencies. This method is now widely used in various branches of chemistry, physics and biology. The main fields of application are: microwave rotation–inversion spectroscopy, electron spin resonance, nuclear magnetic resonance and nuclear quadrupole resonance. Electrical conductivity and dielectric properties of matter, as related to its physical and chemical structure, are also discussed here. The former is sometimes referred to as dielectric spectroscopy or Debye spectroscopy; the latter is being extensively studied now in connection with charge transfer and energy transfer problems.

Radio frequency and microwave spectroscopy deal with the interaction of electromagnetic waves and matter. Emission and absorption of radiation occur at certain frequencies which characterize the energy states of atoms, ions, molecules or chemical groups. The measurements are usually made in absorption, i.e. transitions are induced between the energy levels of the material corresponding to absorption of discrete quanta from the radiation field. In the case of ultraviolet, visible, and infrared spectroscopy the absorbed energy quanta range from 0.1 to 10 eV approximately, according to the relation

$$h\nu = \Delta E$$

where ν is the frequency of the electromagnetic radiation, h is Planck's constant, ΔE is the energy separation between two energy states of the material.

At radio and microwave frequencies the values of $h\nu$ range from 10^{-2} to 10^{-6} eV, indicating that this method can be used for measuring extremely low energy separations or energy splittings.

As well as the difference in the energy separations there is another difference between optical and radio frequency absorption spectroscopy. In optical spectroscopy, mainly electrical dipole transitions are induced and observed; i.e. the interaction of the electric component of the radiation field with atomic or molecular electric dipole moments. Most radio frequency methods, however, induce magnetic dipole transitions. In the case of nuclear magnetic resonance, for example, the magnetic component of the radiation field (radio waves) interacts with the magnetic dipole moments of nuclei. The same happens in electron spin resonance where magnetic dipole transitions between unpaired electron spin energy levels are observed at microwave frequencies. Indeed, radio frequency and microwave spectroscopy have proved to be very powerful methods of investigating magnetic properties of atoms, ions, molecules or groups. This is extremely important in free radical chemistry since free radicals are characterized by their paramagnetic behaviour. Investigation of magnetic properties can be useful in every case where paramagnetic atoms, ions or nuclei are present in the system. Since many nuclei have magnetic moments, one can hardly find a system which is completely free of paramagnetism. The methods based on the magnetic interaction between radio waves and matter are often referred to as magnetic resonance spectroscopy.

In the case of microwave molecular rotation–inversion spectroscopy both electric and magnetic dipole transitions can be induced. In Debye spectroscopy only electric interactions are considered.

The following subsections will present a brief survey of the principles of radio frequency and microwave spectroscopic methods. The experimental technique and the applications are discussed in separate chapters.

1.1 MICROWAVE ROTATION–INVERSION SPECTROSCOPY

Microwave rotation or inversion spectra can only be taken from gaseous materials or vapours, preferably at low pressures and at low temperatures. Transitions between rotation or inversion energy levels can be induced by microwave irradiation of 0.1–10 cm wavelength. The first such measurements were made by Cleeton and Williams[1.1] in 1934. They irradiated ammonia (NH_3) gas with cm-waves and observed sharp absorption peaks near 1.25 cm (22 Gc/s). The method has been largely developed since then; spectra of thousands of molecules have been recorded and interpreted. The usual bands of microwave spectroscopy range from several tenths of a millimetre up to decimetre waves.

Most of the microwave absorption spectra can be interpreted in terms of radiation-induced transitions between certain rotational energy levels of the molecules. The rotational energy of a simple diatomic molecule is

$$E_{rot} = \frac{N^2}{2\theta}$$

where N is the angular momentum, and θ is the moment of inertia of the molecule.

According to quantum mechanics the absolute value of the angular momentum vector N should only have the following discrete values

$$N = \frac{h}{2\pi} \sqrt{J(J+1)}$$

where $J = 0, 1, 2, 3, 4$ is the quantum number of the rotation, and h is Planck's constant.

According to this quantization of the angular momentum the rotational energies become

$$E_{rot} = \frac{h^2}{8\pi^2\theta} J(J+1) \qquad 1.1$$

Each value of the rotational quantum number J corresponds to a definite rotational energy state of the molecule, which is represented by the energy level scheme of Fig. 1.1. The difference between two successive rotational energy levels of a molecule depends in first approximation on its moment of inertia θ. The rotational energy level scheme of a molecule is thus principally determined by its structure, i.e. the mass distribution of the atoms of which it consists. The heavier the atoms and the greater the distance between them, the less the rotational energy levels will be separated. Since absorption of microwaves corresponds to transitions between successive rotational levels, heavy molecules having large θ have low frequency absorption peaks. The

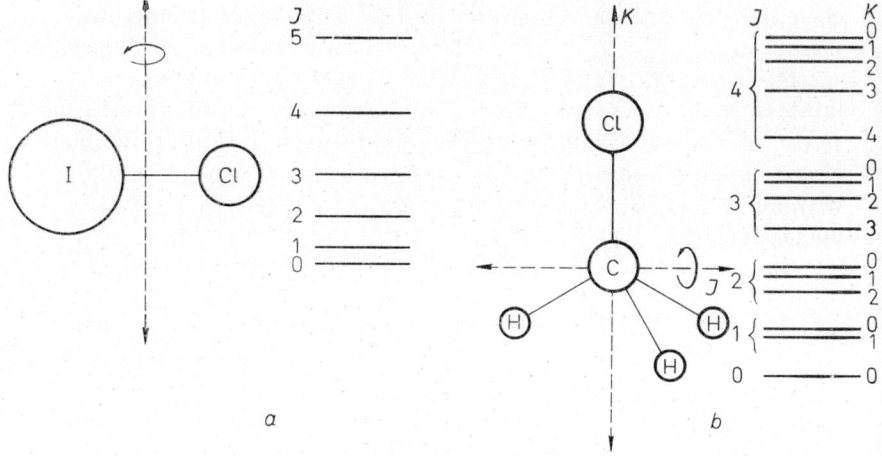

Fig. 1.1 Rotational states of molecules.
a — ICl; b — CH$_3$Cl

rotational absorption peaks of HCl, for example, lie mostly in the infrared region. The heavier ICl molecule, on the other hand, has rotational absorption peaks in the microwave region. Microwave rotation spectroscopy can be regarded in this respect as an extension of infrared spectroscopy to much lower frequencies, although its technique is quite different.

Equation 1.1 is, of course, oversimplified. In general the molecule might have several axes of symmetry, about which it can rotate. This results in several rotational quantum numbers, and corresponding sets of rotational energy levels. Besides rotation, vibrational motion is also present, resulting in a perturbation of the energy levels. The centrifugal force of the rotation will cause distortions of the molecule, and this should also affect the energy level scheme. Finally, interaction between molecules would increase the widths of the spectral lines, resulting in wide absorption bands. This is why microwave absorption spectroscopy can only be applied to the gaseous state, mostly at low pressures. In more condensed phases or at higher pressures the molecular interactions cause a serious drop in resolution and a corresponding decrease in the amount of information derived from the spectra.

Data of microwave rotation–inversion spectra are collected in the tables of Kisliuk and Townes.[1,2]

Effects of electric and magnetic fields. Molecules in general have electric and magnetic dipole moments. The interaction energies between molecular dipole moments and external electric field **E** or magnetic field **H** are

$$E_{\text{electric}} = \mu_{\text{el}}\mathbf{E}$$

$$E_{\text{magnetic}} = \mu_{\text{m}}\mathbf{H}$$

where μ_{el} and μ_{m} are the electric and magnetic dipole moments, respectively (vectors).

These energies result in a splitting of the rotational energy levels, known as Stark splitting in electric fields and Zeeman splitting in magnetic fields. The splitting of the energy levels results in a splitting of the spectral lines. The number of components depends on the angular momentum of the molecule. An energy level corresponding to a rotational quantum number J splits into $2J + 1$ sub-levels. According to quantum mechanics the z-component of the total angular momentum is

$$N_z = m_J \frac{h}{2\pi}$$

where

$$m_J = -J, -J+1, \ldots +J$$

and z is the direction of the magnetic or electric field.

m_J can have $2J + 1$ different values. The energy between two Stark or Zeeman sub-levels is

$$\Delta E_{\text{Stark}} = \Delta m_J \mu_{\text{el}} \mathbf{E}$$

$$\Delta E_{\text{Zeeman}} = \Delta m_J \mu_{\text{m}} \mathbf{H}$$

The values of **E** and **H** being known, the electric and magnetic moments of molecules can be determined from the Stark or Zeeman

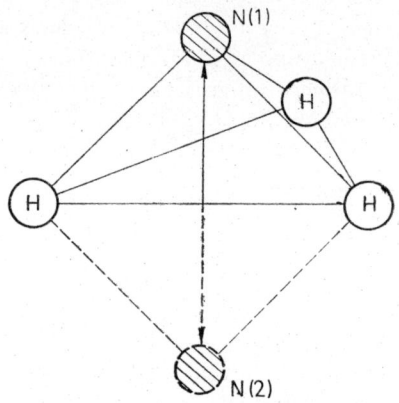

Fig. 1.2 Inversion of the ammonia molecule

splitting of the microwave molecular spectral lines. Many molecular dipole moments have been measured in this way. The results are collected in the tables of Kisliuk and Townes.[1.2] The magnetic dipole moments of molecules are usually small, while their electric dipole moments may be large. This is why Stark splittings can be observed at electric field strengths as low as 50 V/cm, while for the measurement of molecular Zeeman splitting magnetic fields up to several thousand gauss are needed.

Inversion. Besides rotation and vibration, some molecules may have another kind of motion which can result in absorption of microwaves; the molecule NH_3 has for example two different configurations as indicated in Fig. 1.2. The nitrogen atom can be inverted from position (1) into position (2). For inversion (1) → (2) the nitrogen atom has to overcome the potential-barrier formed by the three hydrogen atoms. The energy difference between the two inversion states is of the order of 1.24×10^{-4} eV, which corresponds to 1–2 cm waves. Thus on microwave radiation a selective absorption of radiation occurs. Since the potential-barrier of the hydrogen atoms is strongly affected by the rotational state of the molecule, the transition energy is slightly different for each rotational state. Transitions among these energy levels are observed as rotation–inversion spectral lines.

Microwave rotation–inversion spectroscopy can mainly be applied to determination of molecular parameters: bond lengths, bond angles, molecular electric and magnetic dipole moments. It can also be of use in chemical analysis.

1.2 ELECTRON SPIN RESONANCE

Electron spin resonance or electron paramagnetic resonance is one of the most powerful physical methods in organic chemistry. It offers a very sensitive way to measure electron paramagnetism of atoms, molecules, ions or chemical groups.

Electron paramagnetism is a result of unpaired electron magnetic moments. Electrons in atoms, ions or molecules are usually arranged so that their magnetic moments are compensated. The hydrogen atom for example has but one electron and a corresponding magnetic moment of one Bohr magneton

$$\mu_s = eh/4\pi M_e c$$

where e is the charge, M_e the mass of the electron, h is Planck's constant and c is the velocity of light. The hydrogen atom is thus electron paramagnetic.

In the case of the hydrogen molecule the magnetic moments of the hydrogen atoms compensate each other, resulting in zero net electronic magnetic moment. The hydrogen molecule is therefore not electron paramagnetic, although it has a slight net magnetic moment resulting from the two protons. The magnetic moments of nuclei are expressed in units of nuclear magnetons $\mu_I = eh/4\pi M_p c$, where M_p is the mass of the proton. According to the electron–proton mass ratio the magnetic moments of nuclei are about three orders of magnitude lower than those of electrons.

The most common electron paramagnetic materials are the ions of the iron group elements and the rare earths. These ions have unpaired electrons in the incomplete $3d$ orbits. Metals and semiconductors are also electron paramagnetic because of the uncompensated moments of the conduction electrons and holes.

There are, however, other possibilities for a molecule or a chemical group to have unpaired electrons:

1. Rupture of chemical bonds, which results in formation of free radicals.
2. Exciting the molecules into *triplet states*, where electron magnetic moments are not compensated any more.
3. Capture of electrons by crystalline defects of solids (colour centres).

Free radicals are often formed during the course of reactions, by mechanical or thermal treatment of the material, or by irradiation with ultraviolet light, X-rays or high-energy radiation. The study of free radicals is one of the central problems in chemistry today. The study of triplet states is becoming of interest in connection with charge- and energy-transfer problems. Colour centres in crystalline solids can be formed by mechanical treatment, by electrolysis or by ultraviolet or high-energy irradiation.

The energy of an electron paramagnetic atom, ion or molecule in a magnetic field $\mathbf{H}_0$ is

$$E_s = 2m_s \mu_e \mathbf{H}_0 \qquad (m_s = -s, -s+1, \ldots +s)$$

where m_s is the quantum number corresponding to the z-component (in direction $\mathbf{H}_0$) of the angular momentum (spin) of the electron, and μ_e is the magnetic moment of the molecule. As the electron spin is $1/2$, m_s has just two values, $+1/2$ and $-1/2$, corresponding to the parallel versus antiparallel orientation of the electron spin with respect to the magnetic field H_0. The energy difference between these two states is

$$\Delta E_e = 2\mu_e H_0$$

If the magnetic moment of the atom, ion or molecule were only that of the unpaired electron, μ_e would be one Bohr magneton, μ_s. The electron is, however, never completely free: some of its orbital magnetic moments are always coupled, causing the net magnetic moment to change. The magnetic energy splitting is therefore usually expressed as

$$\Delta E_s = g_s \mu_s H \qquad 1.2$$

where the *splitting factor*, g_s, for free electrons has a value of 2.00229. Any deviations from this *free spin value* indicate the presence of magnetic interactions with the orbital magnetic moments. If the spin–orbit interaction is large, as in the paramagnetic ions, the deviation of g_s from the free spin value is considerable; if it is small, as in most free radicals, the g_s-value is very close to 2.00229.

The magnetic energy splitting ΔE_s has a value of about 10^{-3} to 10^{-4} eV at magnetic fields of 1,000–10,000 gauss. Transitions between spin states $s = +1/2$ and $s = -1/2$ can be induced by microwave radiation according to the equation

$$h\nu_0 = g_s \mu_s H_0 \qquad 1.3$$

The system of electron spins will absorb microwave radiation at ν_0 and H_0, where $s = +1/2 \to -1/2$ magnetic dipole transitions are induced. This resonance absorption is called *electron spin resonance*. The intensity of the electron spin resonance line corresponds to the number of paramagnetic centres, i.e. spins. From the position of the line (ν_0, H_0) the electronic splitting factor g_s can be calculated from Equation 1.3.

The first electron spin resonance measurements were made by Zavoisky in 1944.[1.3] He succeeded in detecting the resonance lines of some salts of the iron group elements.

Hyperfine splitting. Since electrons can only have two orientations with respect to the magnetic field H_0, the electron spin resonance spectrum should always consist of a single line. This is definitely not true, however, for in most cases a number of electron spin resonance lines are detected. In practice the number of the lines can be as high as 100. The reason is the interaction between the unpaired

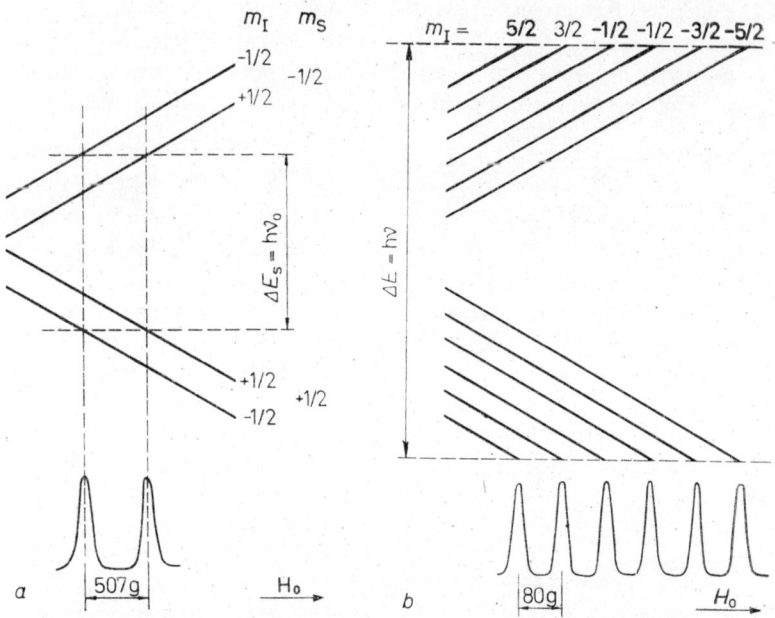

Fig. 1.3 Hyperfine splitting in electron spin resonance spectra. a — Atomic hydrogen; b — Mn^{++} in an aqueous solution of $MnSO_4$

electrons and nuclei having magnetic moments. This interaction would split each electron spin level into several *hyperfine levels*, as shown in Fig. 1.3. The energy differences can be expressed as follows:

$$\Delta E_s = g_s \mu_s H_0 + m_I a_i \qquad 1.4$$

where m_I is the quantum number of the z-component (along H_0) of the nuclear spin I, $m_I = -I, -I+1, \ldots +I$, a_I is the hyperfine splitting, which characterizes the interaction between nuclear and electron spins.

The number of the hyperfine components is thus $2I + 1$. For the hydrogen atom, the proton spin being $I = 1/2$, each electron spin level splits into $2I + 1 = 2$ hyperfine levels resulting in a two-line spectrum as indicated in Fig. 1.3a. The splitting of the energy levels and the spectrum is represented as a function of the magnetic field **H** at constant frequency v_0. For Mn^{+++} ions, the nuclear spin of Mn being $I = 5/2$, the number of the hyperfine lines is $2I + 1 = 6$ (Fig. 1.3b).

The hyperfine interaction of electron spin resonance is extremely useful in the study of free radical structures. It is a unique method of determining the exact position of the free bond in a molecule. It is being extensively used to study the *delocalization* or *resonance* of radicals, which by interacting with a large number of nuclei result in complicated, multiline spectra.

More details and illustrative examples of hyperfine splitting of electron spin resonance spectra are presented in Chapters 3 and 4.

Effects of electric fields. Electric fields should have no effect on the spin states of electrons. However, if the spin–orbit coupling is not negligible they can alter the magnetic spin energies by altering the orbital motion of the electrons. The crystalline electric fields can in this way make the electronic splitting factor g_s anisotropic. Indeed, in single crystals of paramagnetic ions, which usually have large spin–orbit interactions, the g-factor has been found to depend strongly on the orientation of the crystal with respect to the magnetic field direction **H**. In free radicals the spin–orbit coupling is usually small resulting in a small g-anisotropy. The electronic splitting factor g_s should therefore be regarded as being a tensor of symmetry determined by the crystalline field around the paramagnetic centres. An example of studying oriented structures by g-anistropy is presented in Chapter 3.

Temperature dependence. One would expect an electron spin system in magnetic field H_0 to be completely oriented or polarized parallel to the field. The energy difference between the parallel and antiparallel orientations is, however, so small, that at ordinary temperatures both orientations are almost equally populated.

The relative number of electrons with spin 'up' and of those with spin 'down' is according to Boltzmann statistics

$$\frac{N_1}{N_2} = \exp\left(-\frac{\Delta E}{kT}\right) \qquad 1.5$$

where ΔE is the energy difference between states (1) and (2), k is the Boltzmann constant and T is the absolute temperature.

The following types of transitions exist between the spin states (1) and (2).

1. Spontaneous transition from state (2) to state (1) accompanied by spontaneous emission of radiation. For microwave spin systems the probability of this transition is in general negligibly small.
2. Transition (2) → (1) induced by the electromagnetic radiation. By this transition the system emits microwave quanta of $h\nu = \Delta E$.
3. Transition (1) → (2) accompanied by absorption of radiation;
4. Radiationless transition (2) → (1), the energy ΔE being transferred to the surroundings of the spin system, i.e. in a generalized sense to the lattice.

At equilibrium the number of (1) → (2) transitions should be equal to that of (2) → (1) transitions. According to radiation theory, the probability of induced emission equals that of absorption p_a. The equilibrium condition is

$$(p_a + p_r)N_1 = p_r N_2$$

where p_r is the probability of the radiationless transition (2) → (1). The probability of spontaneous emission is neglected.

Using Equation 1.5 one gets for the probability of absorption

$$p_a = p_r \frac{N_2 - N_1}{N_1} = p_r \left[\exp\left(\frac{-\Delta E}{kT}\right) - 1\right] \qquad 1.6$$

For $\Delta E \ll kT$ the intensity of the electron spin resonance line is approximately

$$\text{Intensity} \sim \frac{\Delta E}{kT} \qquad 1.7$$

The intensities of the electron spin resonance lines are thus in first approximation inversely proportional to the absolute temperature. It is advisable therefore to measure electron spin resonance spectra at the lowest possible temperatures. Line intensities can also be in-

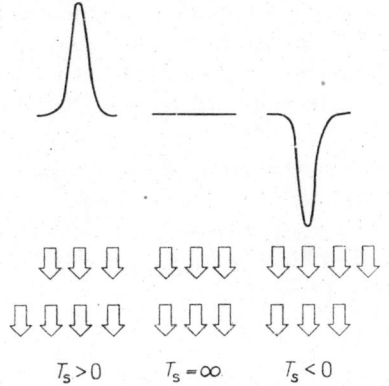

Fig. 1.4 Illustration of spin temperatures

creased by increasing the magnetic energy splitting ΔE. This would require operation at the highest possible frequencies and magnetic fields.

Equation 1.5 is often used to define the temperature of the spin system, or *spin temperature*.

$$T_s = \frac{\Delta E}{k \ln \frac{N_1}{N_2}} \qquad 1.8$$

In ordinary conditions the spin system is in thermal equilibrium with the surrounding lattice, and correspondingly the spin temperature T_s equals the temperature of the lattice T. There are possibilities, however, for a spin system to be cooled down or heated up to temperatures different from that of the lattice. If, for example, high-power microwave radiation is applied, the spin levels will tend to be equally populated ($N_2 = N_1$), resulting in an infinite spin temperature T_s. In this case the electron spin resonance line vanishes; the transition is said to be *saturated*.

If the population of the spin levels is inverted, i.e. $N_2 > N_1$, the spin temperature T_s will be negative. In this case the absorption probability becomes negative (Equation 1.6); the spin system becomes active: it emits microwave radiation instead of absorbing it. Illustrations of the different spin temperature regions are presented in Fig. 1.4. The corresponding resonance signals are also shown. Methods based on spin systems not in equilibrium with the lattice are discussed in more detail in Section 2.4.

Electron spin resonance spectroscopy is widely used in cases where paramagnetic centres are present in the system. It is a basic method in free radical chemistry: in reaction kinetics, in photochemistry, in radiochemistry and in biochemistry. Examples of these fields are presented in Chapters 3 and 5.

1.3 NUCLEAR MAGNETIC RESONANCE

The basic principle of paramagnetic resonance discussed in Section 1.2 can also be applied to paramagnetic nuclei. A large number of nuclei have magnetic moments, expressed in units of nuclear magnetons μ_I. The energy of a paramagnetic nucleus in a magnetic field $\mathbf{H}_0$ is

$$E_I = m_I g_I \mu_I \mathbf{H}_0 \qquad 1.9$$

where the quantum number m_I corresponds to the z-component of the nuclear spin vector $\mathbf{I}$, where z is in the direction of $\mathbf{H}_0$. Thus m_I can have $2I + 1$ values $-I, -I + 1, \ldots +I$; g_I is the nuclear magnetic splitting factor.

Since nuclear magnetic moments are about three orders of magnitude lower than that of the electron, transitions between nuclear spin levels can be induced at much lower frequencies. The usual band for measuring nuclear magnetic resonance spectra is 10–100 Mc/s, corresponding to magnetic fields of 10,000–20,000 gauss.

According to Equation 1.9 the positions of the nuclear magnetic resonance lines are determined by the following equation

$$h\nu_0 = g_1\mu_1\mathbf{H}_0$$

It is in first approximation independent of the atom or molecule in which the nucleus is present. In reality the position of the lines is rather sensitive to the surroundings of the nuclei. Changes in the molecule or group in which the paramagnetic nucleus occurs would cause a shift of the values of ν_0 and $\mathbf{H}_0$. This shift of the nuclear magnetic lines is mainly caused by the diamagnetism of the electrons, i.e. the external field $\mathbf{H}_0$ generates electronic currents in the atom, molecule or group, and the magnetic field of these currents is added to $\mathbf{H}_0$. Thus the position of the line is determined by the following equation:

$$h\nu_0 = g_{\mathrm{I}}\mu_1(\mathbf{H}_0 + \mathbf{H}_l) \qquad 1.10$$

where $\mathbf{H}_l$ is the intensity of the local magnetic field, which depends on the chemical structure around the nucleus whose resonance is being investigated. The shift of the resonance line caused by the local field $\mathbf{H}_l$ is called *chemical shift*.

The method of nuclear magnetic resonance spectroscopy in chemistry essentially involves accurate measurement of local fields $\mathbf{H}_l$ which are characteristic of the chemical structure of matter. A large variety of chemical structures have been analyzed in this way. Examples of chemical structure determination by nuclear magnetic resonance are presented in Chapter 4.

The first nuclear magnetic resonance experiment was made by Rabi[1.4] in 1937. The method which is now being used in chemical structure analysis was discovered independently by Purcell and Bloch[1.5] in 1946. It has been further developed and applied to many chemical problems since then.

Spin–lattice relaxation. As in the case of the electron spin system, the nuclear spin system is generally in equilibrium with the surrounding lattice. The equilibrium populations of the nuclear spin energy levels are also determined by Boltzmann statistics according to Equation 1.5. ΔE corresponds in this case to the nuclear magnetic energy splitting given by Equation 1.9. If a nuclear spin system, which is in equilibrium with the lattice, is irradiated by high-power radio waves of frequency v_0, the difference in spin-level populations will decrease. The radio frequency field thus tends to depolarize the system or heats it up to a higher spin temperature. By disconnecting the r.f. field the spin system will relax until equilibrium population is reached. The equation for this thermal relaxation is

$$\frac{d(\Delta n)}{dt} = 2p_r(\Delta n - \Delta N) \qquad 1.11$$

where Δn is the population difference between the spin levels in the presence of the r.f. field, ΔN is the equilibrium population difference, t is the time, p_r is the probability of radiationless transitions between levels (2) and (1). Since the population difference, or polarization, corresponds to the net magnetization $\mathbf{M}_z$ of the system in the direction of the field $\mathbf{H}_0$, the solution is

$$\mathbf{M}_z = \mathbf{M}_0 [1 - \exp(-2p_r t)] \qquad 1.12$$

The spin–lattice relaxation time is defined as

$$T_1 = \frac{1}{2p_r}$$

T_1 is thus a measure of how quickly the disturbed spin system would reach the equilibrium magnetization $\mathbf{M}_0$. The more the spin system is isolated from the lattice, the longer is T_1. Nuclear spin systems may have T_1 relaxation times of several seconds at room temperatures and several minutes at low temperatures. The spin–lattice relaxation times of electron spin systems are generally short, 10^{-4} to 10^{-6} sec.

Precession model. Spin–spin relaxation. In 1946 Bloch[1.6] developed a semi-classical theory for describing the behaviour of nuclear spin systems in external magnetic fields. The spin system is regarded as an ensemble of classical magnets which, having angular momenta, precess around the direction of the polarizing field $\mathbf{H}_0$. The motion is described by the equations of classical mechanics. The net magnetic moment of the precessing bar magnets along the field $\mathbf{H}_0$ is $\mathbf{M}_z$ as indicated in Fig. 1.5. This longitudinal component of magnetization corresponds to the difference in the spin levels, as discussed above. The change in the net magnetization M_z as a function of time is described by the relaxation Equation 1.12.

The transverse net magnetization components M_y and M_x also change with time. If a radio frequency field is applied to the system

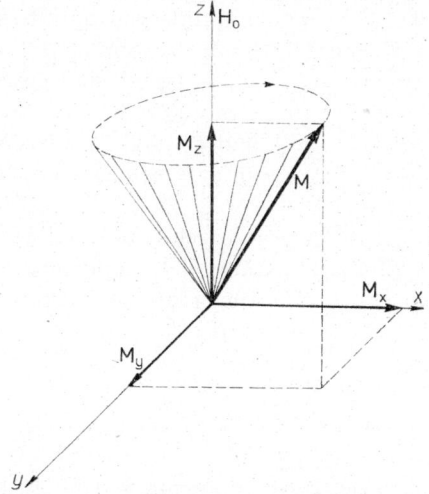

Fig. 1.5 The components of magnetization vector **M**

with a frequency corresponding to the frequency of precession v_0, M_x and M_y will be large, because the individual magnets precess in phase. If, however, the r.f. field is off, the phase coherence among the individual moments is spoiled by the dipole–dipole interaction. The phase coherence will therefore be lost gradually, corresponding to the decrease of the transversal net magnetizations M_x and M_z. The relaxation equations are

$$\frac{dM_x}{dt} = -\frac{M_x}{T_2}$$

$$\frac{dM_y}{dt} = -\frac{M_y}{T_2} \qquad 1.13$$

where T_2 is the *spin–spin relaxation time*, which characterizes the magnetic dipole–dipole interaction among the elements of the spin system.

The spin–lattice and the spin–spin relaxation processes tend to increase the widths of the spin resonance lines in the case of nuclear magnetic resonance and of electron spin resonance as well. The shorter the relaxation times T_1 and T_2, the wider the spin resonance lines are. This suggests that high resolution can only be obtained in systems having long T_1 and T_2. The dipole–dipole interaction, being anisotropic, is usually averaged out in liquids and gases because of the rapid tumbling motion of the molecules. This is why high resolution can only be achieved in the liquid or gaseous phase. The spin–lattice interaction is isotropic, so T_1 can only be decreased by decreasing the temperature.

The measurement of nuclear or electronic relaxation times offers new possibilities of investigating such important physico-chemical processes as adsorption, electron or proton exchange, and catalysis. Examples are discussed in Chapter 5.

1.4 NUCLEAR QUADRUPOLE RESONANCE

Nuclei have no electric dipole moments, but might have quadrupole moments resulting from their asymmetric charge distributions. The charge of one group of nuclei can be regarded as being distributed in a sphere. Owing to this symmetry they have no electric moments at all. The other group has asymmetric charge distribution, which can be taken in first approximation as being ellipsoidal. The electric quadrupole moment of an ellipsoid-like nucleus with an eccentricity of $\varepsilon = (b - a)/(b + a)$ is

$$Q_I = \frac{3}{5} \varepsilon\, ZA^{2/3}\ 10^{-26}\ \text{cm}^2$$

where Z and A are the charge and mass numbers of the nucleus respectively, and b and a are the axes of the ellipsoid.

Nuclei having electric quadrupole moments can be oriented by inhomogeneous electric fields. If an electric field gradient exists along axis z the ellipsoid-like nucleus will tend to be oriented along this direction (Fig. 1.6). The interaction energy is

$$E_Q = \frac{3m_I^2 - I(I+1)}{4I(2I+1)} eQ_1 \left(\frac{\partial^2 V}{\partial z^2}\right)_n \qquad 1.14$$

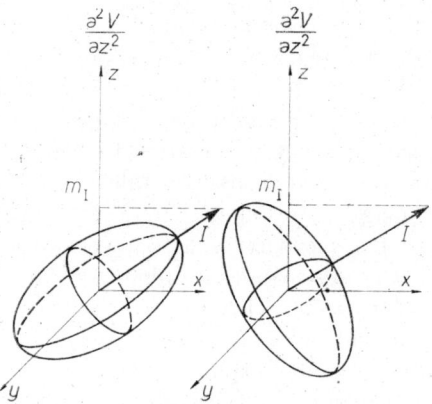

Fig. 1.6 Orientation of ellipsoid-like nuclei in inhomogeneous electric fields

where V is the potential of the electric field, e is the charge of the electron, the derivative is to be taken at the nucleus; I is the nuclear spin, m_I is the spin component along axis z.

Such inhomogeneous electric fields exist in molecules having asymmetric electron distributions, i.e. covalent bonding. The field gradient $(\partial^2 V/\partial z^2)_n$ is therefore characteristic of the electron configuration of the molecule. It is called the quadrupole coupling constant q.

In 1950 Dehmelt and Krüger[1.7] discovered that it is possible to induce magnetic dipole transitions among the nuclear quadrupole energy states. This is a kind of magnetic resonance, because magnetic dipole transitions are induced, but the energy splitting is caused by the electric quadrupole interaction described by Equation 1.14.

Theoretically it would be possible to induce electric quadrupole transitions among the quadrupole levels, but the line intensities are too small to be observed.

For the chemist, nuclear quadrupole resonance is interesting because it offers a possibility of measuring the quadrupole coupling constant q. For molecules having completely ionic bonding, q should be zero because of the symmetry of the ionic electron distribution. In the case of covalent bonding, electrons are shared among nuclei with different electronegativities. This results in an inhomogeneity of the molecular electric field, i.e. $q \neq 0$. So by nuclear quadrupole resonance ionic-covalent bonding percentages can readily be measured provided the nuclei have electric quadrupole moments.

Nuclear quadrupole energies might influence the rotational energies and the nuclear magnetic resonance energies as well. This results in a multiplication of the spectrum lines known as *quadrupole hyperfine splitting*. In this way quadrupole coupling constants can also be measured in the gaseous state, where direct quadrupole resonance lines would be too weak to be observed.

Nuclear quadrupole energy splittings are strongly affected by lattice vibrations. The quadrupole resonance frequencies therefore have a strong temperature dependence. The vibration of the quadrupole coupling constant q as a function of the amplitude of the lattice vibrations ϑ is approximately

$$q = \left(\frac{\partial^2 V}{\partial z^2}\right)_n \left(1 - \frac{3}{2}\overline{\vartheta^2}\right)$$

where $\overline{\vartheta^2}$ is the time average of the lattice vibration amplitudes squared.

1.5 DIELECTRIC SPECTROSCOPY

Dielectric spectroscopy deals with electric interactions between electromagnetic waves and matter. As a result of this interaction the wave passing through a dielectric medium would change its phase and amplitude. It is convenient to describe these changes in terms of complex permittivity

$$\varepsilon = \varepsilon' - i\varepsilon'' \qquad 1.15$$

where the real part ε' corresponds to the change in phase, while the imaginary part ε'' is proportional to the power absorbed from the electromagnetic wave.

Data at low frequencies are very often given in terms of dielectric constant ε and loss factor $\tan \delta$

$$\varepsilon = \varepsilon'$$

$$\tan \delta = \frac{\varepsilon''}{\varepsilon'} \qquad 1.16$$

At optical frequencies the refractive index and the absorption coefficient are used instead of complex permittivity.

The frequency dependence of ε' and ε'' is referred to as the dielectric spectrum of the material. ε'' as a function of the frequency exhibits several maxima or spectral bands corresponding to different electric dipole relaxation mechanisms in the medium. Since the dipole absorption bands are rather wide, $\varepsilon''(\nu)$ is often represented in semilogarithmic scale.

The main factors which determine the position of a dielectric dipole absorption band are as follows.

1. *Dipole orientation.* In polar substances having permanent dipole moments, the electromagnetic field tends to change the equilibrium distribution of the dipole orientations. The power absorbed by the dipole system will be a maximum when the dipole relaxation frequencies equal the frequency of the field. Since the dipole relaxation frequencies are strongly affected by temperature, the maximum of the dipole absorption will shift as temperature is changed:

$$\nu_{max} = \nu_0 \exp\left(\frac{-E_0}{kT}\right) \qquad 1.17$$

where E_0 is the activation energy of the dipole relaxation, k is the Boltzmann constant, T is the absolute temperature and ν_0 is a structural constant. By measuring dipole absorption bands as a function of temperature molecular dipole moments can be calculated.

2. *Absorption due to chemical and physical defects in solids.* In perfect ionic crystals no electric dipole moments exist. Impurity centres and crystalline defects, however, may often form dipoles which exhibit characteristic dielectric absorption bands.

3. *Dipole absorption due to adsorbed polar molecules.* Polar molecules in the sorbed phase exhibit peculiar dipole absorption. Water, for example, in a solid system may be built in the crystals as crystalline water, or may be bonded to the chemical structure, or may be in a sorbed phase. This results in a significant difference in the activation energies of the dipole relaxation, which can readily be calculated from Equation 1.17. Thus by measurement of dielectric dipole absorption shifts as a function of temperature information can be gained about the physical or chemical state of polar liquids in a solid, non-polar network.

4. *Phase transitions.* Dielectric relaxation is strongly influenced by phase transitions since activation energies of dipole orientations change abruptly with first- or second-order phase transitions. The jumps observed in the dielectric properties at phase transition points supply valuable information about the physical structure of matter.

5. *Absorption due to dipoles formed in heterogeneous media.* In heterogeneous media, dipoles are formed in the boundary layers, between domains having different permittivities. These are often referred to as Maxwell–Wagner dipoles.

6. *Induced dipole absorption.* Electric dipole absorption in non-polar compounds corresponds to the dipole moments induced by the radio frequency field. These dipoles are formed by deformation of atomic or molecular charge distributions. Since much more energy is required for this deformation than for orientation of permanent dipoles, this type of dipole absorption band lies at the highest frequency region of the spectrum.

Dielectric spectroscopy has been extensively used for determination of molecular dipole moments in the liquid and solid states where microwave rotation spectroscopy cannot be used. It is being used now for studying chemical and physical defects in solids. It can be applied to investigation of such problems as adsorption, impurity analysis, second-order phase transitions and in similar structural problems of solids and liquids. An illustrative example of the use of dielectric methods in the study of polymer structures is shown in Chapter 5.

1.6 ELECTRICAL CONDUCTION IN ORGANIC COMPOUNDS

Investigation of the electrical conductive properties of organic compounds is of interest to the organic chemist mainly in connection with charge transfer and energy transfer phenomena. Attention has been drawn to the importance of this method in biochemistry by Szent-Györgyi in 1957.[1.8]

Electrical conduction essentially involves charge transfer processes. Charges can be supplied by ions which by moving through the liquid or solid medium result in a net ionic current. The electrochemical processes of ionic conduction are not discussed here.

There are, however, other possibilities for charge transfer among organic molecules. A simplified process of electron–hole transfer is shown in Fig. 1.7. The molecular electrons are represented in this

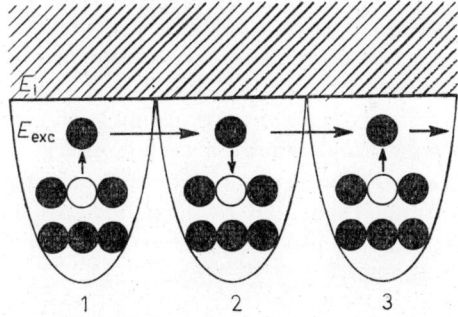

Fig. 1.7 Schematic representation of electron–hole conductivity in organic solids

simplified picture as being in a potential valley represented by the attractive forces of the nuclei. At very low temperatures the electrons cannot penetrate the potential-barriers existing between molecules or groups; the conductivity is correspondingly very small. At elevated temperatures the molecular electrons get excited, and the difference between the ionization energy E_i and the excitation energies E_{exc} will be smaller. According to quantum mechanics there is a probability for an electron to penetrate the potential-barrier E_i even if its energy E_{exc} is less than E_i. This is referred to as the *tunnel effect* in quantum theory. The probability of tunnelling through a potential-barrier E_i is approximately

$$p_t \sim \exp\left[(E_i - E_{\text{exc}})^{1/2} d\right] \qquad 1.18$$

where d is the width of the potential-barrier.

Equation 1.18 indicates that the smaller the difference $E_i - E_{\text{exc}}$, the more probable the charge transfer, i.e. the higher the electron conductivity.

As shown in Fig. 1.7 an electron transferred from molecule 2 to molecule 3 leaves a positively charged defect in its place, which is called a hole. If electron charge transfer occurs from molecule 1 to 2 the hole vanishes in 2 but appears in 1. This results in an apparent movement of holes in the opposite direction to that of electrons. Holes can thus be regarded as being individual charge carriers which can be transferred between molecules in a similar way to electrons. The conductive mechanism based on this charge transfer is referred to as electron–hole conductivity.

Activation of electrons and holes. According to the simplified picture of Fig. 1.7 a system would require a certain energy of activation in order to excite electrons into a state where charge transfer is probable. Electrons can be activated by heat, light or high-energy radiation. The electrical conductivity activated by the thermal energy kT is

$$\sigma = \sigma_0 \exp\left(\frac{-\Delta E}{kT}\right) \qquad 1.19$$

where σ_0 is a structure-dependent constant, ΔE is the activation energy, which generally consists of two terms: the energy of excitation and the energy of transfer between molecules. If molecules are excited by light or by high-energy radiation, Equation 1.19 only determines the energy required for the charge transfer between molecules or groups.

Electron and hole conductivities are very strongly affected by molecular interactions. These result in a multiplication of the excited energy states forming wide energy bands. The stronger the molecular interactions, the wider the energy bands are. Physical structures such as orientation and crystallinity also have a very pronounced effect on conductivities.

Electron–hole conductivities in organic solids range from about 10^{-20} ohm^{-1} cm^{-1} to as high as 10 ohm^{-1}cm^{-1}. This enormous variation in conductivities is explained by the exponential probability of tunnelling between molecules as stated by Equation 1.18.

Some organic compounds are almost as conductive as metals. These very high conductivities cannot be interpreted in terms of the simple electron-hole conductivity. There are other types of interactions which result in very high conductivities. This is the case with compounds known as *charge transfer complexes* which have electron-donor and electron–acceptor groups, ordered in such a way that charge can very easily be transferred. An example of biochemical interest in charge transfer is presented in Section 5.2.

Proton transfer. An important type of reaction involves proton exchange or proton transfer between molecular groups. In certain cases protons are found to be extremely mobile in liquid and solid phases. Proton transfer is very often observed when intermolecular hydrogen bonds are present. It can be studied by the proton magnetic resonance method and also by electrical conductivity measurements. An example of this is presented in Section 5.6.

REFERENCES

1.1 Cleeton, C. E. and Williams, N. H., *Phys. Rev.* **45**, 234 (1934).
1.2 Kisliuk, P. and Townes, C. H., *J. Res. Nat. Bur. Stand.* **44**, 611 (1950).
1.3 Zavoisky, E. K., *J. Phys. U.S.S.R.* **9**, 245 (1945).

1.4 Rabi, I. I., *Phys. Rev.* **51,** 652 (1937).
1.5 Purcell, E. M., Torrey, H. C. and Pound, R. V., *Phys. Rev.* **69,** 37 (1946); Bloch, F., Hansen, W. W. and Packard, M., *Phys. Rev.* **70,** 474 (1946).
1.6 Bloch, F., *Phys. Rev.* **70,** 460 (1946).
1.7 Dehmelt, H. G. and Krüger, H., *Naturwiss.* **37,** 111 (1950).
1.8 Szent-Györgyi, A., *Bioenergetics* (ed. L. G. Augenstine), Academic Press, New York (1957).

2
Experimental Methods

In this chapter the experimental methods for measuring radio frequency and microwave spectra are described briefly. The principles of operation and limits of sensitivity and of resolution are discussed rather than technical details. It is intended that the reader should get a fairly general picture of this technique, which would help him to make use of commercial equipment, or in planning to buy the most suitable one for his research programme.

2.1 MICROWAVE MOLECULAR SPECTROMETERS

As shown in Section 1.1, the rotation and inversion–transition frequencies of gas molecules are in the order of 10^{10} c/s, i.e. 10 Gc/s. These are radio waves with wavelengths in the order of cm, commonly called microwaves. The technique of microwaves has been developed enormously since the discovery of radar. It is extensively used today in different fields of telecommunications.

Some of the microwave bands important in microwave spectroscopy of gases are shown in Fig. 2.1. In the right-hand column some molecules having rotation transitions in the corresponding bands are indicated. It is usually possible to use the same microwave equipment within one band, but one must change the whole apparatus if bands are to be changed.

2.1.1 PRINCIPLE OF OPERATION

Microwave gaseous spectra are generally taken in *absorption*. The microwaves are sent through a cell containing the gas and the amount of absorbed energy is measured as a function of the frequency (or

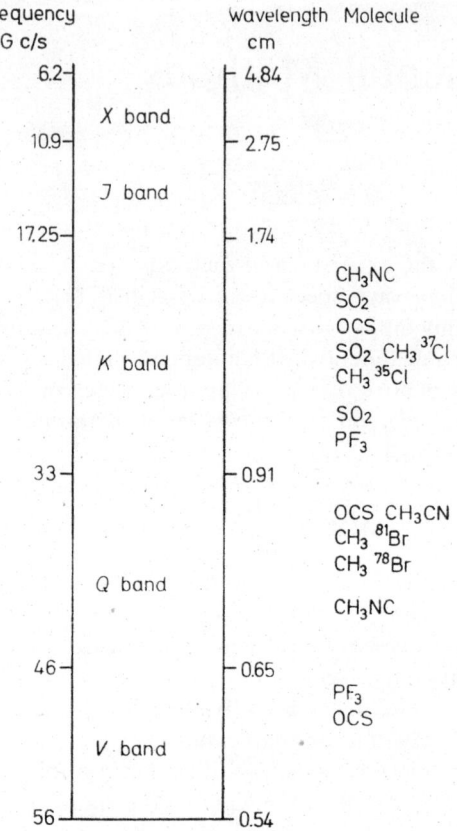

Fig. 2.1 The main bands of microwave molecular spectroscopy

wavelength). According to Equation 1.1 sharp absorption peaks should appear at the transition frequencies.

The simplest way of displaying such absorption spectra is shown in Fig. 2.2. Microwaves are transmitted by a generator through the absorption cell filled with the gas to be investigated. The pressure of the gas is in the order of 0.01 mm Hg. The microwaves, which have

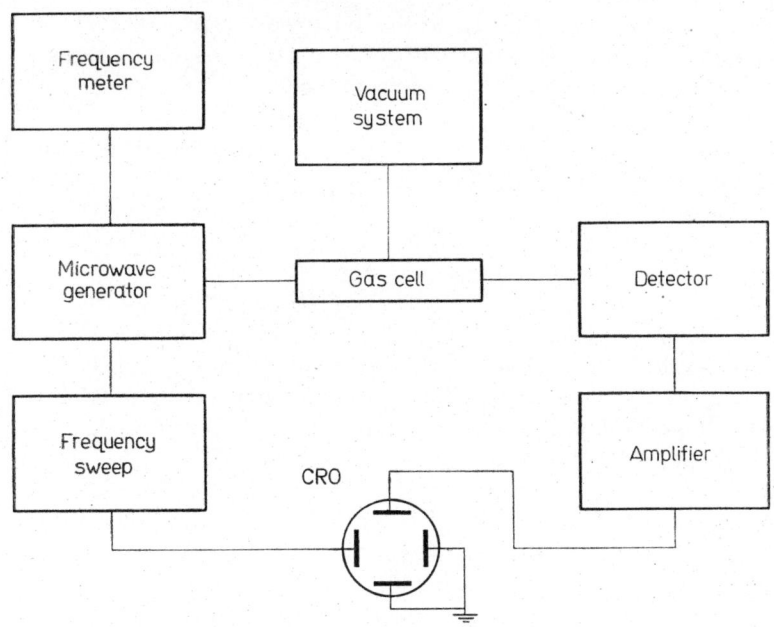

Fig. 2.2 A simple microwave spectrometer scheme

passed through the gas, are picked up by a crystal detector. The signal of the detector is amplified by a simple broad band amplifier and displayed by a cathode ray oscilloscope (CRO). The frequency of the microwave generator is swept through a certain region, where the spectral lines are expected. The same repetitive sweep supplies the horizontal deflection of the CRO. If rotation or inversion transitions are present in the frequency region swept through, the signal level at the detector decreases sharply and the spectrum lines appear on the oscilloscope. The repetition frequency of the sweep is unimportant; it is usually the mains frequency.

The frequency of the spectral lines is measured by a frequency standard, which usually consists of a highly stabilized quartz crystal oscillator of about 10 Mc/s and a frequency multiplier, which multiplies the oscillator frequency up to the microwave band. The beat signals of this frequency standard can also be displayed by the CRO, and so the position of the spectrum lines can be very accurately measured.

This simple method of displaying microwave spectra of gases is called the *video* method.

2.1.2 MICROWAVE COMPONENTS

For *microwave generators* in most cases *reflex klystrons* are used, because they are easy to handle, very stable and can easily be frequency-modulated. Reflex klystrons are commercially available for each important microwave band. For microwave gas spectroscopy low-power (20 mW) klystrons are needed. The frequency of such klystrons can very easily be swept by applying a saw-tooth voltage to their reflector electrodes. It is also very easy to stabilize klystron frequencies electronically. Stabilized signal generators are also available commercially for each band.

Magnetrons are not used in microwave spectroscopy, for they are not stable enough. Some researchers use *travelling-wave tubes*, and probably the recently developed *solid state generators* can also be applied. However, there is but little experience with these systems at the moment.

The *transmission elements* in the centimetre and millimetre bands are hollow pipes, or waveguides, which are also available commercially in standardized sizes. Almost exclusively rectangular waveguides are used with inner measurements of about one half of the operating wavelength. This means that the waveguides in millimetre or submillimetre bands are very small, and must be very precisely made.

The gas cell is usually a piece of waveguide of 1–2 m length. It is sometimes coiled up in order to save place. Such coiled cells can very easily be cooled down even to liquid nitrogen temperatures.

The *detector* is almost invariably a silicon crystal, available commercially for almost every band. The crystals are mounted in a tower-like form for the lower frequency bands (*X*-band) and in a coaxial form for the higher frequency bands (Fig. 2.3).

The *absorption cell* can be made of commercial waveguides. Usually larger sizes are used than those required for the given band. In

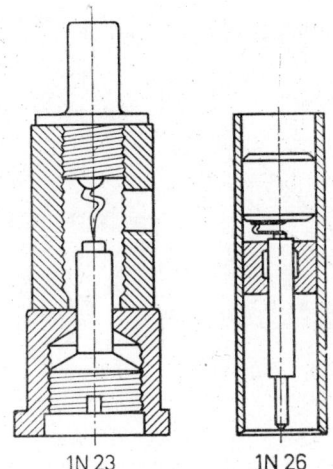

Fig. 2.3 Microwave crystal detectors.
Type 1N23 for *X*-band; Type 1N26 for higher frequencies

the *K*-band (1.25 cm wavelength), e.g., an *X*-band (3 cm) waveguide can be used for an absorption cell. The cell must be vacuum-tight, so it is separated from the other parts by thin mica or plastic windows. The optimal length of the cell depends on the attenuation caused by the walls. In the *J*-, *K*-, and *Q*-bands cell lengths of 2–4 m are usually used. The walls of the cell must be coated by silver and highly polished in order to decrease the attenuation. A possible coupling of the absorption cell to the microwave components is shown in Fig. 2.4. Since the cross-sectional area of the cell is higher than that of the waveguides, a tapered section should be placed between them. A good match must be assured between the generator and cell, and between cell and detector in order to avoid unwanted reflections which would cause similar signals in the CRO to show as spectrum lines do.

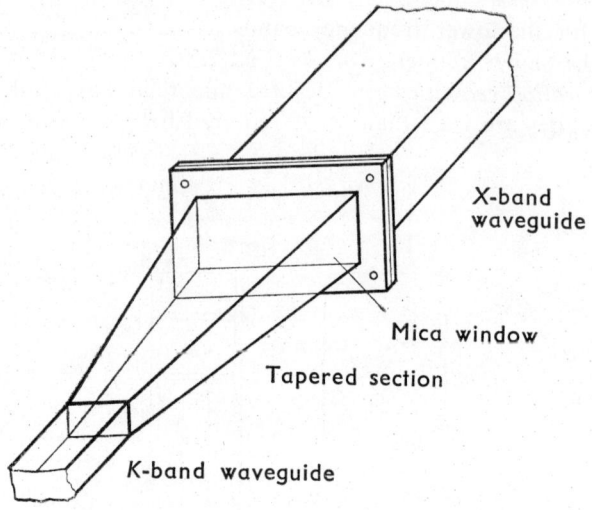

Fig. 2.4 Coupling of the absorption cell

2.1.3 SENSITIVITY

The sensitivity of microwave spectrometers depends on the signal-to-noise ratio which appears at the detector. The signal intensity depends on the microwave power, on the amount of gas in the cell, on the conversion factor of the crystal, and on the temperature. The microwave power cannot be increased at will, because of the saturation of the transitions (see Section 2.2). Saturation would increase the widths of the spectrum lines and correspondingly the sensitivity is decreased. Since the rotation–inversion lines are rather sharp (10–100 kc/s) the highest microwave power is in the order of one milliwatt. The amount of gas in the cell is also limited by the wall losses. This is why an optimal cell length should be used in each band. The pressure of the gas also should not exceed a limit of about 0.1 mm Hg in order to avoid the increase of the line widths due to collisions. The conversion factor of a crystal detector at a certain microwave power level is constant for each crystal. The signal can also be increased by decreasing the temperature, which in most cases can be done easily.

The other important factor determining the spectrometer sensitivity is the *noise level* at the detector. 'Noise' is unwanted statistical fluctuation of the detector current, which is amplified with the signal and displayed at the CRO screen. *Noise sources* can be the following: statistical fluctuation of the amplitude and frequency of the microwave generator, thermal noise of the detector and the amplifier, mechanical vibrations. The most important noise source is the detector itself.

Taking into account the different factors which influence the signal-to-noise ratio of a given spectrometer it is possible to express the minimum detectable absorption coefficient of the gas to be measured as follows[2.1]

$$\alpha_{min} = 2e\alpha_0 \sqrt{\frac{2kTSN}{P_0}} \qquad 2.1$$

where e is the charge of the electron, α_0 is the absorption coefficient of the cell, k is the Boltzmann constant, T is the absolute temperature, S is the bandwidth of the system, N is the noise factor of the crystal detector, P_0 is the microwave power. The length of the cell is assumed to be optimal.

Great sensitivity means small α_{min}. From Equation 2.1, α_{min} can only be decreased by reducing either the operating temperature T, or the bandwidth of the system S. The noise factor N can be reduced somewhat by carefully selecting the crystals. Its order of magnitude in the K-band is 10^3. The bandwidth of a video spectrometer should be at least 4,000 c/s in order to avoid distortion of the lines. So at room temperature the minimum detectable absorption coefficient for video spectrometers is approximately α_{min} 10^{-7} cm^{-1}. This figure can further be reduced to about 10^{-9} cm^{-1} by decreasing the bandwidth of the system which, however, requires a more elaborate technique. An example of this will be given below (Stark modulation spectrometers).

The absorption coefficient of rotation or inversion transitions can be found in tables (Kisliuk and Townes[2.2]). The figures given above enable one to estimate the minimum concentration of the gas to be measured.

2.1.4 RESOLUTION

The resolution of a microwave spectrometer defines how well the individual spectrum lines can be distinguished, i.e. how sharp they are. The line widths are determined by various physical and technical parameters. The most important ones are the following:

1. *Collision broadening*. The gas molecules in the absorption cell during the course of their thermal movement collide with each other or with the walls. Each collision will disturb the rotational or inversional state of the molecules and thus results in a spread of the energy levels. The more frequent the collisions are, the more uncertain the energy levels and the wider the spectral lines will be. This is why no spectra can be taken at atmospheric pressures, where frequent collisions make the lines too broad to be detected. At lower pressures the lines become progressively sharper. At 10^{-2} mm Hg, the line widths due to collisions are in the order of 100 kc/s, which is at 22 Gc/s a relative value of 5×10^{-6}. Further reducing of the pressure would result in a better resolution, but with reduced sensitivity.

2. *Doppler broadening*. The molecules in the absorption cell are moving in an electromagnetic field. The effective frequency on each molecule depends on its velocity with respect to the direction of the wave propagation (Doppler effect). Since the molecules are moving at random, there will be a spread of the effective frequencies, which broadens the spectrum lines to about 40 kc/s.

3. *Saturation broadening*. As shown in Section 1.1 the differences between rotation and inversion energy levels are about the same order as the thermal energy kT. This means that the higher levels are quite highly populated even in the absence of the microwaves. The microwave field present in the cell would induce transitions between the levels. If the power is high enough the rate of 'upwards' transitions exceeds that of the 'downwards' ones, the population of the levels tends to be equalized. Under these conditions, no net upward transitions can be induced, and the absorption tends to be zero. This is illustrated in Fig. 2.5, where a single rotational spectrum line is shown taken at different microwave power levels. At low power of about 1 μW the line is very sharp, the absorption

EXPERIMENTAL METHODS 43

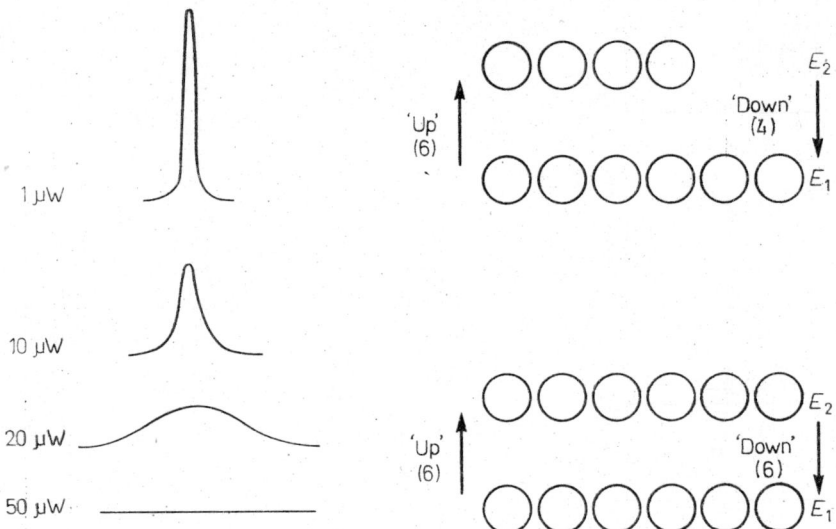

Fig. 2.5 Saturation broadening of the microwave spectral lines

coefficient is high. At about 10 μW the line broadens suddenly and vanishes completely in the noise at 50 μW, indicating that the corresponding energy levels are almost equally populated.

Saturation broadening can thus always be avoided by reducing the microwave power. This would, however, result in a decrease in the sensitivity.

The resolution of microwave gaseous spectrometers is determined by physical parameters rather than technical ones. The resolution is good enough to resolve the splittings caused by the vibrational energies, by the effects of the nuclei and others.

2.1.5 STARK MODULATION SYSTEMS

Since molecules have electric dipole moments, at least at higher excited rotational levels, it is possible to orient them by strong electric fields. The interaction energy of the molecules with the external

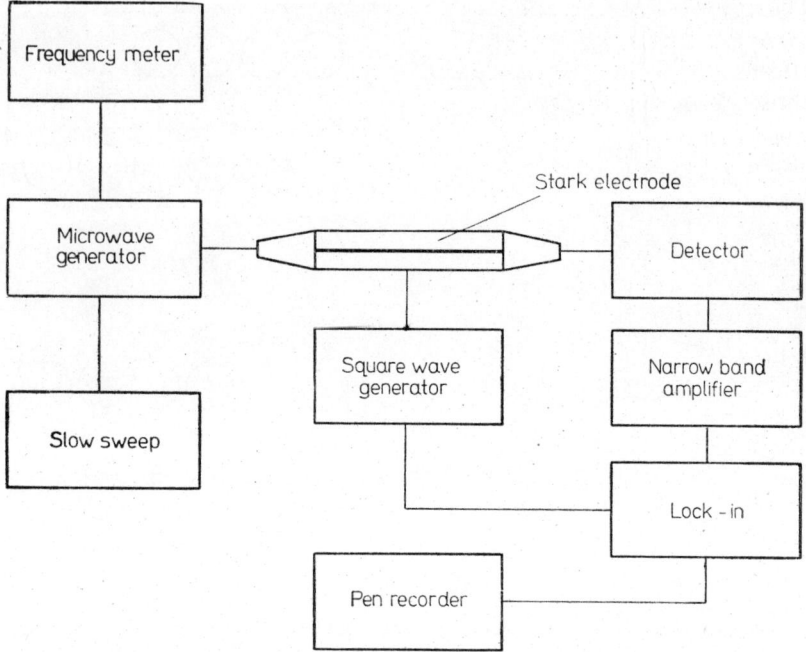

Fig. 2.6 Scheme of a Stark modulation microwave spectrometer

electric field would split the energy levels and the corresponding spectral lines. This is called *Stark splitting*.

By the use of Stark splitting it is possible to construct such spectrometers which can be operated at very narrow bandwidths without distorting the lines. Such a system is shown in Fig. 2.6. It differs from the video system of Fig. 2.2 in two main respects:

1. The cell is supplied with an electrode where a high voltage square wave is produced from a special generator.
2. Instead of using a broad band amplifier it has a narrow band one followed by a phase-sensitive lock-in detector.

The principle of operation is very simple. The high voltage square wave modulates the amplitude of the microwaves through the Stark effect without introducing frequency modulation. The Stark field is chosen to be so high that the line is completely shifted out of the band when the field is on, and is undisturbed when the field is off. Thus an alternating frequency signal appears at the detector only if the generator frequency is set on a spectrum line. This signal is amplified and recorded by a pen recorder. By the use of the lock-in detector it is possible to reduce the bandwidth to as low as 0.1 c/s. The operation of the lock-in will be discussed later. In comparison with the $S = 4,000$ c/s bandwidth of a video spectrometer the gain in sensitivity according to Equation 2.1 is 200. The reduction of the bandwidth requires that the speed of the frequency sweep should also be reduced. In the case of $S = 0.1$ c/s, for example, the time constant of the system is 10 sec. The total sweeping time should be considerably more than that, at least 10 min. At higher sweep rates the lines are distorted (partly integrated). Since the time constant (bandwidth) can very easily be changed, it is possible to choose the most suitable one for each case.

2.1.6 LOCK-IN DETECTOR

The phase-sensitive lock-in detector is widely used in radio frequency and microwave spectroscopy and is available commercially. Its simplified circuit is given in Fig. 2.7. The signal coming to the input has to be modulated by the same generator as the one for the 'reference-in'. At the output a d.c. voltage appears which is proportional to the amplitude and phase of the signal with respect to the reference. The amplitude of the reference signal is constant, the phase can be shifted at will. If this phase is properly adjusted, the recorder will follow the change in amplitude and phase of the input signal. The bandwidth of this system is determined by the capacitor C and resistor R

$$S = \frac{1}{RC} \qquad 2.2$$

where R is given in ohms, C in farads, S, the bandwidth, in c/s.

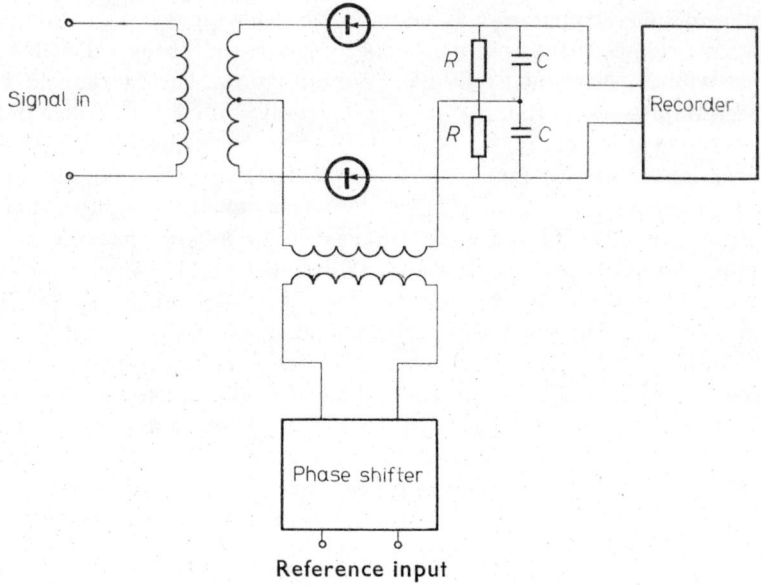

Fig. 2.7 Simplified circuit of a phase-sensitive lock-in detector

Thus the bandwidth can simply be changed by changing either R or C in the lock-in by a switch. The product RC (seconds) is called the time constant of the system. It determines how fast is the response of the system to amplitude or phase changes of the signal. The response to a voltage step at the input is described as follows

$$A(t) = A_0 e^{-t/RC} \qquad 2.3$$

where A is the reading of the recorder, t is the time.

This is why the frequency sweep is to be slow enough for a given time constant RC in order to ensure distortionless display. The usual values of time constants are given by $RC = 1\text{--}10$ sec, i.e., bandwidths, $1/RC = 1\text{--}0.1$ c/s and sweep rates, 0.1–1 min/cycle. Lock-in detectors are available commercially.

The laboratories doing research work in microwave spectroscopy of gases use home-made instruments. Recently the Japan Electron

Optics Laboratory (JEOL) has developed a commercial spectrometer for the *K*- and *J*- bands. To the authors' knowledge this is the only commercially available spectrometer at present.

2.1.7 EMISSION SPECTRA

Molecules excited at higher rotational–vibrational energy levels would tend to return to their ground state with simultaneous emission of radiation. Usually at equilibrium more molecules are in the lower

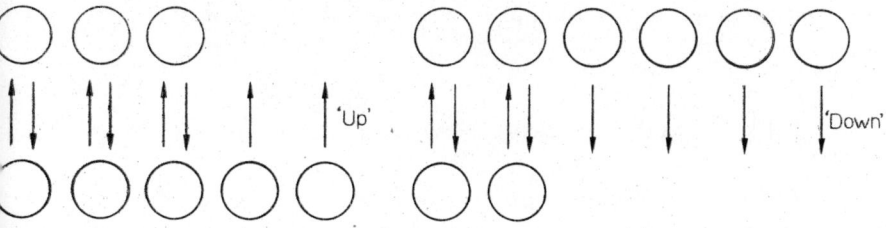

Fig. 2.8 The principle of maser action

energy levels than in the higher ones. The events of 'up' transitions (absorption) are therefore more frequent than the 'down' ones. In systems where most of the molecules are in higher-energy levels the 'down' transitions would predominate. 'Down' transitions would occur at random or can be induced by microwaves as well as 'up' ones (Fig. 2.8). Thus the emission of radiation coming from the 'down' transitions can be synchronized by the electromagnetic field and the result is emission of a coherent electromagnetic radiation. This is the principle of the maser (Microwave Amplification by Stimulated Emission of Radiation) action. A system of molecules can be brought into the state of maser action by filtering out the molecules which are in the ground state. The beam enriched in molecules at higher rotation–vibration or inversion levels enters a cavity resonator. The electromagnetic field (thermal radiation) in the cavity induces 'down' transitions at the transition frequencies determined by Equation 1.1. The first molecular maser system was realized by Gordon, Zeiger and Townes,[2,3] in

1954. The technique has been extensively developed since then. Paramagnetic solids and semiconductors can be brought into maser action and also coherent light can be generated and amplified by this principle (laser – Light Amplification by Stimulated Emission of Radiation). Masers and lasers are now being widely used in telecommunication, physics, chemistry and even in surgery.

Maser action in microwave molecular spectroscopy essentially means taking rotation–vibration or inversion spectra *in emission*. The emission lines are extremely sharp and thus the resolution is tremendously improved. However, gaseous masers are difficult to operate and the improved resolution is of limited use to chemists.

2.2 ELECTRON SPIN RESONANCE TECHNIQUE

Electron spin resonance spectroscopy has recently become one of the most useful tools in investigating free radicals and reactions Its technique has been highly developed. Several types of ESR spectrometers are commercially available with very high sensitivity and resolution. They are relatively easy to operate at wide temperature ranges. Different kinds of accessories are available for observing ESR spectra during the course of reactions, under ultraviolet irradiation and during electrolysis. It is possible to observe spectra in the solid, liquid and gaseous phase as well.

2.2.1 PRINCIPLE OF OPERATION

Electron spin resonance involves radiation-induced transitions between magnetic energy levels of unpaired electrons (see Section 1.2). The magnetic energy splitting is created by the field of a big electromagnet. According to Equation 1.3 the transition frequency v_θ depends linearly on the magnetic field intensity H_0. It is desirable to work at the highest possible H_0- and v_0-values, because this would result in greater difference in the population of the energy levels. It can be shown from Equations 1.3 and 1.5 that the relative difference

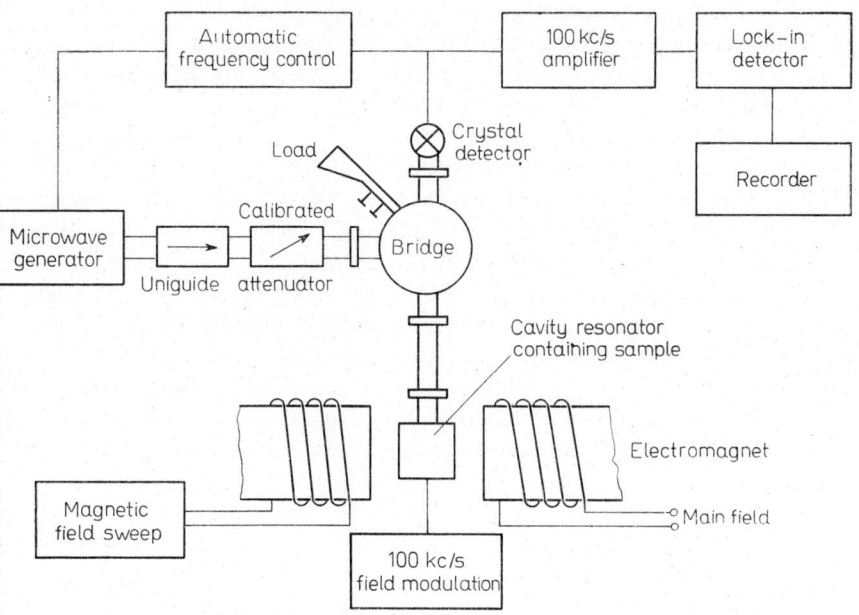

Fig. 2.9 Block diagram of a conventional ESR spectrometer

in the population is approximately

$$\frac{\Delta N}{N} \approx \frac{h v_0}{kT} = \frac{g_s \mu_s \mathbf{H}_0}{kT} \qquad 2.4$$

where ΔN is the difference in the population of the levels, N is the total number of unpaired electrons, h is Planck's constant, k is the Boltzmann constant, T is the absolute temperature, g_s is the electronic splitting factor, μ_s is the Bohr magneton, and v_0 and $\mathbf{H}_0$ are the frequency and magnetic field strength at resonance, respectively.

According to Equation 2.4 the electron spin resonance absorption is strong if the values of v_0 and $\mathbf{H}_0$ are high and the temperature is low. This is why ESR spectrometers are operated at microwave frequencies, where the resonant frequency v_0 is sufficiently high, and the resonant magnetic field H_0 is not too high to cause too many technical difficulties. The most useful band for ESR spectrometers is the X-band (3 cm), where the resonant magnetic field is around 3,000 gauss. Some of the spectrometers are operated in the K- and J-bands with correspondingly higher magnetic fields. As follows from the principle of ESR, described briefly in Section 1.2, ESR spectra can be taken in a fixed microwave band by changing the magnetic field strength only. Instead of sweeping the frequency, ESR lines are searched by sweeping the magnetic field, which is technically much easier.

A very commonly used scheme is shown in Fig. 2.9. Microwaves are usually generated by a reflex klystron oscillator. The waves are passed through an isolator or uniguide, a piece of ferrite placed in a homogeneous magnetic field of a small permanent magnet. The microwaves can pass the uniguide in the forward direction almost unattenuated, but for the reflected wave an attenuation of about 30–40 dB is effective. The generator is this way isolated from the rest of the system. The microwave power level can be regulated by a calibrated attenuator. The wave is fed into a microwave bridge, which separates it into two channels. In one is placed the measuring cell, in the other a matched load. The sample is placed in the cavity resonator, which is between the pole-pieces of a large electromagnet. The wave reflected from the cavity arm of the bridge is mixed with that reflected from the load and the residual signal is detected by a crystal detector. In the absence of resonance the bridge is balanced in such a way that a relatively small constant microwave power level is present at the detector. If the magnetic field is swept through resonance, additional losses are introduced in the cavity arm which upset the balance of the bridge. If the bridge is correctly set, the change in the d.c. level of the crystal is proportional to the energy absorbed by the sample.

D.C. signals are difficult to amplify. This is why usually a 100 kc/s magnetic field modulation is introduced by auxiliary coils, which would transform the direct current into an alternating one. This is illustrated in Fig. 2.10. Curve a is a typical ESR absorption spectrum line taken as a function of the magnetic field H. Absorption is maximum at the resonance field H_0. If the magnetic field is slightly modulated with 100 kc/s and swept slowly through the spectrum line, an alternating current of 100 kc/s appears in the detector. The a.c. signal level varies as a function of the magnetic field **H**. It can be shown that the signal appearing after a second, phase-sensitive (lock-in) detection (Fig. 2.10*b*) is the derivative of the absorption curve *a*. The reso-

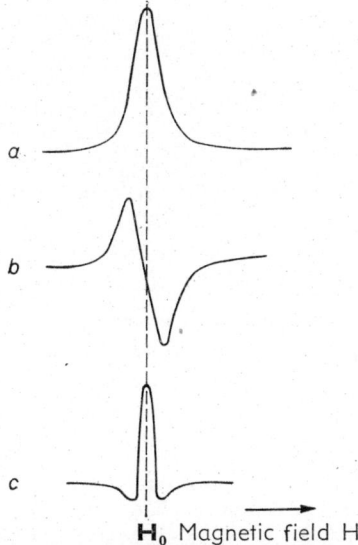

Fig. 2.10 Typical ESR spectrum line representations.
a — Absorption line; *b* — First derivative; *c* — Second derivative

nance field H_0 corresponds to the zero intersection of the derivative curve. This is the 'position' of the spectrum line. The width of the line is often defined as the peak-to-peak distance of the derivative curve measured in gauss. The peak-to-peak width $\varDelta H_0$ of the derivative curve is not the same as the half-width of the absorption curve. The difference depends on the line shape. It is also possible to record the second derivative curve shown in Fig. 2.10*c*, as will be discussed later.

The introduction of field modulation makes it possible to apply very narrow band detection. The high-gain narrow band 100 kc/s amplifier followed by a lock-in detector ensures a bandwidth of about 0.1–1 c/s, and a correspondingly low noise level. The spectrum is recorded by a pen recorder driven synchronously with the slow magnetic field sweep.

2.2.2 CAVITY RESONATORS

The heart of an ESR spectrometer is the cavity where the sample is placed. Cavity resonators are metallic boxes, in which the microwave energy is present in the form of standing waves. The simplest cavity resonator is an X-band waveguide section closed at both ends, as shown in Fig. 2.11a. The microwave energy can be coupled into the cavity by a hole as indicated in the figure. The distribution of the microwave magnetic and electric fields within the cavity is also shown. This is important because the sample is to be placed at that part of the cavity where the microwave magnetic field is maximum. As shown in Section 1.2, electron spin resonance involves induced magnetic

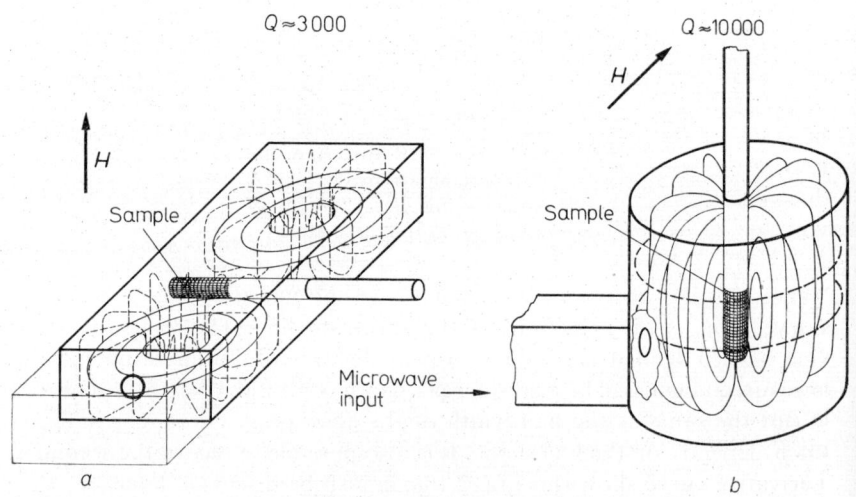

Fig. 2.11 ESR cavities.
a — Rectangular; b — Cylindrical

dipole transitions. This can only be done by the magnetic component of the microwave field. It is also necessary that the microwave field be perpendicular to the direction of the polarizing d.c. magnetic field.

The losses in the cavity are characterized by the Q-factor defined as

$$Q = \frac{2\pi v_9\, E_{\text{stored}}}{P_{\text{loss}}} \qquad 2.5$$

where v_0 is the resonant frequency of the cavity, E_{stored} is the microwave energy present in the cavity, P_{loss} is the power dissipated.

Losses within the cavity are due to the microwave energy transformed into heat in the cavity walls and the dielectric losses of the sample holder and of the sample itself. Some additional loss is introduced by the coupling hole. Since the microwave bridge is sensitive to the variation of the Q-factor of the cavity it would be desirable to have Q-values as high as possible. In order to minimize dielectric losses sample holders are usually made of quartz. If samples having high dielectric loss (aqueous solutions) are to be investigated, only thin capillary tubes can be introduced in the cavity.

A typical cylindrical cavity resonator most commonly used in ESR is shown in Fig. 2.11b. This type has the highest Q-value (about 10,000) in the smallest volume. The microwave magnetic field is a maximum along the axis of the cavity, so the sample holder tubes can easily be introduced. The minimum diameter of the cavity in the X-band is about 40 mm. Quartz tubes of about 6 mm diameter can be introduced axially without serious drop in the Q-value. The effective sample volume for low dielectric loss samples is about 0.1 cm^3.

ESR cavities are mostly made of brass with the highly polished inner walls coated with very pure silver. A protecting layer of gold is also often used. Cavities can also be made of silver-coated glass, quartz, ceramics and plastics. In the case of metallic cavities the 100 kc/s modulation coils are placed inside the cavity, for otherwise the modulating field could not penetrate the thick metal walls. With the cavities made of insulating materials external modulation coils can be used. The thin silver wall has practically no effect on the 100 kc/s modulating frequency because the skin depth at this frequency is much greater than the wall thickness.

2.2.3 AUTOMATIC FREQUENCY CONTROL

Cavities resonate at definite frequencies determined by their dimensions. The higher the Q-value of a resonator, the sharper the resonant frequency band is. So the cavity should always be kept exactly tuned

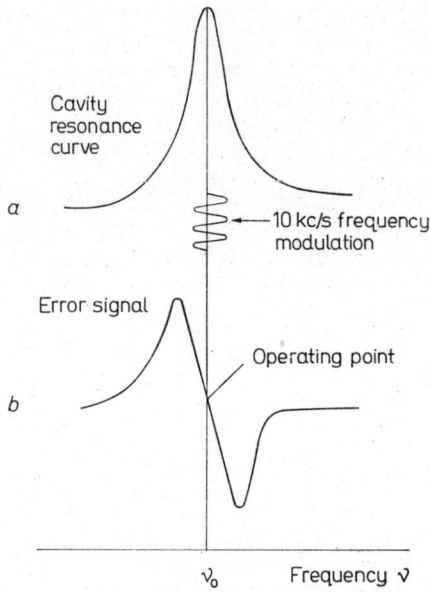

Fig. 2.12 The principle of automatic frequency control.
a — Cavity resonance curve; b — Error signal to klystron

to the microwave generator, for otherwise frequency drifts would turn into random amplitude modulation at the detector, increasing the noise level seriously. Automatic tuning is done by the automatic frequency control (AFC) unit, which consists of a similar narrow band amplifier and lock-in detector to the 100 kc/s signal channel, but it is operated at lower frequency (1–10 kc/s). The frequency of the microwave generator is modulated slightly by this frequency (usually 10 kc/s), and the signal appearing at the detector is amplified, detected and fed back to the reflector electrode of the klystron. If the cavity is correctly tuned, no 10 kc/s signal appears at the detector. If the frequency drifts, an error signal appears with a phase dependent on the direction of the drift. The phase-sensitive detection converts this a.c. signal into a positive or negative d.c. error signal, depending on the direction of the drift (Fig. 2.12). This system, when correctly set, would always keep the cavity on tune during the course of the measurement. A good AFC system having small response time can thus very effectively reduce the noise level of the spectrometer.

2.2.4 SENSITIVITY

Radicals are very seldom present in chemical systems in very high concentrations. In many important cases, as for example in reactions, radical concentrations are not higher than 10^{-7}–10^{-8} mole/l, which generally cannot be detected by ESR. Increasing the sensitivity is therefore a basic problem in the ESR technique.

Sensitivity is usually defined as the minimum number of paramagnetic centres which can be detected. The number of the paramagnetic centres, commonly called 'spins', is proportional to the area under the absorption signal. Thus the amplitude of the derivative line recorded by the spectrometer depends on the line width. The signal-to-noise ratio, however, depends on the amplitude and correspondingly broader lines have poorer signal-to-noise ratios than sharper ones corresponding to the same number of spins. This is why spectrometer sensitivities are usually given in terms of spins/gauss, which refers to a signal having a signal-to-noise ratio of 1 : 1 and a line width of 1 gauss. The signal-to-noise ratio is, of course, determined by many other technical parameters, such as the magnetic field modulation amplitude, temperature, the dielectric loss of the sample, the microwave power level and so on.

The usual sensitivity value of X-band ESR spectrometers is about 10^{11} spins/gauss. The line widths of the radicals are in the order of 10 gauss. The sample volume is about 0.1 cm^3 for low dielectric loss materials. This would mean that the sample should contain at least 10^{13} radicals/cm^3 in order to see the line (signal/noise = 1). For structure determinations or other measurements at least a signal-to-noise ratio of 10 : 1 is required; for quantitative analysis even more. This corresponds to 10^{14} radicals/cm^3, i.e. a concentration of about 10^{-6} mole/litre. This is about the lowest radical concentration which can be quantitatively measured by simple ESR technique in the X-band.

Technical parameters. The sensitivity of the ESR spectrometers as a function of technical parameters has been discussed extensively by G. Feher[2,4] in 1956. According to his arguments the minimum detectable number of spins is determined by the following main technical factors

$$N_{min} \propto \frac{V_c}{V_s} \frac{C}{Q} \left(\frac{TS}{P_0}\right)^{1/2} \qquad 2.6$$

where V_c, V_s are the volumes of the cavity and of the sample respectively, Q is the Q-factor of the cavity filled with the sample, T is the absolute temperature, S is the bandwidth of the system, P_0 is the microwave power. C is a technical factor characterizing the noise level-to-signal conversion of the detector crystal. C depends upon the quality of the crystal, upon the modulation frequency and upon the microwave power converted by the crystal.

As seen from Equation 2.6 the sensitivity of a given spectrometer can be increased by decreasing the temperature of the sample. Measuring at the temperature of liquid nitrogen (77 °K) would increase the nominal sensitivity of the spectrometer by a factor of 2. In some special cases temperatures down to 4 °K are used (the temperature of liquid helium). This, however, requires a more elaborate cooling technique. Measurement at very low temperatures has also the advantage that radicals are not reactive; they are frozen in the system. This problem will be discussed in detail in Chapter 5.

The other possibility for increasing the sensitivity of an ESR spectrometer is to increase the microwave power level P_0. However, the effect of saturation discussed in Section 1.2 sets a limit to the microwave power. Sharper spectrum lines would saturate at lower microwave power levels than broader ones. The usual power level for radicals in the liquid state is 1–10 mW. In the solid state, where the lines are broader, the power level can be increased up to about 500 mW. Thus the loss in sensitivity due to the line widths can be partially compensated by increasing the microwave power level.

This is indicated in Fig. 2.13, where the minimum detectable number of spins are plotted against the microwave power for different line

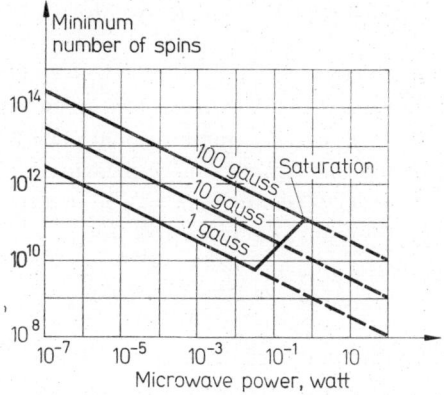

Fig. 2.13 Diagram for estimating ESR spectrometer sensitivities. X-band, room temperature, response time 0.1 sec. After Feher[2.4]

widths. The chart corresponds to a typical X-band spectrometer, as that shown in Fig. 2.9, with main technical parameters as follows

$$Q = 5,000$$
$$S = 0.1 \text{ c/s}$$
$$V_c = 200 \text{ cm}^3$$
$$V_s = 0.1 \text{ cm}^3$$
$$T = 300 \text{ °K}$$

It should be stressed that the figures given in Fig. 2.13 correspond to a signal-to-noise ratio of 1. For any structure determination or quantitative measurement at least 10–50 times more spins should be present in the sample.

The crystal noise factor C in Equation 2.6 cannot be improved significantly. The spectrometers are designed in such a way as to ensure optimal conditions for C. This is why the modulation frequency is chosen as high as 100–500 kc/s for the conversion loss is much higher at low frequencies. The optimum microwave power converted by the crystal is adjusted by adjusting the matched load of Fig. 2.9. The only way of reducing the crystal noise factor C is to select crystals experimentally by measuring the signal-to-noise ratio of a standard sample.

As for the optimal volume of the sample V_s, the choice of sample holder diameter is essential. The Q-value of the cavity should not drop more than about 30% with the sample present. The change in the Q-value upon introducing the sample can be measured by observing the cavity resonance curve. The Q-value is expressed in terms of the parameters of the cavity resonance curve as follows

$$Q = \frac{v_0}{\Delta v} \qquad 2.7$$

where Δv is the width of the cavity resonance curve at half-power points expressed in Mc/s, v_0 is the operating frequency in Mc/s. In the X-band $v_0 \approx 9{,}000$ Mc/s, for $Q = 10{,}000$ the width $\Delta v = 0.9$ Mc/s. Upon introducing the sample this width should not exceed 1.5 Mc/s. If it does, the diameter of the sample tube should be decreased.

The resonance curve of the cavity can be displayed in CRO by applying a saw-tooth voltage to the reflector of the klystron. This would modulate the klystron frequency over a wide range, say about 50 Mc/s. The power reflected from the cavity is minimal at resonance v_0. A typical picture seen at the CRO screen is shown in Fig. 2.14a. It is called the 'klystron mode' with the cavity resonance curve in it. The variation of the width can easily be measured.

Displaying the klystron mode and cavity resonance makes it very easy to adjust the correct coupling of the cavity resonator. If coupling is correct, the cavity resonance peak should reach the zero power level, indicating that at the resonance frequency v_0, all the power has been dissipated by the cavity. If the coupling is not correct, the peak is not so deep as indicated in Fig. 2.14b.

Sensitivity can be increased somewhat by applying higher microwave frequencies. In the Q-band (8 mm) for example, a sensitivity of

5×10^9 spins/gauss has been achieved in commercial spectrometers (Varian Associates, Palo Alto, California). This increased sensitivity can, however, be realized only in those cases, where small amounts of samples are to be used. In the millimetre bands the dimensions of

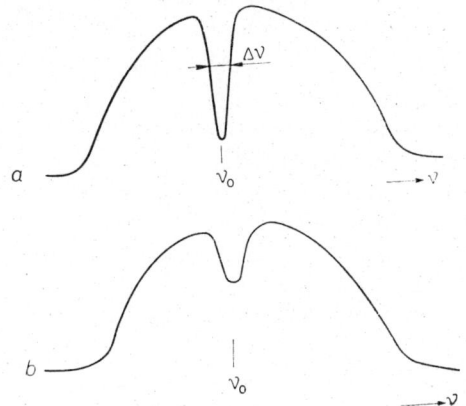

Fig. 2.14 A klystron mode with a cavity resonance curve.
a — Correct coupling; *b* — Incorrect coupling

the cavity resonator are smaller, and thus much smaller amounts of sample can be used than in the centimetre bands.

Superheterodyne detection. The sensitivity of the ESR spectrometers can in principle be increased by a factor of about 5 by using microwave superheterodyne detection. A superheterodyne receiver consists of a microwave local oscillator whose frequency is mixed with the signal. The intermediate frequency of 30–60 Mc/s formed in this way is amplified and detected. The field modulation frequency in this case is low (30–60 c/s). The superheterodyne scheme usually consists of two lock-in detectors, one for the intermediate frequency, the other for the field modulation frequency. Instead of the automatic frequency control, other frequency stabilization systems are used. The main advantage of the superheterodyne scheme is that its sensitivity is high even at very low microwave power levels. So it is chiefly used in cases where very sharp lines are to be detected (e.g. in metals and semiconductors).

Recently such superheterodyne spectrometers have also been developed in which only a single klystron is used.

Double modulation. In cases when the temperature or the dielectric loss of the sample is rapidly changing during the course of the measurement the double modulation technique is used. In this scheme the magnetic field is modulated by a low frequency, e.g. 60 c/s, and by a high frequency of 100 kc/s. The receiver system is also doubled. First the usual 100 kc/s narrow-band amplifier and lock-in detector are coupled to the crystal detector as shown in Fig. 2.9, then a low-frequency (60 c/s) amplifier and lock-in detector. As a result the second derivative of the absorption curve is recorded (Fig. 2.10c). The double modulation system prevents serious drifts in the base line due to drastic changes in the cavity parameters. The sensitivity is also slightly increased. This scheme is very useful in such cases, where only the presence of radicals is to be indicated and their structure is to be determined. It is not practical for determination of radical concentrations, because the original absorption line can only be obtained from the recorded curve by double integration.

Accumulation of spectra. In the last few years a very effective electronic method has been developed for increasing the sensitivity of ESR spectrometers and for making interpretation of complicated spectra easier. The principle of spectrum accumulation lies in the fact that upon repetitive sweeping through a spectrum the signal peak will always be the same, but the peaks due to noise will fluctuate statistically. If the same spectrum is repeated n times and the spectrum amplitudes are added, signal amplitudes will be added linearly but not the statistical noise amplitudes. The signal thus will be proportional to n, the noise to $\sqrt{n}$, and thus the signal-to-noise ratio

$$(S/N)_{\text{sum}} = (S/N)_0 \sqrt{n} \qquad 2.8$$

where $(S/N)_{\text{sum}}$ is the signal-to-noise ratio of the sum of spectra repeated n times, $(S/N)_0$ is that obtained by sweeping once.

For addition of spectra, electronic computer circuits are used. The conventional spectrum, which is a variation of electric current amplitudes in time, is first converted into frequency variations which can easily be stored in magnetic memory units. The spectra stored in such a way can be summed rather easily and the sum is transformed back into current amplitude variations. If, for example, a spectrum is swept through 100 times, the signal-to-noise ratio of the sum produced and recorded by the computer will be improved by a factor of 10. Using this technique it is possible to detect less than 10^{10} spins/gauss at room temperature by a conventional X-band ESR spectrometer.

A compact commercial spectrum accumulator has recently been developed by the Japan Electron Optics Laboratory, Tokyo. It is a small electronic computer capable of storing 65,536 information units (bits). It can be connected to any ESR or NMR spectrometer via the recorder output (0–10 mV). The accumulated spectrum is also recorded on chart paper.

Besides increasing the sensitivity, other problems can also be solved by the computer. It is possible to add and subtract spectra, to integrate, i.e. to convert derivative curves into absorption ones, to analyze line widths and shapes and to simulate complicated spectra by introducing hyperfine splitting constants.

2.2.5 RESOLUTION

As shown in Section 1.2 the hyperfine splitting of the ESR spectrum lines makes possible the determination of radical structures. In many cases the radical electron is delocalized so that it is coupled to a number of magnetic nuclei. Most ESR spectra should then consist of many lines. If they are well resolved, the hyperfine interaction constants and thus the structure of the radicals can readily be determined. However, the ESR line shapes and line widths, and thus the resolution, are limited by technical and physical factors.

1. *Dipole–dipole interaction.* The spin systems considered in ESR spectroscopy consist of electronic and nuclear magnetic moments. As shown in Chapter 1, energy can be exchanged between elements of the spin system through¹ the local magnetic fields of the dipoles.

This interaction would destroy the phase coherence of the spins and result in a broadening of the spectrum lines. The dipole–dipole broadening is the most important physical factor limiting the resolution of ESR spectra.

The variation of the line width due to the dipole–dipole coupling is expressed as follows

$$\delta H = \mu_i \mu_j \frac{3\cos^2 \vartheta_{ij} - 1}{r_{ij}^3} \qquad 2.9$$

where μ_i, μ_j are the magnetic moments of the dipoles i and j, respectively, ϑ_{ij} is the angle, r_{ij} is the distance between them. Since a large number of dipoles are present in the sample, the relative angle ϑ_{ij} and distance r_{ij} are to be averaged.

As follows from Equation 2.9 the largest dipole–dipole broadening of the spectrum lines is expected in the polycrystalline or amorphous solid phase, where the dipoles are randomly oriented and the average distance between them is small. This is why the resolution of the ESR spectra is usually poor at low temperatures. The dipole–dipole broadening in the amorphous or polycrystalline state is in the order of 20–30 gauss, which is about the same as or even larger than the hyperfine splittings in most radicals.

At higher temperatures the random tumbling motion of the molecules would decrease the dipole–dipole broadening very effectively. In this case the effect of orientation is averaged out by the rapid motion of the molecules. In solutions, where the average distance between the paramagnetic centres is reduced and the orientation effect is averaged out, the line widths can be as small as 0.1 gauss. Therefore, good resolution in ESR spectra can only be expected in the liquid or in the gaseous phase.

In some cases fair resolution can also be obtained in the solid phase. If, for example, a relatively small concentration of paramagnetic centres is present in a diamagnetic single crystal, the dipole–dipole broadening will be small. Thus, a relatively high resolution can be obtained in single crystals doped with paramagnetic ions, such as Mn^{++} in ZnS, Cr^{++} in MgO and others. Similarly the resolution is rather good in irradiated organic or inorganic *single crystals* even at very low temperatures. Some interesting examples for chemists are given in Chapters 3 and 5.

The shape of the spectrum lines, provided that only dipole–dipole interaction is present, is Gaussian. The amplitude G of the line is expressed as a function of the frequency as follows

$$G(v - v_0) = \frac{T_2}{\pi} \exp\left[-4\pi (v - v_0)^2 T_2^2\right] \qquad 2.10$$

where v_0 is the resonance frequency, T_2 is the spin–spin relaxation time characterizing the dipole–dipole interaction. (See Section 1.3 for the definition.)

The width at the half-power points is

$$(\Delta v)_{1/2}^G = \frac{1.182}{T_2} \qquad 2.11$$

A Gaussian spectrum line is shown in Fig. 2.15a.

2. *Exchange interaction.* In most paramagnetic materials, especially in free radicals, line widths are considerably smaller than those expected from the dipole–dipole interaction. Line shapes are also different from Gaussian. This is caused by the so-called exchange interaction existing between delocalized electrons provided their spin densities overlap. The situation is similar to the contact interaction between electrons and nuclei, discussed in Section 1.2. It can be shown that exchange interaction is isotropic and makes the lines sharper. The ESR line shape in the case of strong exchange interaction is Lorentzian. The amplitude of a Lorentzian line as a function of the frequency is given as follows

$$L(v - v_0) = \frac{1}{\pi} \frac{T_2}{1 + 4\pi^2 (v - v_0)^2 T_2^2} \qquad 2.12$$

where v_0 is the resonance frequency, T_2 is the spin–spin relaxation time.

As shown in Fig. 2.15 the Lorentzian line is much sharper at the centre than the Gaussian one. The wings are, however, farther extended.

The half-width of a Lorentzian line is

$$(\Delta v)^L_{1/2} = \frac{1}{T_2} \qquad 2.13$$

Since the unpaired electrons of free radicals are always somewhat delocalized in the solid state, the corresponding ESR lines are Lorentzian.

The characteristic line widths of the main groups of paramagnetic materials are shown in Table 2.1. As seen there, the more the delocalization of the unpaired electron is, the less the line widths are. In the paramagnetic ions such as those of the rare earth group and of the iron group the unpaired electrons are located in an inner electronic shell and correspondingly are strongly localized. In metals and semi-

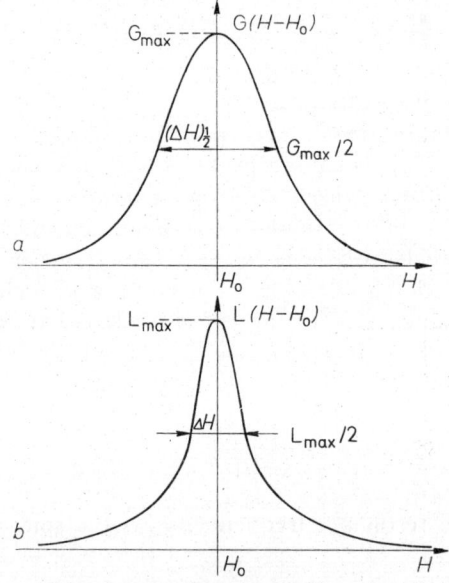

Fig. 2.15 ESR spectrum line shapes.
a — Gaussian; b — Lorentzian

Table 2.1

ESR Line Widths

Material	Widths (gauss)	
	Solid	Liquid
Paramagnetic ions	300–1,000	50– 100
Radicals with localized electron (e.g. ROȮ, RṠ)	10– 100	1– 10
Radicals with strongly delocalized electron (e.g. aromatics)	3– 10	0.1– 1
Conduction electrons in metals and in semiconductors	< 0.1	–

conductors, on the other hand, electrons are fully delocalized and therefore exchange interaction is very strong. Radicals in the solid state form an intermediate case between paramagnetic ions and metals. Radical electrons are usually delocalized to their nearest neighbours. There are, however, highly conjugated systems with electrons delocalized to macroscopic distances. In these systems the ESR lines are very sharp.

3. *Inhomogeneous broadening.* The resolution of ESR spectra is very often limited by unresolved structures and by the inhomogeneities of the polarizing magnetic field. In the case of unresolved structures the spectrum line is an envelope of several lines smeared out by orientation or by other effects. This is a physical limitation of the resolution and thus can only be improved by altering the physical conditions of the measurement. The inhomogeneities of the polarizing magnetic field would cause similar effects, but it can be improved technically. For investigation of radical structures in the liquid or in the gaseous state, the inhomogeneity of the ESR magnets should not exceed 1 part in 10^5. This is why the electromagnet is the most delicate and expensive part of an ESR spectrometer.

The inhomogeneous broadening due to unresolved hyperfine structure is illustrated in Fig. 2.16. Curve *a* represents a well-resolved hyperfine spectrum arising from the interaction of the unpaired electron with 4 equivalent protons. The spectrum consists of 5 lines with the binomial intensity ratio of 1 : 4 : 6 : 4 : 1. Such spectra can easily be recorded from dilute solutions of free radicals. If, however, line widths are increased by dipole–dipole broadening, as in concentrated so-

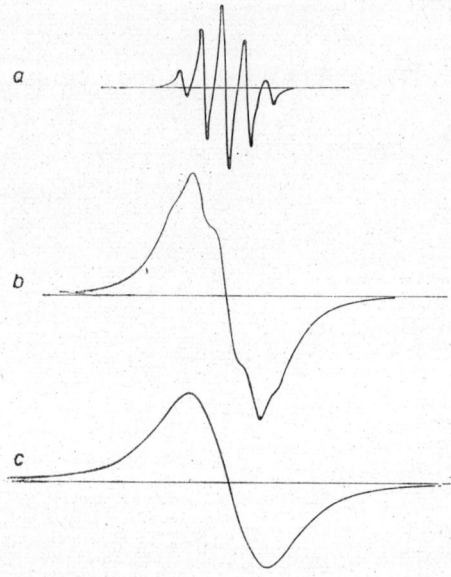

Fig. 2.16 Inhomogeneous broadening of the ESR spectral lines due to unresolved hyperfine splitting

lutions or in solids, the spectrum is poorly resolved or even not resolved at all (Fig. 2.16b and c). The resulting broad line is symmetrical. In this case the total width of the line in Fig. 2.16c is certainly not due to the spin–spin coupling and correspondingly Equation 2.11 does not hold. In favourable cases it is possible to reconstruct the hyperfine splitting and the line widths from the shape of the inhomogeneously broadened line. Standard theoretical spectra for this can be found in 'Atlas of ESR Spectra' edited by Voevodsky.[2.5] For computation of these theoretical spectra an electronic computing machine was used.

Inhomogeneously broadened lines can be distinguished from homogeneously broadened ones by *saturation*. As shown in Section 1.2 by application of high microwave power the population of the electron spin levels tend to equalize. This corresponds to the decrease of the ESR line amplitudes. In the case of homogeneously broadened lines, the line shape also changes by saturation. The higher the microwave power is, the broader the lines will be. The width of an inhomogeneously broadened line is, however, not affected appreciably, for this line is the envelope of several homogeneous lines with equally decreasing amplitudes. The intensity of the spectral lines is

$$I \propto \begin{cases} \dfrac{1}{P} & \text{(homogeneous)} \\ \dfrac{1}{P^{1/2}} & \text{(inhomogeneous)} \end{cases} \qquad 2.14$$

where P is the microwave power, and I is the intensity of the spectrum line proportional to the area under the absorption curve.

Thus, by recording the line at increasing microwave power level it can be easily decided if the line is inhomogeneously broadened or not.

Distortion of the lines by the anisotropy of the splitting factor. In crystals and in oriented radicals the position of the ESR spectral lines may be influenced by the strong electric fields of the crystal. As mentioned in Section 1.2 the crystalline electric fields affect the spins through the spin–orbit coupling and make the splitting factors g anisotropic. In the case of polycrystalline powders or randomly oriented systems the anisotropy of the g-factor would cause a distortion of the spectrum lines. The anisotropy of the g-factor depends on the spin–orbit coupling and on the symmetry of the crystalline electric field. In the case of axial symmetry the splitting factor is given by the following equation

$$g^2 = g_\perp^2 \sin^2\theta + g_\parallel^2 \cos^2\theta \qquad 2.15$$

where $g_\parallel$ and $g_\perp$ correspond to orientations parallel and perpendicular to the polarizing magnetic field, respectively.

In polycrystalline substances or other randomly oriented systems the g-factors are averaged over all possible orientations to result in typical spectra shown in Fig. 2.17. Spectrum a corresponds to the case when $g_\perp \gg g_\parallel$. Such spectra have been measured, e.g. in systems

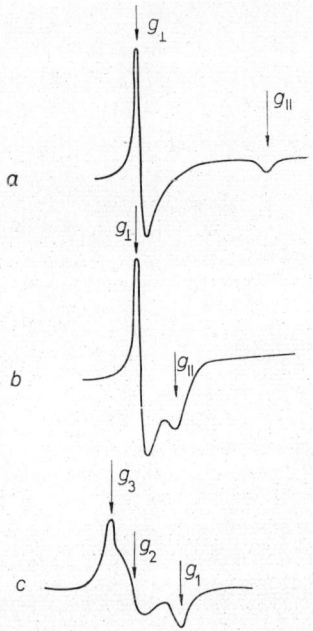

Fig. 2.17 ESR lines of randomly oriented polycrystalline substances
$a - g_\perp \gg g_\parallel$; $b - g_\perp \approx g_\parallel$; $c - g_1 \neq g_2 \neq g_3$

of *haemin*, and *haemoglobin*. Spectrum b corresponds to the case when $g_\perp$ and $g_\parallel$ are nearly equal. Such spectra can be observed in many irradiated organic crystals and polymers. Spectrum c corresponds to a case when the crystalline field has no axial symmetry and thus three g-values are to be given g_1, g_2 and g_3. Such spectra can be observed, e.g., in polycrystalline $CuCl_2.2H_2O$.

ESR lines broadened by the effect of g-anisotropy are usually asymmetrical and thus can be easily recognized.

EXPERIMENTAL METHODS

Modulation broadening. In most of the commercial ESR spectrometers high-frequency field modulation of 100–500 kc/s is used. Low-frequency modulation alone is not favourable because of the increased crystal noise factor. However, at higher resolution, lines may be

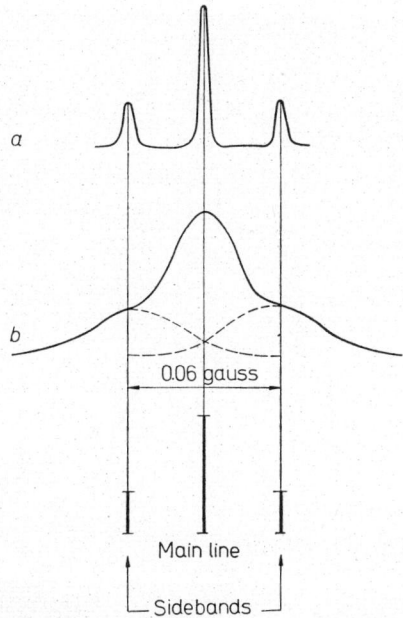

Fig. 2.18 Modulation broadening of the ESR spectrum lines

broadened by the effect of the high-frequency modulation. If the lines are very sharp, the high-frequency field modulation will cause sideband lines to appear at both sides of the main line at a distance corresponding to the modulation frequency (Fig. 2.18a). If it is 100 kc/s, the relative deviation of the spectrum line at 10,000 Mc/s is 10^{-5}. The deviation of the sideband lines from the main line in the magnetic field scale is $3{,}000 \times 10^{-5} = 0.03$ gauss. The effect of line broadening is illustrated in Fig. 2.18. As can be seen there, modulation sidebands affect the apparent line widths only in those cases where they are in the order of 0.06 gauss. This would set an upper limit of 2 parts in 10^{-5} for the resolution of the 100 kc/s field modulation systems.

In such cases when physical factors permit higher resolution a superheterodyne system should be used which can be modulated at

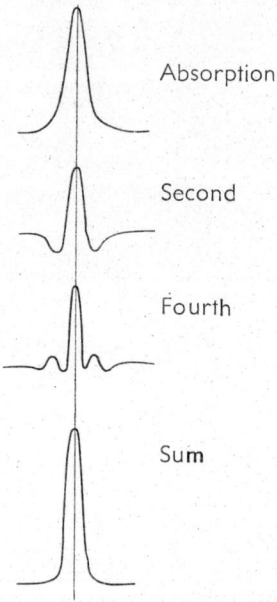

Fig. 2.19 The improvement of resolution by addition of derivative spectra[2,6]

low frequencies without increasing the noise level. In the case of very sharp lines there is another reason for using a superheterodyne system; i.e. saturation. To avoid saturation broadening very low microwave power levels should be used. The field modulation systems do not work well at low power level, because the crystal conversion is very unfavourable there. Superheterodyne spectrometers can be operated at very low microwave power levels without a drop in sensitivity.

Addition of derivative spectra. For improving the resolution of ESR spectra often the second derivative representation is chosen, because the second derivative lines are sharper than absorption or first derivative ones. Based on this fact, a new method has been developed recently[2.6] for identifying the position of spectrum lines. If, for example, a set of even derivatives is taken and the resulting spectra are added with definite amplitude ratios and correct phases, very sharp lines will appear at the place of the original line. Derivative spectra can be taken simultaneously by using a multi-channel lock-in detector system driven with harmonics of the field modulation frequency. In Fig. 2.19a a set of derivative lines is shown to explain why the resolution is improved by addition of such spectra. It can be seen that the side lines of the multiple derivatives are in the opposite phase and thus are cancelled by addition.

The method is a kind of electronic spectrum analyzer which is capable of giving the exact positions (splittings) of not quite resolved complex spectra and thus makes radical structure determination easier.

2.2.6 OPERATION AT UNUSUAL CONDITIONS

Conventional ESR spectrometers are constructed for operation at ordinary conditions, i.e. near atmospheric pressures and at not too low nor too high temperatures. Usually liquid or solid samples can be measured. However, in chemistry, measurements at extreme conditions are often required. For the study of reaction kinetics e.g. ESR spectra are to be taken at the temperature and pressure of the reaction. In the study of photochemical reactions the sample is to be illuminated during the course of the measurement.

Commercial spectrometers can be equipped with various attachments for operation in unusual conditions. Some of them are briefly described below.

Operation at various temperatures. The most frequently used temperature range for ESR measurements in chemistry is 77–400 °K. A variable temperature accessory for this range is a necessity. Sometimes it is advisable to get the temperature to as low as 4.2 °K (the boiling point of helium). A simple liquid helium cooling system is shown in Fig. 2.20. The sample is placed in a rectangular cavity resonator (see

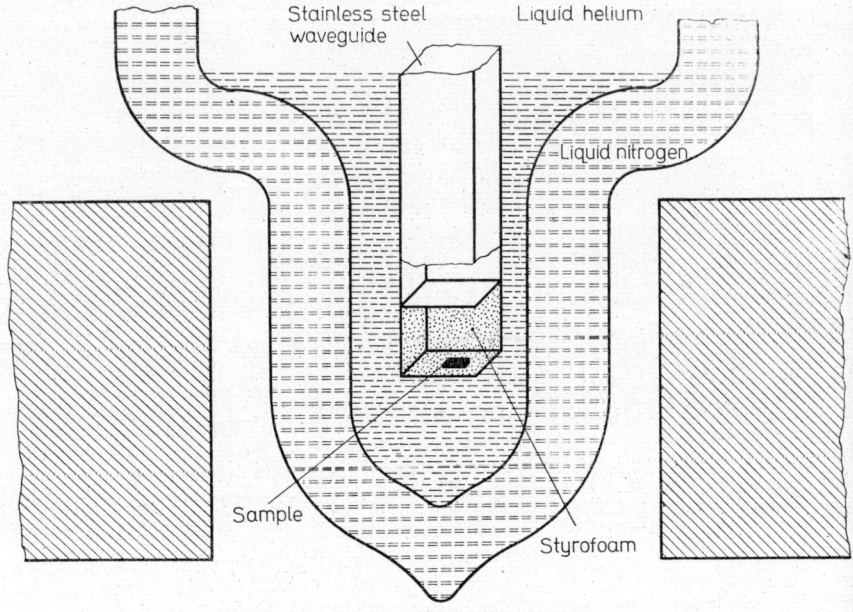

Fig. 2.20 Liquid helium Dewar for ESR

Fig. 2.12a) which is immersed in a liquid helium bath. The helium Dewar is placed in another Dewar vessel containing liquid nitrogen. The waveguide leading to the cavity is made of stainless steel. The cavity and the waveguide are usually filled with styrofoam in order to minimize the vibration caused by the bubbles of the boiling helium. The double Dewar vessel is placed between the pole pieces of the electromagnet. The distance between the pole pieces should be at least 60 mm.

Operation at liquid nitrogen temperatures (77 °K) is much easier. Special quartz Dewar flasks are available for circular cavities. The sample is placed in the Dewar which is filled with liquid nitrogen. The whole set-up is placed in the circular cavity as shown in Fig. 2.21. The outer diameter of the quartz Dewar neck should not exceed 7 mm.

A common variable temperature set-up is shown in Fig. 2.22. It consists of a metallic or glass Dewar flask and a heater for evaporating the liquid nitrogen. The cold nitrogen gas flows through the cavity in a quartz tube placed axially. By varying the heating power the gas

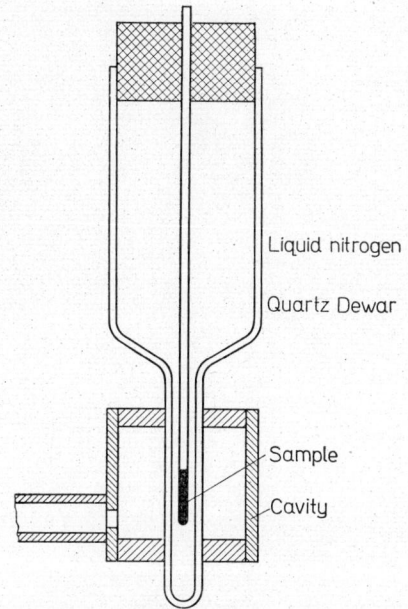

Fig. 2.21 Insertion-type quartz liquid nitrogen Dewar for ESR

pressure is increased within the Dewar and thus the flow rate increases. This results in a lower temperature of the sample. This technique can cover the temperature range of 100–270 °K; the temperature settles down quite quickly and remains constant within ± 1 °C. Some setups use automatic temperature control.

It is usually necessary to let dry nitrogen gas flow through the cavity (outside the quartz tube) to prevent moisture condensation or frost within the cavity which would cause a serious drop in sensitivity.

ESR measurements at elevated temperatures can be made by using the arrangement shown in Fig. 2.23. The sample is heated by hot nitrogen or air flow. The required temperature is reached by setting the flow rate and the heating current. Equilibrium is attained easily and

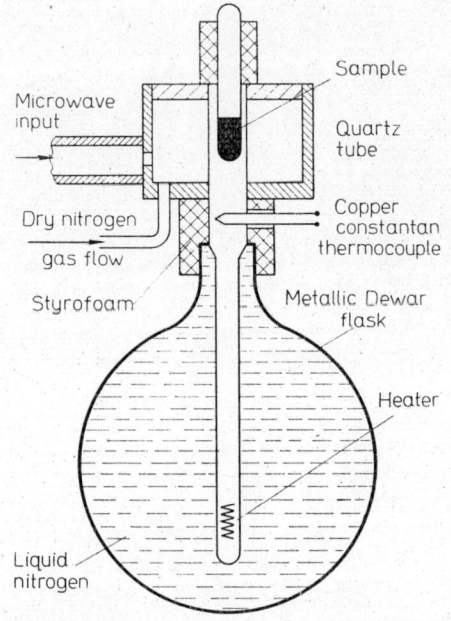

Fig. 2.22 Variable temperature equipment for ESR in the range of 100–300 °K

rapidly. The temperature can be controlled automatically by regulating either the heating current or the gas flow rate. It is possible to measure at temperatures up to 300 °C by simple technique, or even higher, up to 1,000 °C by using special cavities.

Measurement during irradiation. By studying photochemical reactions or photosynthesis in biochemistry, measurements *in situ* are often needed. It is possible to irradiate the samples during the course of the measurement by ultraviolet light, by X-rays and by accelerated electrons. Ultraviolet irradiation is the simplest; it can be done in both rectangular and circular cavities. A possible arrangement is shown in Fig. 2.24. The light of a high-pressure mercury lamp is focused into a hole bored in the side walls of the cavity. The beam

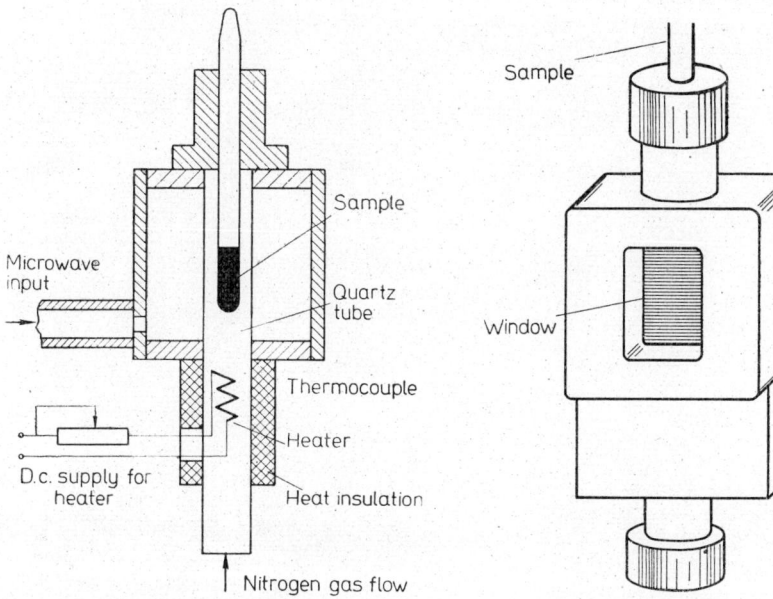

Fig. 2.23 Heating of ESR samples by hot air or nitrogen flow

Fig. 2.24 Cavity resonator for ESR measurements during ultraviolet illumination

can reach the sample through the quartz walls of the heater (cooler) tube and that of the sample tube. Ordinary high-intensity ultraviolet lamps can produce radical concentrations high enough for measurement.

Flow technique. In the study of reactions the lifetime of intermediate radicals is usually too small to be measured by the ordinary 'static' technique. Recently a very powerful dynamic technique has been developed[2.7] to study short-lived intermediate radicals in the liquid state. Instead of using fixed samples the fluid sample is made to flow through the cavity, as shown in Fig. 2.25. The reagents are mixed just before the flow enters the cavity. If the flow rate is high enough in comparison with the lifetime of the radicals formed upon mixing

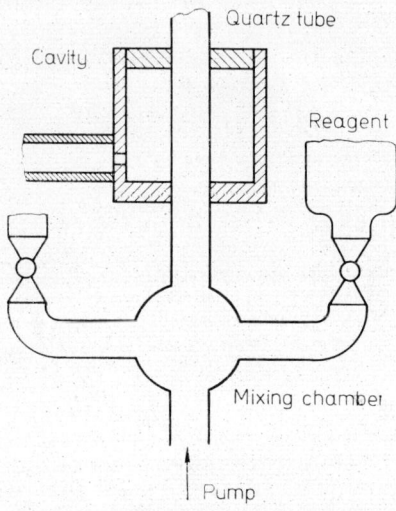

Fig. 2.25 Simplified scheme of the flow technique to study transient radicals[2-7]

the reagents, the effective radical concentration in the cavity can be kept constant. In this way a kind of stationary concentration of short-lived radicals can be achieved within the cavity, and thus spectra can be taken by sweeping slowly at a correspondingly high sensitivity level. Being in the liquid state the resolution is excellent, and owing to the 'steady state' the sensitivity is also quite good.

Measurement during the course of electrolysis. The free radicals formed during the course of electrolysis of organic materials can be measured by using the cavity resonator shown in Fig. 2.26.

The sample is first deoxygenated by nitrogen flow. A small voltage of 2–3 V is applied to the electrodes and the formation of radicals is followed by successive sweeping of the magnetic field. Combined with polarography ESR thus offers an excellent way of studying electrolytic reactions.

The cavity resonator equipped with the electrolytic cell is easy to operate. The temperature can be controlled in the way described above.

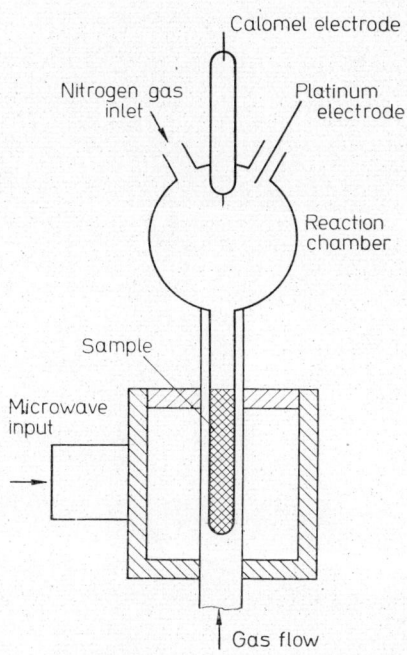

Fig. 2.26 Cell for ESR measurement during the course of electrolysis

ESR of gases. Paramagnetic gases or gaseous radicals can simply be measured by letting the gas flow through the cavity resonator of the spectrometer. Only a standard diffusion pump equipped with the gas inlet accessories is needed for this. It is also possible to take ESR spectra of excited gas molecules. Excitation can be made by using a high-frequency discharge section just before the gas enters the cavity.

2.2.7 ERRORS IN OPERATION

In operating an ESR spectrometer there are some typical errors which can easily be avoided by choosing suitable technical parameters. Some of these are summarized as follows.

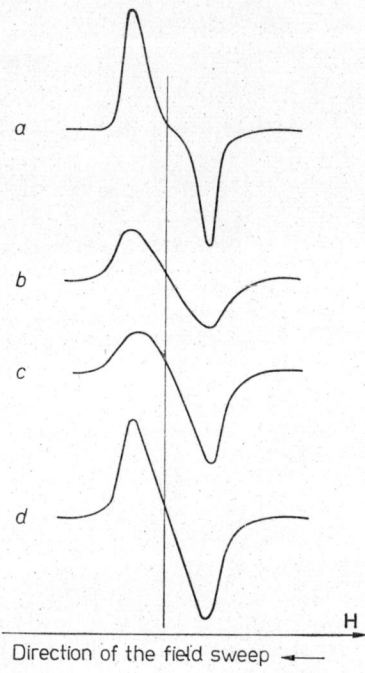

Fig. 2.27 Erroneous operation of an ESR spectrometer.
a — Field modulation is too high; *b* — Field modulation is too low;
c — Scanning rate is too high or response time too long;
d — Correct operation

Field modulation is too high or too low. The amplitude of the field modulation must be in the order of the smallest line width in the spectrum. If field modulation is too high, the lines will be distorted as shown in Fig. 2.27*a*. If it is too low in comparison with the line width, the signal amplitude and thus the sensitivity will be reduced (Fig. 2.27b).

Scanning rate is too high. The scanning rate of the field sweep must be adjusted to the response time of the lock-in detector. If scanning rate is too high for the response time used, the line will be distorted as shown in Fig. 2.27c. The field is swept from high to low fields and thus the low field side of the signal is distorted most. If the response time is too long, the line will be partly integrated.

Sweep rate is given in gauss/min. It can be adjusted either by choosing a wider scanning range or by sweeping faster. It is a common practice to choose scanning rates of the order of $\Delta H/60 t_0$ where ΔH is the width of the narrowest line in the spectrum and t_0 is the response time of the lock-in. If the linewidths are of the order of 1 gauss and the time constant $t_0 = 1$ sec, the sweep rate will be 1 gauss/min. The total time required for scanning a range of 20 gauss would then be 20 min.

Phase of reference signal is not correctly set. The d.c. signal at the output of the lock-in detector depends on the amplitude and phase of the signal with respect to the reference. Correct representation of the spectrum can only be expected if the phase is properly set. This is done by setting the field by hand to the maximum of a resonance line and adjusting the phase shifter until maximum reading is obtained on the recorder. The signal amplitude is shown at the output of the narrow band amplifier by a CRO. Incorrect phase setting would result in a drop in sensitivity and serious distortion of the line.

Microwave power is too high or too low. If microwave power is too high the narrowest lines will begin to saturate and thus resolution and sensitivity are reduced. At too low microwave power level the sensitivity will be too low according to Equation 2.6.

Dielectric loss of sample is too high. This will result in a serious drop in the Q-value of the resonator and a corresponding drop in sensitivity. It is advisable to check the drop in Q-value before measuring high dielectric loss samples, e.g. aqueous solutions. In some special cases the dielectric loss of the samples is continuously changing during the course of the measurement. This is the case when reactions of highly polar materials are studied which result in nonpolar products. The consumption of the polar reagent would in this case increase the Q-factor of the cavity and correspondingly the signal level would be increased too.

2.3 NUCLEAR RESONANCE TECHNIQUE

Nuclear magnetic resonance has become a very important, widely used tool for investigation of chemical structures. Its technique is highly developed. Compact, easy to operate commercial NMR spectrometers are available with a resolution as high as some parts in 10^9. Besides determination of complex chemical structures they are now being used for routine analysis and for the study of such physicochemical problems as rotation of groups, proton exchange phenomena and hydrogen bonding.

The other magnetic resonance technique, nuclear quadrupole resonance (NQR), is mostly used for the study of chemical bonding. Since NQR can only be measured in covalent compounds having nuclei with quadrupole moments, its use in chemistry is rather limited.

2.3.1 PRINCIPLE OF OPERATION

Nuclear magnetic resonance involves induced transitions between nuclear magnetic energy levels. As in electron spin resonance, magnetic levels are created by placing the spin system in a strong homogeneous polarizing magnetic field. As shown in Section 1.3, the separation of the nuclear spin levels is proportional to the polarizing magnetic field **H**. These are the following main differences between ESR and NMR:

1. The separation of the nuclear magnetic spin levels is about 10^{-3} times that of electron spin levels. This factor is due to the difference between electronic and nuclear magnetic moments. As a result of this the resonant frequencies are much lower at the same field **H** and the sensitivity in NMR is reduced.
2. The widths of the NMR lines are much smaller than those of ESR lines. In the liquid state the usual NMR line widths are of the

order of 10^{-3} gauss. This means that NMR spectra can supply much more information than ESR.

3. ESR can only be measured in samples having paramagnetic centres: NMR spectra can be taken of every material having magnetic nuclei. As shown in Table 6.1 the number of magnetic nuclei is so great that practically any compound can be investigated by NMR.

As mentioned in Section 1.3 NMR is used in chemistry as a precise method of measuring local magnetic fields inside the molecules. The magnetic nuclei act as probes for this local magnetic field measurement.

The distribution of the local magnetic fields inside the molecules is directly connected with the molecular structure. Therefore it is irrelevant which magnetic nuclei are used. In the compound to be investigated there are usually several magnetic nuclei, as e.g. proton, fluorine, carbon ^{13}C etc. Taking NMR spectra from all of them will provide so much information that molecular structures can be determined very accurately even in the case of very complex organic compounds.

Methods of detection. For detection of nuclear magnetic resonance spectra the sample is placed in a strong d.c. polarizing magnetic field **H**; it must also be placed in a radio frequency field H_1 in order to induce magnetic dipole transitions between nuclear magnetic spin levels. For illustrating nuclear spin resonance the vector model of Fig. 1.5 is used again. The nuclear moment representing the nuclear spin system is precessing around axis z, the direction of the polarizing field H_0 with an angular frequency:

$$\omega_p = \gamma_p H_0 \qquad 2.16$$

where γ_p is the 'gyromagnetic ratio' of the spin system. According to Equation 1.10, $\gamma_p = (2\pi/h)g_1\mu_1$.

If a circularly polarized radio frequency field of the same frequency ω_p is applied in plane xy, a constant force will act on the nuclear magnet and cause its orientation to be changed (Fig. 2.28). Instead of a circularly polarized field a simple linearly polarized field of a coil can be used in a direction perpendicular to H_0. Such fields will always have the circular polarized component required.

According to this picture, nuclear magnetic resonance means change in orientation of nuclear magnets with respect to the polarizing field

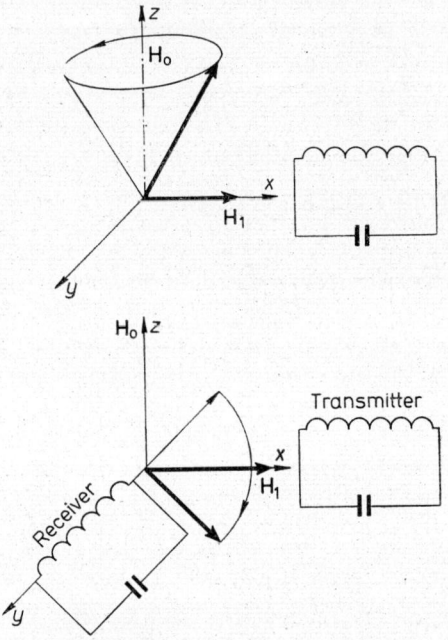

Fig. 2.28 The principle of nuclear magnetic resonance absorption and induction

H_0. This can be detected in two ways. One is to detect the energy absorbed from the radio frequency field, as done in the case of ESR. The other is to pick up the field generated by the nuclear magnet with a receiver coil placed perpendicular to both H_0 and H_1, as shown in Fig. 2.28. At resonance the nuclear magnets changing orientation will induce an r.f. field in the receiver coil which can be amplified and detected in the conventional way. This induction method of detecting nuclear magnetic resonances was introduced by F. Bloch in 1946.[1,5]

A simplified block scheme of an absorption NMR spectrometer is shown in Fig. 2.29. The energy absorbed by the spin system is usually measured by a radio frequency bridge. Depending on the adjustment of the bridge the change in the resistive component (absorption)

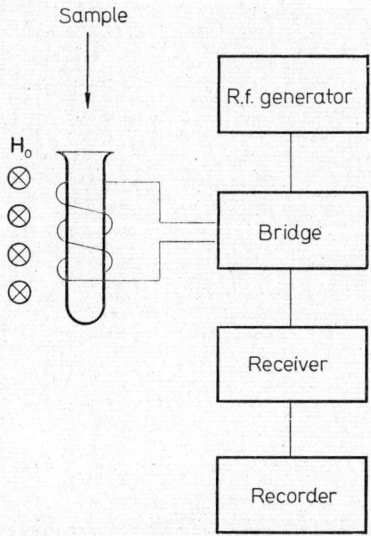

Fig. 2.29 Simplified scheme of an absorption NMR spectrometer

and that of the reactive component (dispersion) can be displayed. The r.f. voltage of the bridge is amplified and detected by a narrow-band lock-in detector described in Section 2.1. The absorption or dispersion spectrum line is recorded by a pen-recorder.

Besides the bridge method there is another very simple way of measuring NMR absorption spectra, the *marginal oscillator method*. This is widely used for magnetic field measurements and for measuring wide-line NMR spectra in the solid state. A simple reactive circuit is given in Fig. 2.30. The sample is placed directly in the coil of the oscillator, which is very loosely coupled. The losses produced in the coil make the collector current change. The d.c. signal appearing

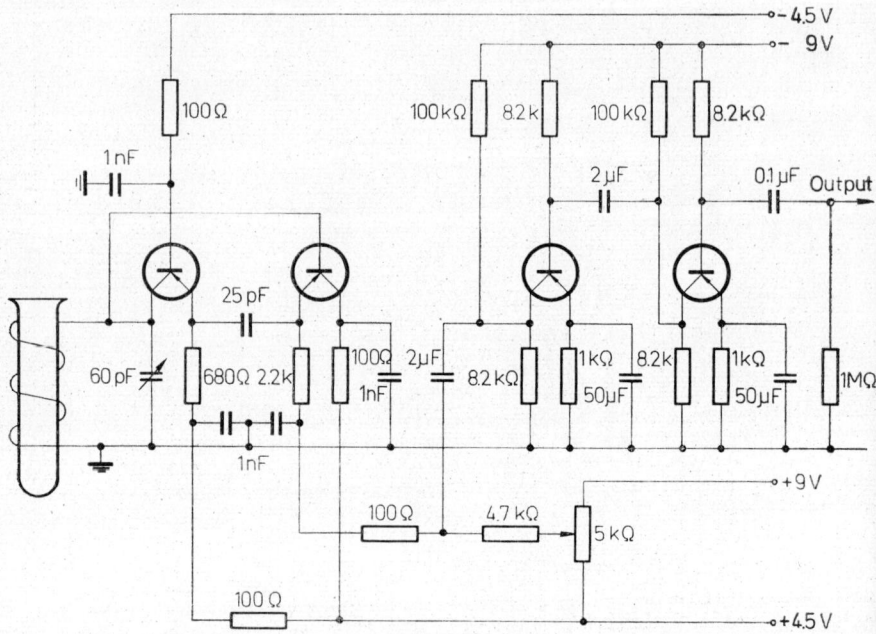

Fig. 2.30 Marginal oscillator for proton resonance (proton field meter) with OC 45 transistors. After Robinson[2.24]

at the collector resistor is thus roughly proportional to the energy loss introduced by the sample at resonance. In this case the magnetic field must be modulated in order to get an a.c. signal. This signal can be either directly displayed on a cathode ray oscillograph or recorded by using exactly the same technique as in ESR, described in Section 1.2 (narrow band amplifier, lock-in detector and recorder).

The simplified scheme of an induction NMR spectrometer is shown in Fig. 2.31. The head contains two coils perpendicular to each other. It is necessary that the receiver coil be completely uncoupled from the transmitter and thus no signal should appear in it provided that the spin system is not at resonance. Decoupling is achieved by additional resistive and reactive paddles which are to be adjusted carefully.

As the main field is passing through a resonance line, an r.f. voltage appears at the receiver coil which is amplified and recorded.

In high resolution NMR highly stabilized quartz crystal-controlled oscillators are used. This is why spectra are usually taken at fixed

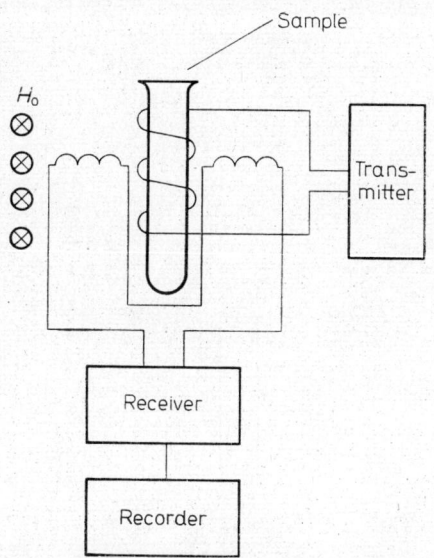

Fig. 2.31 Simplified scheme of an induction NMR spectrometer

frequency by sweeping the magnetic field. Since the chemical shifts are increased by increasing the magnetic field and the operating frequency, it is desirable to work at the highest possible frequencies. The usual bands for proton resonance are 60 Mc/s (15 kG), 100 Mc/s (25 kG) and 220 Mc/s (55 kG). 100 Mc/s is the highest frequency commonly used in NMR as yet because of the technical difficulties in designing high field electromagnets with the very high homogeneity and stability required.

Detection of nuclear quadrupole spectra. Nuclear quadrupole spectroscopy involves magnetic dipole transitions between nuclear electric quadrupole levels split by the inhomogeneous molecular electric fields. Since no external magnetic field is required, these spectrometers are much simpler and much less expensive than NMR spectrometers. They usually consist of marginal oscillators similar to

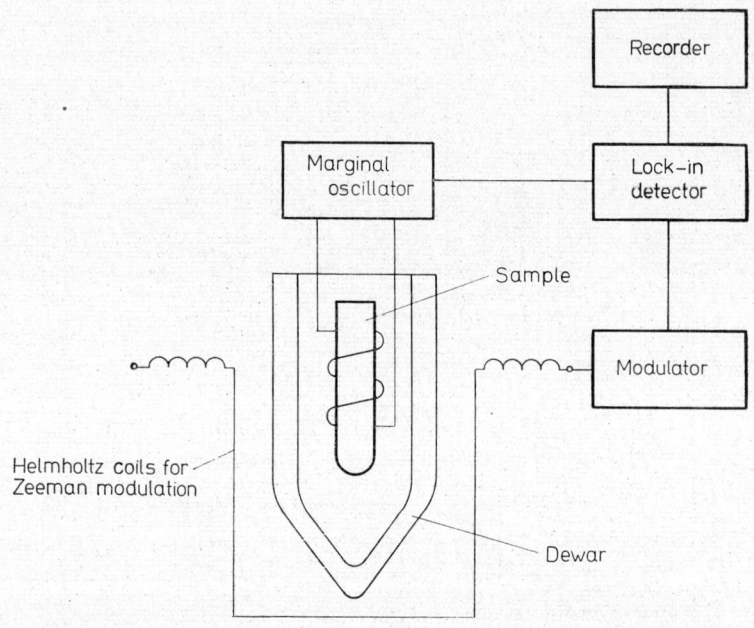

Fig. 2.32 Scheme of a nuclear quadrupole resonance spectrometer

that of Fig. 2.30 with frequencies swept continuously through the spectrum. The block scheme of an NQR spectrometer is shown in Fig. 2.32. Besides the frequency sweep, a small frequency modulation is also introduced by reactance tubes or reactance diodes. This frequency modulation would turn into amplitude modulation when the sweep is passing through an absorption line.

The amplitude-modulated signal appearing at the output of the reactive oscillator is amplified, detected by a lock-in system and recorded in the usual way.

As in the case of microwave molecular spectroscopy (see Section 2.1) it is possible to use Zeeman modulation by using an alternating magnetic field. The quadrupole spectrum lines are split in weak external magnetic fields (Zeeman effect); in higher fields they vanish completely. It is therefore possible to modulate the signal by using alternating magnetic fields and, thus, the narrow-band lock-in detection can be used without introducing frequency modulation.

In most of the laboratories doing NQR work home-made spectrometers are used since the technique is rather simple. The operating frequencies are determined by the nuclear quadrupole moments Q_I and quadrupole coupling constants q (see Section 1.3). Some characteristic frequencies are given in Table 2.1. As seen there, NQR spectrometers are to be operated in the frequency range of 10 Mc/s to 1,000 Mc/s, which would require several set-ups. However only the oscillators must be changed when changing frequency bands; the display system remains the same.

Measurement of chemical shifts and spin–spin couplings. As shown in Section 1.3 the position of the NMR lines are determined by the local magnetic fields which depend on the environment of the nuclei. Local fields can be generated by the diamagnetism and paramagnetism of the electrons surrounding the nuclei and by the magnetic coupling between them. The shifts of the resonance lines caused by the former effects are called *chemical shifts*; the splittings caused by the latter are called *spin–spin splittings* (for detailed definition of these concepts see Sections 4.1.1 and 4.1.2). Technically in both cases small differences in the resonance fields are to be measured accurately. Chemical shifts are usually measured with respect to a standard signal and given as relative values defined as follows

$$\delta = \frac{H_{sa} - H_{st}}{H_{st}} 10^6 = \frac{\nu_{st} - \nu_{sa}}{\nu_{st}} 10^6 \qquad 2.17$$

where ν is the chemical shift in parts per million (ppm); (H_{sa}, ν_{sa}) is the position of the spectrum line in magnetic field and frequency, respectively; (H_{st}, ν_{st}) is the position of the standard line in magnetic field and frequency, respectively.

Technically it is more convenient to measure the shifts in frequency units. Although the spectrum is often displayed as a function of the magnetic field, the splittings of the lines are usually given in terms of c/s. The reason for this is that line shifts can very easily and accurately

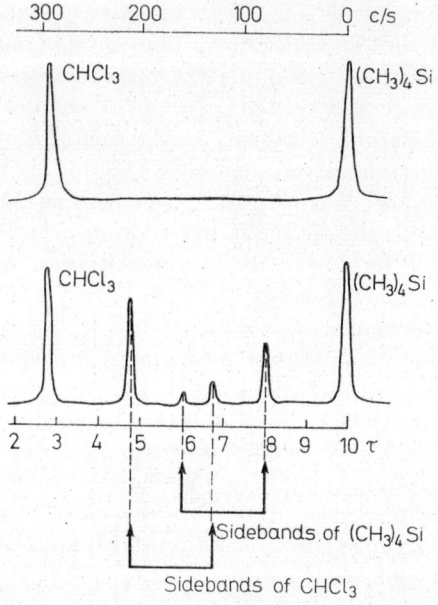

Fig. 2.33 Sideband technique for measuring NMR chemical shifts and spin–spin couplings

be measured by modulating the magnetic field with a frequency of v_m. This would cause the appearance of sideband lines as shown in Fig. 2.33. The position of the sideband lines in the frequency scale is directly equal to the modulation frequency v_m. So it is very easy to calibrate the range in terms of v, or directly in terms of δ

$$\delta = \frac{\Delta v}{v_0} 10^6 \qquad 2.18$$

where ν_0 is the fixed operating frequency, $\Delta\nu$ is the difference in frequency between the standard and signal line, measured directly by the sideband technique.

$\Delta\nu$ can simply be measured by adjusting the modulation frequency so that the first sideband line of the standard coincides with the line to be measured. The frequency reading of the modulator will then be exactly $\Delta\nu$. If the magnetic field sweep is exactly linear, it is possible to precalibrate charts in terms of δ. In this case only the position of the standard is to be adjusted and the calibration controlled from time to time. Most of the commercial NMR spectrometers use precalibrated charts.

Spin–spin splittings are given directly in c/s, as measured by the sideband technique. The position of the NMR lines can be determined in this way with a probable error of about ± 0.1 c/s.

Chemical shifts and spin–spin splittings are relative values. For reference in most cases tetramethylsilane $(CH_3)_4Si$ is used mixed in small amounts with the sample. It exhibits a single sharp line. The shifts from this line are expressed as follows

$$\tau = 10.00 - \delta \qquad 2.19$$

where δ is the relative shift from the $(CH_3)_4Si$ signal defined by Equation 2.18.

Equation 2.19 for defining chemical shifts with respect to the internal standard $(CH_3)_4Si$ is convenient because for all important groups the τ-values are positive. With precalibrated charts usually such τ-values are used. The internal standard line is often used as reference for the magnetic field stabilization system, too. Such highly stabilized spectrometers produce exactly coincident spectra upon repetitive sweeping.

Resolution. The great success of the NMR technique in chemical analysis and structure determination is mainly due to the tremendously improved resolution now obtainable. Fortunately in practice the physical limits of the resolution, the relaxation phenomena, are not too strong. The spin–lattice relaxation time T_1 and the spin–spin relaxation time T_2 defined in Section 1.3 are of the order of seconds in the liquid and gaseous states. This sets a physical limit of resolution at 100 Mc/s of about one part in 10^9.

In viscous solutions or in the solid state natural line widths are increased strongly as a result of the incomplete averaging of the local magnetic fields. The physical limit of the resolution in such systems might be less than a few parts in 10^5.

Magnet. The main technical factor determining the resolution of an NMR spectrometer is the homogeneity of the magnetic field. The resolution requirement of 1 part in 10^8 would require a homogeneity of 0.15 mG at 15 kG over the sample volume. This is technically not possible to realize directly. The inhomogeneity of the most carefully designed magnets is in the order of 10^{-7}, which means that field gradients in the order of milligauss exist in every direction. Further improvement of the resolution can be made either by compensating the field gradients, or averaging them by spinning the sample. In practice both methods are used: the gradients are compensated to the highest possible extent and the sample is spun at a rate high enough in comparison with the broadening caused by the field gradient.

The effect of sample spinning on the NMR line width is illustrated in Fig. 2.34.

A signal broadened and distorted by the field inhomogeneities (Fig. 2.34a) becomes quite sharp upon spinning the sample about an axis perpendicular to the field direction. The sample is spun by an air turbine with a rate of approximately 100 r.p.m.

The effect of sample spinning in averaging out magnetic field inhomogeneities is essentially the same as averaging out anisotropic spin–spin couplings by the rapid tumbling motions of the molecules mentioned in Section 2.2.

Spinning of the sample along an axis would only average the field gradients in a perpendicular plane. The gradient along the spinning axis must be reduced by other means. The field contour along the axis of spinning can be flattened by successively raising the magnetizing current above the operating value for a few minutes, and then returning to the original value. As a result of the inertia of the magnetization this procedure (cycling) can result in a flat field contour. The corresponding spectrum line is sharp and symmetrical.

Magnetic field inhomogeneities along the axis of spinning can also be reduced by small auxiliary coils (called magnetic shims or Golay coils) placed flat against the pole pieces. They consist essentially of concentric loops fed with d.c. current. Upon adjusting the current of the loops a fairly straight field contour can be achieved.

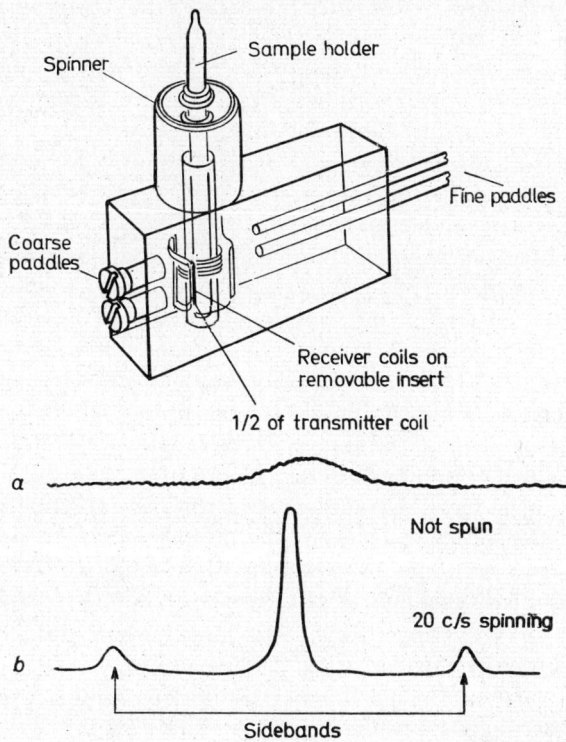

Fig. 2.34 Effect of sample spinning on the resolution of NMR spectra.[2.27]. The NMR head with the spinning sample is also shown (Courtesy of Varian Associates)

The homogeneity of a high resolution NMR magnet once adjusted remains unchanged only when it is operated at constant temperature. Usually there is a heat insulation between the yoke and the coils.

Quadrupole coupling. In some cases resolution and sensitivity of NMR spectra are strongly influenced by the quadrupole energies. As shown in Section 1.4, nuclei having quadrupole moments can be oriented in inhomogeneous electric fields which are present inside molecules bound covalently. In such cases the nuclear magnetic energy levels are perturbed by the quadrupole energies, resulting in either a splitting or a broadening of the nuclear magnetic resonance lines. Therefore it is very difficult to get NMR spectra from nuclei having large quadrupole moments if they are in a molecule with some covalent bonding. The natural chlorine isotopes, e.g. ^{35}Cl, ^{37}Cl, have fairly large nuclear quadrupole moments. Therefore it is very difficult to get ^{35}Cl, ^{37}Cl NMR spectra except in a few cases of pure covalent coupling as for example in HCl.

Nuclear quadrupole interaction can act on the resolution of NMR spectra by altering the spin-lattice relaxation time T_1 and thus make the lines broader. Instead of the spin–spin splittings between nuclei having quadrupole moments, broad bands can be observed because the partial averaging of the spin–spin coupling depends on the lifetime of nuclear magnetic states. If the relaxation time T_1 is long in comparison with the splitting, the spin–spin interaction will not be averaged and the line will be split nicely. By decreasing the lifetime of the magnetic spin states, the spin–spin coupling will be more and more averaged out. This process is called *spin decoupling*. Quadrupole interaction tends to decouple nuclear spins by making spin–lattice relaxation times shorter. Incomplete decoupling (averaging) results in broad lines. As will be shown below, incomplete spin–spin decoupling can be made complete by using a second radio frequency resonant field. This will result in sharp lines without spin–spin splitting.

Line broadening by hindered rotation. Nuclei in different magnetic environments do not always exhibit separate NMR spectrum lines. Rotation of groups can average out the local magnetic fields to result in a single line from each nucleus of the group. The three protons of the methyl group, for example, in most cases exhibit a single sharp line, due to the rapid rotation of the group along the C−C axis. Hindering the rotation would result in incomplete averaging and a corresponding broadening of the line. In the case of rotational isomers the rotation is prevented, and the line is split according to the local fields present at the nuclei. For good resolution a good averaging is required, i.e. the frequency of the rotation must be much higher than the chemical shift. Examples of line broadening due to hindered rotation will be given in Chapter 4.

Chemical exchange. NMR line widths are strongly influenced by inter- or intramolecular exchange of atoms. If the exchange rate is high in comparison with the chemical shifts and spin–spin splittings, the local fields seen from the nucleus of the exchanging atom will be averaged out to result in a single sharp line. A classical example of this is shown in Fig. 5.6 where the effect of the proton exchange rate on the proton resonance spectrum is illustrated for the ethanol–water system. Rapid exchange results in a single sharp line from the water and ethanol hydroxyl protons; the exchanging hydrogens cannot be distinguished. At lower exchange rates, in the neutral ethanol–water system, the averaging of the local fields is incomplete; the resulting hydroxyl line is broad. If no proton exchange is present the lines are sharp again and the spin–spin splitting appears, as in highly purified ethanol in Fig. 5.6a. More examples of chemical exchange are given in Section 5.2.

Rapid chemical exchange and rotation of groups will thus decrease the information obtained from the spectra by averaging out spin–spin splittings and chemical shifts. However, it supplies valuable information about these two important physico-chemical properties of the molecules.

Resolution in the solid state. High resolution NMR spectra cannot be taken in the solid state because of the line broadening due to the anisotropy of the dipole–dipole interaction. However, in some cases, especially with polymers, the compound cannot be dissolved for the NMR measurements. Sometimes, NMR is used to measure parameters characterizing the solid state. In such cases the so-called wide-line NMR technique is used with a resolution of about 1 part in 10^5. Since the line widths chiefly depend on the averaging of the dipole–dipole interactions, it is advisable to take wide-line spectra at the *highest possible temperature* where at least rotation of groups is less hindered.

Wide-line NMR spectrometers are operated by field modulation and lock-in detection similarly to ESR spectrometers. Derivative spectra are recorded. Since the resolution is poor, most of the information is derived from the line shapes. Except for the peak-to-peak width of the derivative line or the width at half amplitude of the maximum of the absorption line the second moments are measured. Second moments are defined as follows

$$M_2 = \int (v - v_0)^2 \, G(v - v_0) \, \mathrm{d}(v - v_0) \qquad 2.20$$

where v is the variable frequency, v_0 is the frequency at resonance, $G(v - v_0)$ is the Gaussian line shape function (see Equation 2.10). Since spectrum lines are represented as a function of magnetic field $\mathbf{H}$ for practical calculations the following expression can be used:

$$M_2 = \frac{-\dfrac{1}{3} \int\limits_{-\infty}^{+\infty} (\mathbf{H}_0 - \mathbf{H})^3 \dfrac{\mathrm{d}G}{\mathrm{d}\mathbf{H}} \, \mathrm{d}\mathbf{H}}{\int\limits_{-\infty}^{+\infty} (\mathbf{H}_0 - \mathbf{H}) \dfrac{\mathrm{d}G}{\mathrm{d}\mathbf{H}} \, \mathrm{d}\mathbf{H}} \qquad 2.21$$

The calculation can be made electronically by using the spectrum accumulator described in a previous section.

Technique of spin–spin decoupling. The most important experimental parameters of NMR spectra from which information about the chemical structure can be derived are the chemical shifts and spin–spin splittings. However, in many cases the lines of spin–spin splitting are mixed with those of chemical shifts due to other groups of nuclei. Some groups cannot even be detected by the usual technique because of the broadening effects due to unresolved spin–spin splittings. The method of spin–spin decoupling offers an excellent way of separating spin–spin coupling from chemical shift. Decoupling involves averaging the local fields caused by the nuclear dipoles by making their lifetimes short. This can be done by introducing a second radio frequency field in order to force transitions in one group of nuclei at the same time when the other group is at resonance. Spin–spin couplings can be decoupled between like and unlike nuclei as well. In the case of unlike nuclei, as e.g. between ^{19}F and ^{1}H, separate radio frequency generators are used. If the main field is swept at the ^{1}H

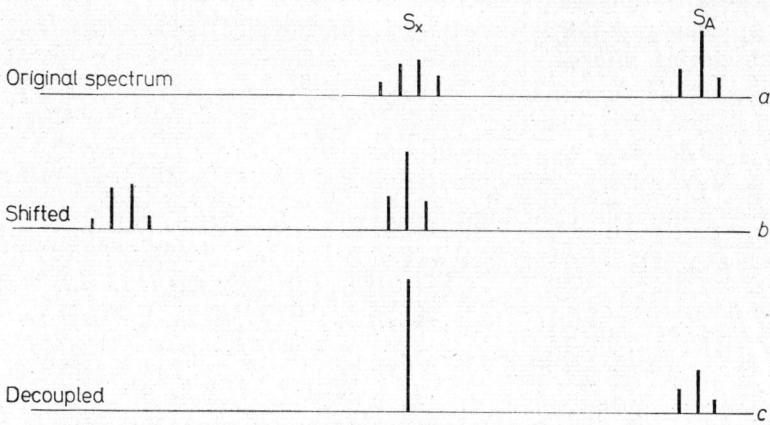

Fig. 2.35 The sideband method of spin–spin decoupling[2.8]

resonance line-group coupled with ^{19}F nuclei, the spin–spin splitting appears unless the second radio frequency field is exactly set to the resonance of the ^{19}F group coupled to the proton group. Upon setting the frequency correctly, the spin–spin splitting of the proton group vanishes and the chemical shifts can be easily measured. Examples of this are given in Chapter 4.

Sideband method. Spin–spin decoupling between like nuclei can be made by using the sidebands of the main r.f. generator. This method was first introduced by Freeman and Whiffen in 1961.[2.8] The procedure is as follows.

1. The r.f. generator of the spectrometer is modulated with a generator of fixed frequency v_1, to produce sideband spectra, as shown in Fig. 2.35a. The centre band in the figure corresponds to two sets of lines with a chemical shift of Δv. The spin–spin coupling of the lines is not shown.

2. Another set of sidebands is produced with a variable frequency generator and the second frequency sideband of group S_2X is made to coincide with the first frequency sideband of group S_1A, as shown in Fig. 2.35b. If the difference in the sidebands $v_0 - v_1$ is set equal to the chemical shift $\varDelta v$ between groups A and X, the spin–spin coupling is averaged and group X is reduced to a single line. In this case, when the field sweep is passing through group X, group A is also in resonance as a result of the second sideband system. Group A can similarly be reduced into a single line by setting the variable generator sideband to group X.

The optimum decoupling frequency shift $v_0 - v_1$ can always be determined experimentally by varying frequency v_1 until the structure collapses. According to Anderson and Freeman[2.9] the true chemical shift δ can be determined by the following equation

$$v_0 - v_1 = \varDelta v - 20 \frac{n^2 J^2}{\varDelta v} \qquad 2.22$$

where n is the number of the components collapsed, J is the spin–spin coupling in c/s, $\varDelta v$ is the chemical shift in frequency in c/s.

Frequency sweep method. High resolution NMR spectrometers of the latest type are stabilized to the internal standard line of $(CH_3)_4Si$. These spectrometers can be run by sweeping the magnetic field or the operating frequency as well.[2.10] For the spin decoupling technique it is more convenient to work at fixed magnetic fields and scan the frequency. In this case the second frequency can be set to a resonance line continuously, and thus the decoupling will be effective throughout the whole sweep. It is also possible to introduce more decoupling frequencies (multiple irradiation) in order to decouple more spin–spin couplings.

2.3.2 SENSITIVITY

NMR spectrometers are much less sensitive than ESR ones. The number of magnetic nuclei in the system is, however, in most cases quite large. Sensitivity is only important in cases where isotopes in low concentration or extremely dilute solutions are to be measured. Analysis and structure determination are based in most cases on proton resonance spectra and the systems usually contain ample protons. Commercial high resolution NMR spectrometers can detect about 10^{-2} mole/litre proton concentrations with a signal-to-noise ratio of about 5 : 1. This sensitivity can be improved by using the spectrum accumulating technique described in a previous section. The upper limit of sensitivity for protons is, for the time being, about 10^{-3} mole/litre. This figure is considerably reduced in other nuclei. In Table 6.1 (in the appendix) the important nuclei are collected with their NMR resonant frequencies and relative sensitivities with respect to that of the proton. Since the sensitivity depends on the operating frequency, the figures given for constant magnetic field differ from those of constant frequency. The reduction of the sensitivity is mainly due to the differences in the nuclear magnetic moments. Nuclei having nuclear spins greater than 1/2 may have quadrupole moments. In these cases part of the sensitivity reduction is caused by the line broadening effect of the nuclear quadrupole interaction.

Integration of spectra. High resolution NMR is widely used in chemistry for quantitative analysis. The number of paramagnetic nuclei giving a specific spectrum is proportional to the intensity of the lines, i.e. the area under the line. So the number of nuclear spins, and of the corresponding chemical groups, can be obtained from the NMR spectra by integration. Most of the up-to-date spectrometers are equipped with a built-in electronic integrator which supplies to the recorder a signal directly proportional to the integral of the lines. The spectrum is usually swept through first by the ordinary technique in order to identify the chemical groups (qualitative analysis). Then it is swept through again by using the electronic integrator and its output voltage is recorded. The voltage drops expressed in volts are proportional to the integrals of the corresponding groups no matter how they are split. A simple example of this is given in Fig. 2.36, where the CH_3 and CH_2 groups in a dilute solution of ethylbenzene are shown. The splitting of groups is the result of the spin–spin coupling between the methyl and methylene protons. The intensity ratio of these groups should be 1.500. The output of the electronic integrator is also recorded on the same chart. According to this the ratio of the voltage drops is $2.02/3.00 = 1.485 \pm 0.03$; this is the intensity ratio of the groups. The deviation from the value of 1.500 is within one per cent.

NMR spectra can also be integrated by the spectrum accumulator

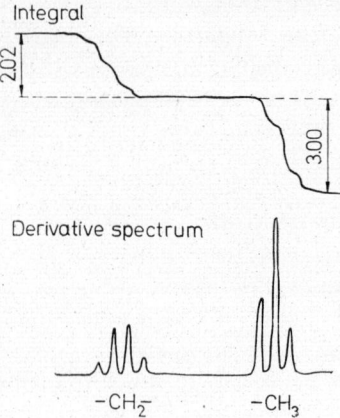

Fig. 2.36 Quantitative analysis by NMR. Integrated spectra of CH_3 and CH_2 groups of ethylbenzene in chloroform solution

described in an earlier section. Further examples on quantitative analysis by NMR are given in Chapter 4.

2.3.3 TYPICAL NMR SPECTROMETER

As an example the block scheme of a typical NMR spectrometer is shown in Fig. 2.37. The operating frequency of the spectrometer is 100 Mc/s for protons and 94.077 Mc/s for ^{19}F nuclei. The highly stabilized oscillator frequency is fed to a bridge-type detector. The signal is fed to a high gain 100 Mc/s (94.077 Mc/s) receiver. For base line stabilization a 4 kc/s field modulation is employed with a corresponding 4 kc/s lock-in detector shown as 'base-line stabilizer' in the figure. The spectrometer is equipped with a spin decoupler having a frequency scaler in order to read the decoupling frequency sidebands directly.

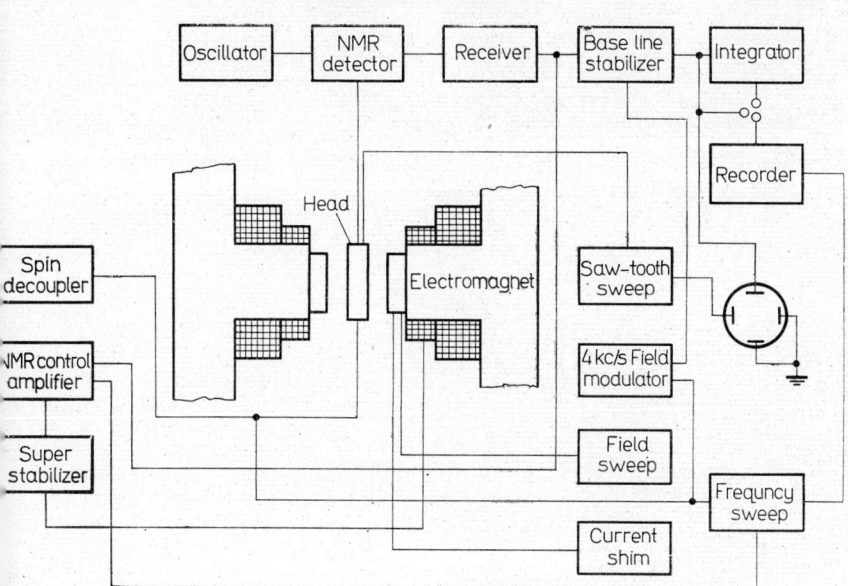

Fig. 2.37 Complete scheme of a typical high resolution NMR spectrometer

The magnetic field is highly stabilized by a current stabilizer and by a superstabilizing system which use the internal standard NMR line (tetramethylsilane) as a reference for stabilization. The chemical shifts can be read directly from precalibrated charts. There is also a frequency counter built in to measure chemical shifts and spin–spin couplings directly by the sideband technique.

2.4 DYNAMIC METHODS

Most of the ESR and NMR work is done at thermal equilibrium, when the population between the spin levels is determined by the Boltzmann distribution described by Equation 1.5. The *degree of polarization* of a spin system is defined as follows

$$p = \frac{\Delta n}{N_0} \qquad 2.23$$

where Δn is the population difference between the spin levels and N_0 is the total concentration of spins. At Boltzmann equilibrium, as follows from Equations 1.2 and 1.5

$$p_0 = \frac{\Delta n_0}{N_0} = \frac{\gamma h \mathbf{H}_0}{4\pi k T} \qquad 2.24$$

where γ is the gyromagnetic factor defined by Equation 2.16, h is Planck's constant, $\mathbf{H}_0$ is the polarizing magnetic field, kT is the thermal energy.

The macroscopic magnetization corresponding to the equilibrium polarization p_0 is

$$M_0 = N_0 \frac{\gamma h}{4\pi} p_0 = \frac{N_0 \mu^2 \mathbf{H}_0}{3kT} \qquad 2.25$$

where μ is the magnetic moment. Equation 2.25 is the well-known Curie law of static magnetization.

The polarization of a spin system p can be altered by various means known as dynamic methods. The intensity of the spin resonance lines corresponding to transition between the spin levels in question is proportional to the degree of polarization p. At positive spin temperatures (see Equation 1.8) p is positive, at saturation it is zero, at negative spin temperatures it is negative.

In this section methods involving operation not at thermal equilibrium are discussed.

2.4.1 ADIABATIC FAST PASSAGE

In ordinary ESR and NMR spectroscopic work, spectra are swept through slowly in comparison with the relaxation times. In the case of NMR, however, in liquids, relaxation times are of the order of seconds. By sweeping through a line faster the spin system has no time to relax to the Boltzmann equilibrium polarization value: a dynamic polarization is reached. It can be shown that at fast passage conditions the population of the spin levels is reversed, the polarization and the spin temperature will be negative. If the spin system is left

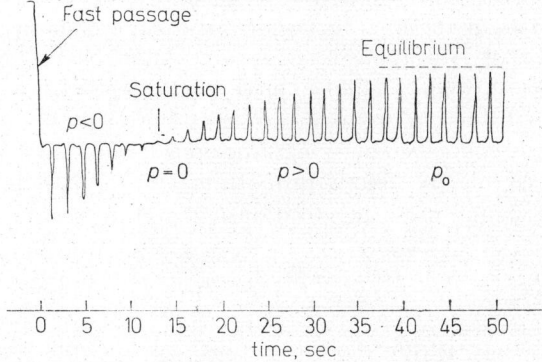

Fig. 2.38 Negative spin polarization produced by adiabatic fast passage. After Drain[2,26]

alone after the fast passage, the Boltzmann equilibrium will be restored. This process can be followed by successively sweeping through the spectrum at slow passage conditions. The intensity of the spectrum lines will then vary in time as polarization $p(t)$ does.

The conditions for adiabatic fast passage are the following:

$$\frac{1}{T_2} \ll \frac{1}{H_1}\left|\frac{dH}{dt}\right| \ll \gamma H_1 \qquad 2.26$$

where T_2 is the spin–spin relaxation time, H_1 is the amplitude of the radio frequency field, H is the main field, γ is the gyromagnetic ratio.

According to Equation 2.25, the condition of adiabatic fast passage can be reached either by increasing the sweep rate dH/dt or by increasing the radio frequency field strength H_1.

An example of adiabatic fast passage is shown in Fig. 2.38. The spin system is protons in benzene. Adiabatic fast passage is done at high radio frequency power level (i.e. high H_1) at point 0. The radio frequency power level is then lowered below the fast passage limit and the proton resonance line is swept through repeatedly. The restoration of Boltzmann equilibrium is clearly seen. First negative signals are obtained corresponding to negative polarization, $p < 0$. In this region the spin system emits radio waves; it is operated as a maser. The amplitude of the line decays to zero in a few minutes, reaching the condition of saturation, where $p = 0$. After that it begins to rise again until Boltzmann equilibrium is reached, $p = p_0$.

It follows from the discussion in Section 1.3 that restoration of equilibrium is controlled by spin–lattice relaxation. The amplitude of the resonance lines is directly proportional to the population difference. The longitudinal magnetization is therefore

$$p \propto M_z = M_0 \left[1 - \exp\left(-\frac{t}{T_1}\right) \right] \qquad 2.27$$

Thus from the experimental curve of Fig. 2.38 the spin–lattice relaxation time T_1 can readily be determined.

The condition of adiabatic fast passage can in principle always be fulfilled providing that the r.f. power and the sweep rate are high enough. However, in ESR the spin–spin relaxation time is of the order of 10^{-5}–10^{-7} sec. Adiabatic fast passage can therefore only be made at very low temperatures, usually below 4.2 °K. In NMR the condition of adiabatic fast passage is easy to fulfil even at room temperature since in the liquid state both relaxation times are long.

Measurement of T_2 by adiabatic fast passage. The spin–spin relaxation time T_2 can be measured by adiabatic fast passage using the dispersion 'u-mode' signals. For NMR dispersion signals at resonance

$$\frac{\mathrm{d}u}{\mathrm{d}t} + \frac{u}{T_2} = 0 \qquad 2.28$$

The method of measurement is the following:

1. Run at adiabatic fast passage conditions until resonance is reached.
2. At resonance the polarizing magnetic field is cut off leaving the system to relax according to Equation 2.28. Recording the u-

mode signal as a function of time provides a curve from which T_2 can easily be calculated.

Besides measuring relaxation times, the method of adiabatic fast passage can be used for detecting very small concentrations of nuclei, e.g. ^{13}C isotopes in natural abundance (1.2%). Running at adiabatic fast passage results in an increase of the signal-to-noise ratio with a corresponding distortion of the lines.

2.4.2 SATURATION

As discussed earlier, saturation means equalization of the population of the energy levels in question. It corresponds to a state where the polarization of the spin system is zero and the corresponding spin temperature is infinite. Saturation is then a dynamic technique, in which the spin system is operated away from equilibrium. If a saturated spin system is left alone it will relax until equilibrium polarization is reached, as shown in the $p > 0$ part of the curve of Fig. 2.38.

Saturation method of measuring T_1. Saturation offers a possibility of measuring the spin–lattice relaxation time T_1, defined by Equation 1.12. As shown in Section 1.3 the population of nuclear spin levels can be equalized, i.e. the state where $p = 0$ can be reached by applying high radio frequency power, i.e. high radio frequency magnetic field inten- sity $\mathbf{H_1}$. Spin–lattice relaxation times of the order of seconds can simply be measured as follows: the spin levels are completely saturated by application of high r.f. power, then the r.f. power is lowered and the NMR resonance line is recorded successively in time. Since the intensity of the spectrum line is proportional to the polarization of the spin system the recorded line amplitudes will be increased until equilibrium polarization is reached (see Fig. 2.38).

If spin–lattice relaxation times are short it is easier to record NMR spectrum lines as a function of the r.f. field intensity $\mathbf{H_1}$. Approaching saturation the intensity begins to decrease as follows

$$\text{Intensity} \propto \frac{1}{1 + (\gamma \mathbf{H_1})^2 T_1 T_2} \qquad 2.29$$

where γ is the gyromagnetic ratio, H_1 is the intensity of the r.f. field. Since the spin–spin relaxation time is unchanged by saturation, from the experimental curve of intensity against H_1, T_1 can be calculated.

2.4.3 SPIN-ECHO TECHNIQUE

Direct measurement of T_1 and T_2 can be made very accurately by the spin-echo method developed by Hahn[2.11] in 1950. The principle of this method is illustrated in Fig. 2.39. The nuclear spin system is represented by a magnetic dipole **M** precessing around the polarizing magnetic field **H** with an angular frequency ω. A rotating coordinate system is used with a z-axis in the direction of **M**. System xyz rotates around axis z with an angular frequency ω. There is a fixed transmitter coil in the X'-, and a receiver coil in the Y'-direction. The situation is similar to the case of nuclear induction discussed previously. The difference is that in the transmitter coil, r.f. pulses with frequency ω are applied in order to make the moment **M** rotate with an angle of 90° in the xy-plane (see Fig. 2.39b). This results in a pulse induced at the receiver coil. However, as a result of spin–spin relaxation, the phase of the individual moments rotating in the xy-plane will be spread, resulting in a decay of the signal at the receiver coil. The next step is to apply a second pulse in the transmitter coil with a duration twice as long as the first one. This reverses the direction of rotation of the moments. Phase dispersion leads to a situation where the moments get in phase again. This after a certain time induces a pulse in the receiver coil called the 'echo-pulse' (see Fig. 2.39b). The amplitude of the echo-pulse depends on the time elapsed between the two driving pulses and also on T_2. By changing the time between the driving pulses a set of echo-pulses can be displayed on the CRO with an envelope characterizing the spin–spin relaxation time T_2. This method essentially gives the absolute value of T_2 in seconds.

For measuring T_1 the second driving pulse is applied for the same duration as the first one. This results in a 90° rotation of the nuclear moment as shown in Fig. 2.39c and thus a voltage corresponding to the longitudinal magnetization M_z is induced in the receiver coil.

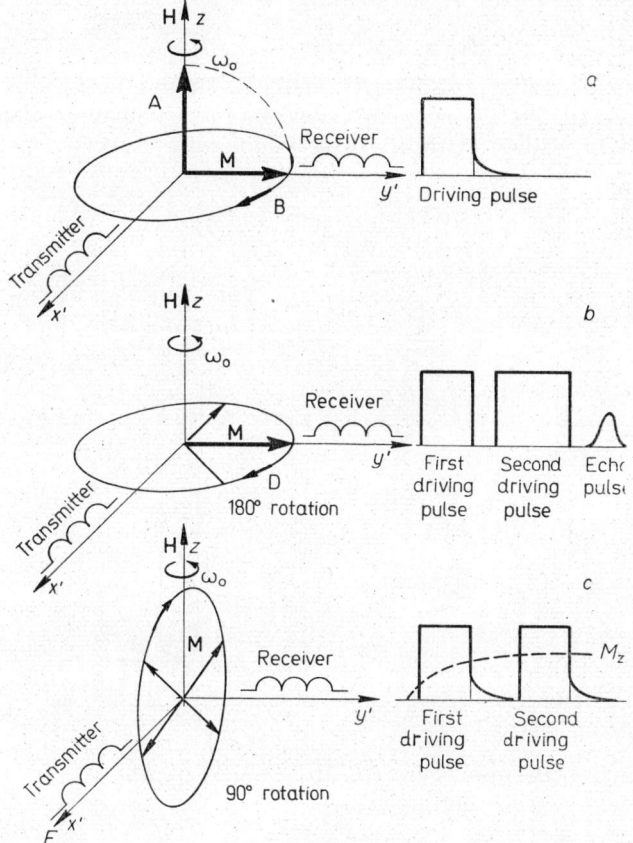

Fig. 2.39 The principle of the spin–echo method

According to Equation 1.12 the time dependence of M_z is determined by the spin–lattice relaxation time T_1.

Spin–echo experiments can also be done in absorption. In the simplest scheme, as in Fig. 2.4, incoherent r.f. pulses are used, i.e. the phase of the pulses is randomly distributed. The r.f. field of the pulses can be expressed as follows

$$\mathbf{H}_1(t) = \sum_k A_k(t) \cos(\omega t + \phi_k) \qquad 2.30$$

where ϕ_k are the phases. With randomly distributed ϕ_k phases the orientation of the r.f. field is also random with respect to the preces-

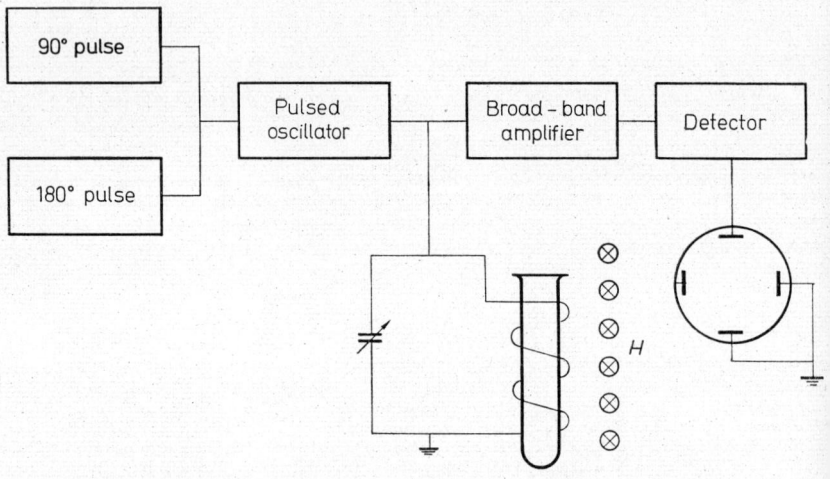

Fig. 2.40 Spin–echo spectrometer. Incoherent pulse system[2.27]

sing spins. If the spin–spin relaxation times are short ($T_2 < 0.1$ sec) the echo envelope will decay as

$$\text{envelope} \propto \exp\left(-\frac{t^3}{4T_2^*}\right) \qquad 2.31$$

where T_2^* is related to the spin–spin relaxation time T_2. The actual measurement is very simple: the sequence of 90° and 180° pulses is started and stopped by a timer and the echo-signals recorded by a simple broad-band amplifier at a cathode-ray oscilloscope. A typical incoherent spin–echo sequence is shown in Fig. 2.41a for protons in ordinary water.

The scheme of coherent r.f. pulses is somewhat more complicated. In this case, phase coherent r.f. pulses are applied, i.e. ϕ_k are the same for each driving pulse. This method produces pure exponential decay of spin–echo pulse amplitudes as shown in Fig. 2.41b

$$\text{envelope} \propto \exp\left(-\frac{t}{T_2^*}\right) \qquad 2.32$$

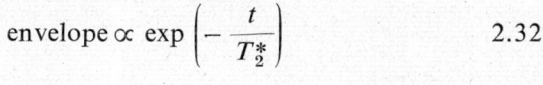

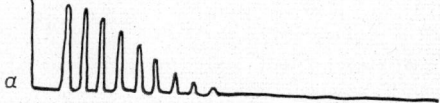

Fig. 2.41 Typical spin–echos in ordinary water.[2.27]
a — Incoherent; b — Coherent

The term T_2^* appearing in Equations 2.31 and 2.32 differs from the spin–spin relaxation time T_2 because of the effects of main field inhomogeneities. Inhomogeneities act the same way as the dipole–dipole interaction among nuclear spins by dephasing the precession of the nuclear magnets. Besides that, during the time elapsed between the 90° and 180° pulses molecular motions would bring the spins into positions where the main field, and thus resonance, is somewhat different; this also results in a spread in the phase of precession. The field inhomogeneity effects can be described by a magnet spin–spin relaxation time T_2^M, which is an instrumental constant. The time T_2 is thus determined by

$$\frac{1}{T_2} = \frac{1}{T_2^*} - \frac{1}{T_2^M} \qquad 2.33$$

The theory and technique of the spin–echo method are fairly well developed. The examples above are only given to illustrate the basic problems. Recently the spin–echo method has been successfully applied to electron spin resonance and nuclear quadrupole resonance as well. For the chemist it has proved to be very useful for studying adsorption problems, chemical exchange, and for investigating intramolecular motions.

2.4.4 DOUBLE RESONANCE METHODS

Chemical compounds usually contain many spin systems as paramagnetic (unpaired) electrons, and paramagnetic nuclei. Nuclei in different local fields form separate spin systems which are connected by various interaction mechanisms. An example of this interaction is the contact interaction existing between electrons and nuclei discussed in Section 1.2, which causes hyperfine splitting of the ESR spectra. Systems of like nuclei being in different chemical surroundings are connected by the electron coupled spin–spin interactions which split nuclear magnetic resonance lines.

Double resonance methods are based on the fact that the polarization state of a spin system can be influenced by altering the polarization of the other with which it is connected. The first experimental observation of this was made by Carver and Slichter[2.12] in 1956 on the basis of the theory developed by A. Overhauser.[2.13] The experiment was made on metallic ^{7}Li magnetic resonance at low magnetic fields. The resonance line was enhanced dramatically as the corresponding electron spin levels were saturated. This experiment shows that the degree of polarization of the ^{7}Li nuclear spin system can be remarkably enhanced by saturating the electron spin levels of the system.

For explaining the dynamic interactions existing between electron spin and nuclear spin systems a simple energy level scheme is shown in Fig. 2.42. It corresponds to paramagnetic electrons (spin = 1/2) and nuclei with spin = 1/2 (e.g. protons). The magnetic energy level scheme determined by the hyperfine interaction consists of four levels. Magnetic dipole transitions can only be induced between levels 1–2 and 3–4 (these are the nuclear magnetic transitions) and between

2–4 and 1–3 (these are the electron spin transitions). Transitions 1–4 and 2–3 are forbidden, i.e. the probability of inducing magnetic dipole transitions between them is small.

In the case of static interaction the ESR spectra of such systems

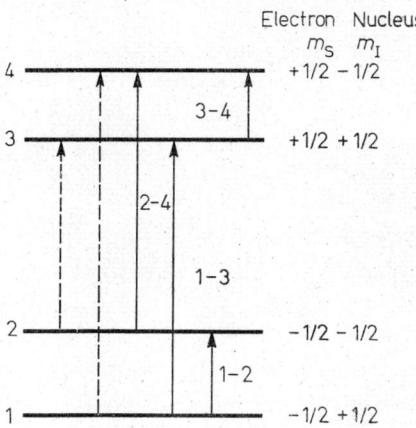

Fig. 2.42 Energy levels of the electron–proton–spin system

exhibit a doublet, according to transitions 2–4 and 1–3. The nuclear magnetic resonance spectrum consists of a single line.

The problem of the dynamic methods is, how will the disturbance of one spin system act on the polarization of the other?

Simple Overhauser effect. In specific spin systems, especially in metals and in some free radicals, the following conditions are fulfilled:

1. The nuclear spin system is not directly connected with the lattice, but through the electron spin system. Thus nuclei cannot relax through ways 2–1 and 4–3 directly.

2. The interaction between the electronic and nuclear spin system makes transition 1–4 allowed, but transition 2–3 remains forbidden.

The double resonance method in such cases involves saturation of transitions 1–3 or 2–4 by a high power radio frequency or microwave field. This will increase the polarization of the nuclear spin system to

$$p_n = \frac{|\gamma_e|}{|\gamma_n|} p_0 \qquad 2.34$$

where p_0 is the equilibrium polarization, γ_e, γ_n are the electronic and nuclear gyromagnetic ratios, respectively.

In the case of protons

$$\left|\frac{\gamma_e}{\gamma_n}\right| = 660$$

upon saturation of the electron spin levels, the intensity of the proton resonance signal should be enhanced by a factor of 660.

Since conditions 1 and 2 are not completely fulfilled the experimental factors are somewhat lower than the theoretical value of 660. As well as in several metals, simple Overhauser effect can be detected in some organic compounds containing stable radicals.

Inverted Overhauser effect. In non-viscous solutions of paramagnetic ions or free radicals, saturation of the electron spin levels results in an inverted polarization of the nuclear levels

$$p_n = -\frac{1}{2} \frac{\gamma_e}{\gamma_n} p_0 \qquad 2.35$$

As the microwave power is increased to saturate the electronic levels, the amplitude of the nuclear resonance line is decreased first, then reversed in phase. In this case the system is operated in maser action.

Saturation of 'forbidden' transitions. In the solid state, especially in solid solutions of paramagnetic centres, the probability of 'forbidden' transitions 1–4 and 2–3 is increased because of the intermixing of the states. In these cases it is possible to saturate states 1–4 or 2–3 by applying high power resonance r.f. fields. The nuclear polarization is

$$p_n = -\frac{\gamma_e}{\gamma_m} p_0 \quad 1\text{--}4 \text{ saturated}$$

2.36

$$p_n = \frac{\gamma_e}{\gamma_m} p_0 \quad 2\text{--}3 \text{ saturated}$$

As shown by Equation 2.36 it is possible to obtain inverted population of the nuclear energy levels, i.e. negative nuclear spin temperature, by saturating one of the 'forbidden' electron spin transitions. This phenomenon is often referred to as 'double effect' or 'the solid state effect'.

Electron nuclear double resonance ENDOR. Feher[2.14] in 1956 introduced a new technique for investigating hyperfine interactions. It is done in the following way:

1. High microwave power is applied to the specimen to bring electron spin transitions 1–3 or 2–4 near saturation. The main field is set on the ESR spectral line. A superheterodyne-type ESR setup is used.

2. A hih-power r.f. field is applied to induce nuclear transitions between levels 3 and 4. This results in a change in the polarization of the electron spin system and a corresponding change in the signal level of the ESR spectrometer.

By sweeping the frequency of the r.f. field, sharp spectral lines appear at the output of the ESR spectrometer.

The block diagram of an ENDOR setup is shown in Fig. 2.43. The ESR spectrum is first taken by a conventional superheterodyne spectrometer. The field is set to a line and the microwave power is increased until the line begins to saturate. Now the second r.f. field is swept by the frequency sweep and the variation of the ESR signal level is recorded as a function of frequency.

The main advantage of this method is that the widths of the double resonance lines are determined by the nuclear relaxation times and thus the resolution is tremendously improved. Theoretically the

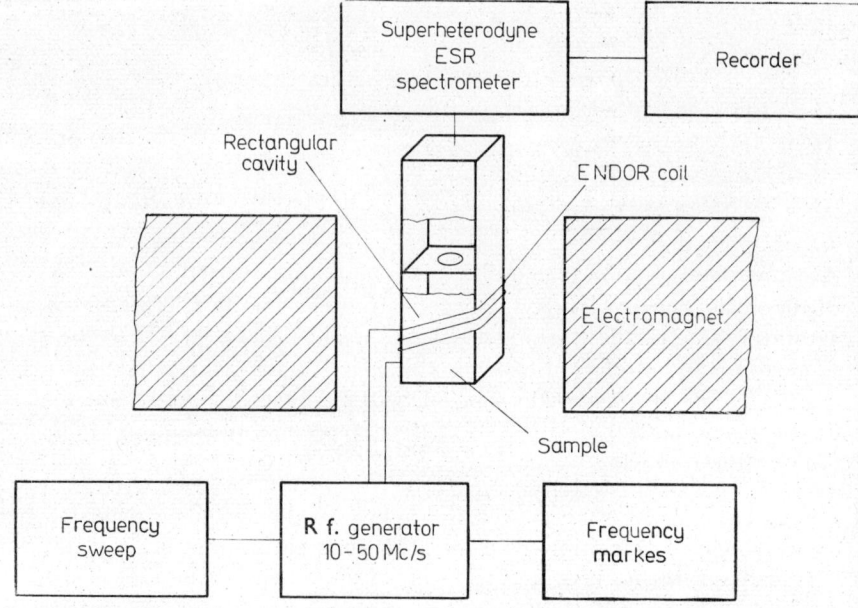

Fig. 2.43 Block diagram of an ENDOR apparatus

resolution can be improved by a factor of 10^4. ENDOR is thus a technique which has the sensitivity of ESR and the resolution of NMR.

The ENDOR technique has been successfully applied to study colour-centres in irradiated alkali halide crystals. Ordinary ESR produces broad unresolved lines from the paramagnetic centers. ENDOR can detect the interaction of the centres with the surrounding nuclei present in the crystal near the defect. Recently the ENDOR technique has been applied to study radicals in organic and inorganic single crystals and also radicals in the liquid state. Examples of this are given in Chapter 3.

Earlier ENDOR measurements have been made at very low temperatures (4.2 °K). By using high r.f. power, however, it is possible to detect ENDOR at higher temperatures, even at room temperature.

In this case, a very high intensity r.f. field is needed. R.f. fields can be applied in the form of pulses and a lock-in technique is used for detection. This results in derivative ENDOR lines. Pulse technique can be applied to separate broad background lines from sharp ones. For this, r.f. pulses of alternating frequency are applied to the specimen and the signal is amplified and detected by the lock-in technique at a frequency which is half the repetition rate of the pulses. This results in derivative ENDOR lines with the broad background line cancelled. Broad background signals often appearing in ENDOR spectra can also be separated by using the spectrum accumulator (computer) described in Section 1.2.

2.4.5 NUCLEAR–NUCLEAR DOUBLE RESONANCES

The principle of double irradiation can be applied if both spin systems are nuclear. As mentioned in Section 2.3, the spin–spin interaction between nuclear spin systems can be decoupled by simultaneous irradiation of both systems, with resonant r.f. fields. The condition for complete decoupling is

$$\gamma \mathbf{H}_2 \gg 2\pi |J_{12}| \qquad 2.37$$

where γ is the gyromagnetic ratio, $\mathbf{H}_2$ is the amplitude of the second radio frequency field, $|J_{12}|$ is the absolute value of the spin–spin coupling constant between spin systems 1 and 2. This method is very useful for determination of chemical shifts, as by successive decoupling, spectra can be made very simple. It is also useful for removing unwanted spin–spin coupling with nuclei having large quadrupole moments. In this way lines broadened by quadrupole effects and thus hidden in the noise level can be detected (see the example in Fig. 4.14 of Chapter 4).

If the second spin system is irradiated with low r.f. field $\mathbf{H}_2$, i.e.

$$\gamma \mathbf{H}_2 \leq 2\pi |J_{12}| \qquad 2.38$$

there will be a transfer of modulation from system 1 to system 2 via the electron-coupled spin–spin interaction. This results in the

114 MICROWAVE STUDY OF CHEMICAL STRUCTURES AND REACTIONS

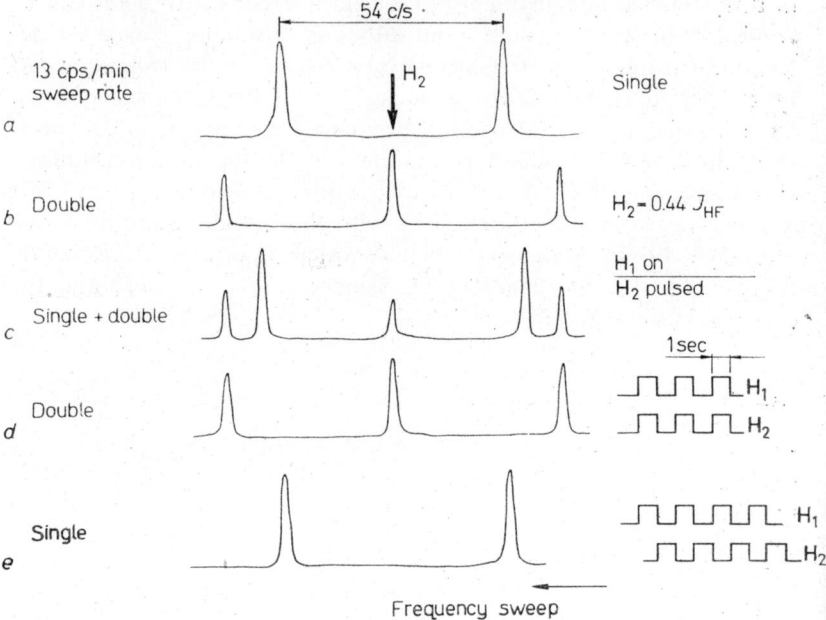

Fig. 2.44 Nuclear–nuclear double resonance in $CHFCl_2$. After Gordon and Baldeschwieler[2.16]

appearance of satellites in those lines of the spectrum which are coupled to the irradiated spin system. This method, developed by Anderson and Freeman in 1962,[2.15] can be used for separating groups in complex spectra which correspond to a single energy level.

In the samples investigated by high resolution NMR, in most cases many spin systems are present. It is possible to use a set of resonance fields $H_2, H_3 \ldots H_i$ simultaneously to decouple the corresponding spin–spin splittings (nuclear–nuclear multiple resonances). Multiple resonance experiments are usually made by the field sweep method mentioned in Section 2.3, for it is more convenient technically and the results are easier to interpret.

Nuclear Overhauser effect. Irradiation of a nuclear spin system results in a redistribution of the energy level populations. The effect of this redistribution on the line intensities is known as nuclear Overhauser effect.[2.16] If a nuclear magnetic resonance signal is observed

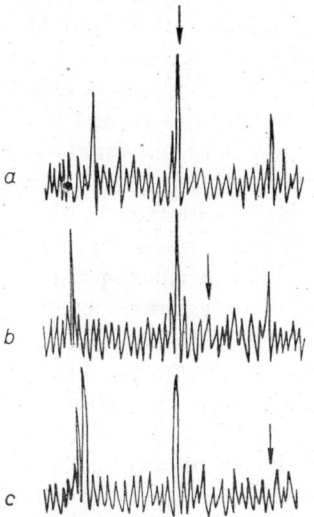

Fig. 2.45 Nuclear Overhauser effect in CHF_2Cl. After Gordon and Baldeschwieler[2.16]

at weak r.f. field H_1 with the simultaneous presence of a second field H_2, the intensity of line 1 should change when 2 is at resonance. These changes can be measured by a pulse method in order to avoid ordinary double resonance features.

The principle of the pulse method for observing nuclear double resonances is illustrated in Fig. 2.44 by the example of $CHFCl_2$. In this molecule the proton resonance spectrum exhibits a doublet with a splitting of 54 c/s as a result of the spin–spin coupling with the ^{19}F nucleus (curve *a*). Irradiation of the system with a constant r.f. field H_2 at the centre of the spectrum results in the appearance of ordinary double resonance pattern (spectrum b). The intensity of the r.f. field in this case is $H_2 = 0.44\ J_{HF}$ where J_{HF} is the proton-fluorine splitting constant. If the sampling field H_1 is on and the second field H_2 is pulsed both the single and double resonance lines appear (spectrum c). Pulsing H_1 and H_2 both in phase results in the disappearance of the single resonance lines (spectrum d). Pulsing H_1 and H_2 both in opposite phase results in disappearance of the double resonance lines. Thus, it is possible to select easily between double and single resonance patterns by pulsing both r.f. fields H_1 and H_2.

Nuclear Overhauser effect as measured by the pulse technique is shown in Fig. 2.45. The sample is CHF_2Cl measured at 40 c/s.[2.16] H_1 and H_2 are pulsed in phase with 0.5 c/s. The intensity of the second field is $H_2 = 0.1\ J_{HF}/\gamma_F$ where γ_F is the fluorine gyromagnetic ratio and J_{HF} is the proton-fluorine splitting constant. Spectrum a corresponds to H_2-irradiation at the centre (see the arrow), spectra b and c are obtained by shifting the H_2-resonance toward the low frequency line. A remarkable intensity change is observed as a result of population redistribution, i.e. nuclear Overhauser effect.

Increasing the amplitude of H_2 does not result in an increase of the effect; at high amplitudes the population distribution is maximalized and the population differences approach thermal equilibrium and so the line intensity changes.

2.4.6 DOUBLE RESONANCE IN MICROWAVE MOLECULAR SPECTROSCOPY

Double resonance in microwave molecular spectroscopy involves changing the intensities of certain lines by using a strong second microwave radiation to alter the population of the energy levels. As an example in Fig. 2.46 a part of the rotational energy levels of the molecule $CH_3C^{35}Cl:CH_2$ is shown at 16,000 Mc/s (transition 1–2) split by the quadrupole interaction with the ^{35}Cl nucleus. Among the many possible lines only the 1/2–1/2, 5/2–7/2, 3/2–3/2 triplet is considered. Upon irradiating the corresponding 0–1 transitions (shown by big arrows in the figure), the population of level 1 is increased and correspondingly the population differences between levels 1 and 2 are decreased. This results in a decrease of the corresponding signal. A set of spectra are shown in Fig. 2.47 after Unland, Weiss and Flygare.[2.17] Spectrum a corresponds to the normal spectrum taken at 16,000 Mc/s without double irradiation. Spectrum b corresponds to the case when

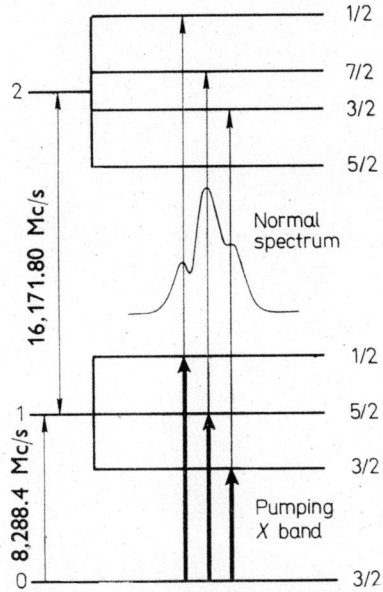

Fig. 2.46 Microwave molecular double resonance energy levels in 2-chloropropene. After Unland, Weiss and Flygare[2.17]

the 3/2–3/2 level of the 0–1 transition is irradiated by a second microwave frequency around 8,000 Mc/s (X band). This 'pumping' results in the decrease of the intensity of the corresponding 3/2–5/2 transition between levels 1 and 2. Similarly pumping transitions 3/2–5/2 and 3/2–1/2 between levels 0 and 1 makes the corresponding 5/2–7/2, and 1/2–1/2 lines of 1–2 transition vanish (spectra c and d).

Generally it is possible to reverse the roles of the pumping and sampling frequencies. Pumping at 16,000 Mc/s of transition 1–2 results in a decrease of the population of level 1, i.e. an increase of the population difference between levels 0 and 1. This results in an increase of the signal intensities of transitions 1–2.

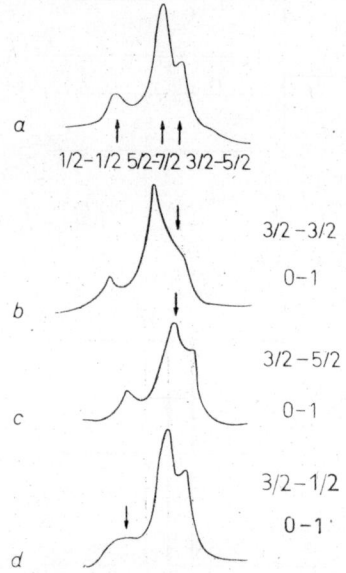

Fig. 2.47 Microwave molecular double resonance in 2-chloropropene. After Unland, Weiss and Flygare[2.17]

As in the case of spin decoupling in NMR it is possible to do double resonance at the same energy levels. Saturation of the central 3/2–5/2 transition between levels 0 and 1 results in a decrease of the intensities of the 3/2–1/2 and 3/2–3/2 side-lines of the same transition. In this experiment one frequency is set to the 3/2–5/2 resonance at high-power level and the side-lines are displayed by sweeping the second frequency through the spectrum.

Technically, microwave double resonance involves sending two microwave signals of different frequencies and different power levels through the absorption cell and detecting only one of them. The first scheme has been realized by Battaglia, Gozzini and Polacco[2.18] for observing double resonance in OCS.

2.5 DIELECTRIC SPECTROSCOPY TECHNIQUE

Dielectric spectroscopy involves electric interactions between radio waves and atomic or molecular electric dipole moments. Instead of sharp lines, dielectric spectra exhibit broad absorption bands as a function of frequency. The main bands are spread over a very wide frequency range. In the high-frequency end, at mm-waves, there is usually a characteristic band associated with vibrational motion of molecules or groups. The classical molecular dipole absorption band (Debye absorption) is observed at the radio frequency region, about 0.1–100 Mc/s. At very low frequencies, below 1 c/s, in some cases, especially in polymers, an extremely slow dielectric polarization can be observed, associated with slow motions of segments or groups in the solid material.

The full range of dielectric spectroscopy extends from about 10^{-4} c/s to 10^{10} c/s. It is evident that this enormously wide range can only be covered by several experimental set-ups differing not only in size but in principle of operation as well. This is why dielectric spectra cannot be taken as easily as ESR or NMR spectra and coherent results in the whole frequency range are rather difficult to obtain. However, in many important cases it is possible to measure dielectric spectra as a function of the temperature instead of frequency, which is technically much easier.

2.5.1 GENERAL PROBLEMS

Radio waves passing through a dielectric medium will interact with the electric dipoles of the material. As a result of this the phase velocity of the wave is decreased and a certain amount of power is transmitted to the system of dipoles and turned into heat. The change in the phase velocity of the wave is described by the refractive index n which is defined as the phase velocity of the wave divided by the freespace velocity c. The absorption of the wave is described by the coefficient α.

At microwave and radio frequencies the complex permittivity is more often used

$$\varepsilon = \varepsilon' - i\varepsilon'' \qquad 2.39$$

Components ε' and ε'' are related to n and α as follows:

$$\varepsilon' = n^2(1 - \alpha^2)$$
$$\varepsilon'' = 2n^2\alpha \qquad 2.40$$

At low frequencies the terms dielectric constant ε_0 and conductivity σ are sometimes used. Their relation with ε' and ε'' is the following

$$\varepsilon' = \varepsilon_0$$
$$\varepsilon'' = \frac{\sigma}{2\pi v} \qquad 2.41$$

where v is the operating frequency, σ is the conductivity in ohm^{-1} cm^{-1} units.

In dielectric spectroscopy, mainly terms ε' and ε'' are used. The total power absorbed from the radio wave is

$$W = \frac{v}{4} E_0^2 \varepsilon'' \qquad 2.42$$

where v is the operating frequency, and E_0 is the amplitude of the dielectric field of the wave.

Plotting values of ε'' against frequency the absorption spectrum is obtained. $\varepsilon'(v)$ is referred to as the dispersion spectrum. In some cases absorption spectra are given in terms of loss angle tangent

$$\tan \delta = \frac{\varepsilon''}{\varepsilon'} \qquad 2.43$$

Technically it is sometimes more convenient to express dielectric absorption and dispersion directly in terms of complex admittance of the capacitor containing the sample

$$Y = G + iB \qquad 2.44$$

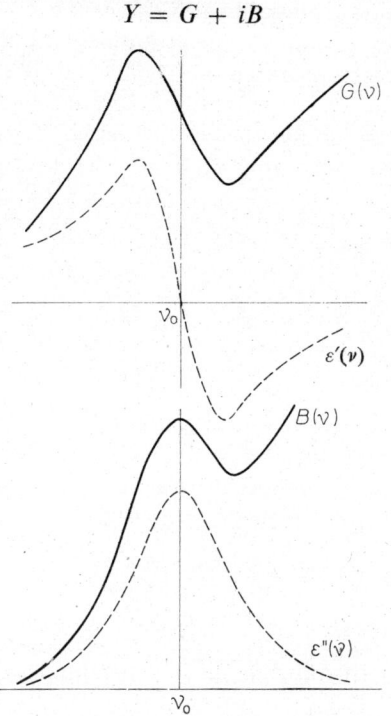

Fig. 2.48 Dielectric spectra in permittivity and admittance representation

where G is the electrical conductance of the material, B is the capacitive admittance

$$B = 2\pi\nu C \qquad 2.45$$

where ν is the frequency, C is the capacity of the sample. The shapes of the spectral bands in $G(\nu)$ and $B(\nu)$ representation are different from those in $\varepsilon'(\nu)$ and $\varepsilon''(\nu)$ representation (Fig. 2.48).

2.5.2 MOLECULAR RELAXATION TIMES

A system of electric dipoles oriented by an electric field will seek equilibrium if the field is switched off. The process of reaching equilibrium (random) orientations is described by the molecular relaxation time τ. The sample usually contains a set of dipoles in different chemical or physical surroundings resulting in a whole set of relaxation times. τ is an average value.

According to the simple theory of Debye[2.19] the components of complex permittivity are simply related to the dipole relaxation time τ

$$\varepsilon' = \varepsilon_\infty + \frac{\varepsilon_0 - \varepsilon_\infty}{1 + \omega^2 \tau^2}$$

2.46

$$\varepsilon'' = (\varepsilon - \varepsilon_\infty) \frac{\omega \tau}{1 + \omega^2 \tau^2}$$

ε_0 is here the static dielectric constant of the material, ε_∞ is the permittivity at very high (optical) frequencies, $\omega = 2\pi$ frequency.

The temperature dependence of relaxation time τ is given according to the theory of Bauer–Fröhlich[2.20] as follows

$$\tau = \tau_0 \exp\left(\frac{E_d}{kT}\right)$$

2.47

where E_d is the activation energy of the dipole orientations, factor τ_0 is expressed as

$$\tau_0 = \frac{h}{kT} \exp\left(\frac{-\Delta S}{k}\right)$$

2.48

Here ΔS is the entropy change of the process. It can be shown that the temperature dependence of the frequency corresponding to absorption maximum is

$$\nu_{max} = \frac{1}{2\pi \tau_0} \frac{\varepsilon_\infty + 2}{\varepsilon_0 + 2} \exp\left(-\frac{E_d}{kT}\right)$$

2.49

Increasing the temperature should thus shift the dipole absorption bands to higher frequencies.

As follows from Equations 2.46 and 2.47 it is possible to display dielectric absorption bands as a function of frequency at fixed tem-

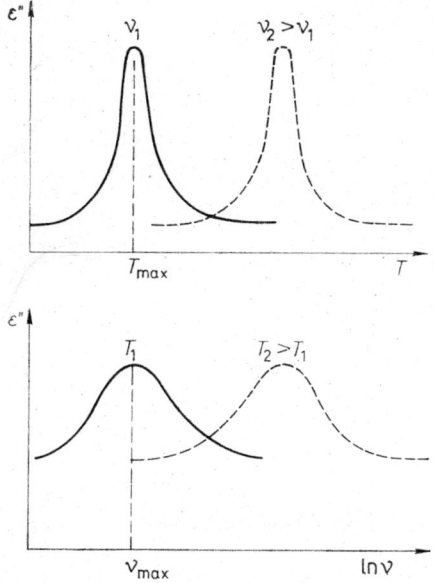

Fig. 2.49 Illustration of dielectric spectrum bands

perature or as a function of temperature at fixed frequency. The corresponding spectral bands are shown in Fig. 2.49. From the measured shift of the absorption frequency as a function of temperature the activation energy E_a can be determined from Equation 2.47.

In most cases it is more convenient to measure dielectric absorption spectra as a function of temperature. Using this representation variations caused by phase transitions can be observed directly. Examples on this are given in Section 4.5.

Cole–Cole diagram. According to Equations 2.46 there is a definite relationship between the components of the complex permittivity

$$\left(\varepsilon' - \frac{\varepsilon_0 + \varepsilon_\infty}{2}\right)^2 + \varepsilon''^2 = \left(\frac{\varepsilon_0 - \varepsilon_\infty}{2}\right)^2 \qquad 2.50$$

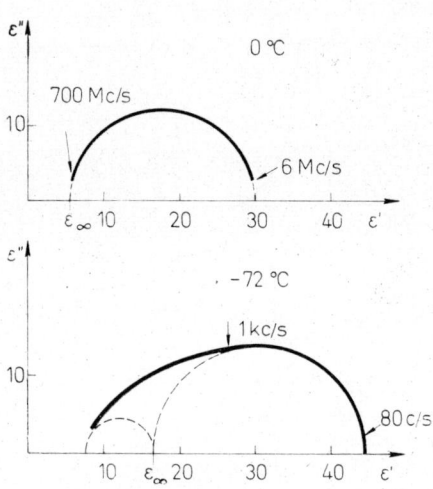

Fig. 2.50 The Cole–Cole plot of methylpentanediol. After Moriamez and Arnoult[2.21]

Thus by plotting ε' against ε'' at various frequencies a semicircle should be obtained, provided that the process can be described by the single relaxation time τ. Deviations from the arc-plot suggest that multiple relaxation processes are present.

Instead of frequency, temperature can also be used as a parameter for displaying Cole–Cole diagrams.

An example of Cole–Cole representation of dielectric data is shown in Fig. 2.50. Measured values of ε' and ε'' are plotted at various frequencies. The material is methylpentanediol.[2.21] The plot taken at 0 °C shows no deviation from Equation 2.50; a nice semicircle is obtained, indicating that the dipole relaxation process can fairly well be described by a single relaxation time. At lower temperatures an appreciable deviation from the semicircle is observed. The experimental plot can be reconstructed by supposing that two entirely different relaxation processes are present with separate relaxation times and corresponding separate Cole–Cole circles (dotted lines in Fig. 2.50).

2.5.3 MEASUREMENT AT INTERMEDIATE FREQUENCIES

There are many types of commercial devices for measuring ε' and ε'' at intermediate frequencies. The frequency range covered by these devices is about 10 c/s–300 Mc/s. For each value of ε' and ε'' a separate measurement is needed, the spectrum usually cannot be recorded automatically. These point-by-point measurements are rather tedious, for displaying a spectrum band at least 20–30 experimental points are needed. It is always advisable to repeat the measurement over a certain temperature range, which would increase the total number of measurements up to about 200 for each material.

According to the principle of operation the devices can be divided into two main groups as follows.

1. *Bridge methods.* The cell containing the sample causes a resistive and capacitive imbalance in a radio frequency bridge. By restoring the balance with known resistances and capacitances the complex permittivity can easily be calculated.

2. *Resonance methods* (Q-meters). The sample cell is a part of a resonant circuit. Upon introducing the sample the resonance frequency and the Q-factor of the circuit are changed. From the measured resonance frequency shifts and Q-values the complex permittivity can be calculated.

Most of the commercial bridges or Q-meters are directly calibrated in ε' and tan δ or these values can be easily calculated from the measured parameters. Details of these techniques will not be discussed here.

Admittance comparators. At intermediate frequencies the comparator method of Fig. 2.51 can be used for measuring complex admittances Y, as a function of the frequency. The device is fed by a sinusoidal r.f. generator. The admittances of arms 1 and 2 are compared by a sensitive receiver. The admittance of the reference arm is determined by resistances R_2 and R_0 and capacitor C_0. By adjusting

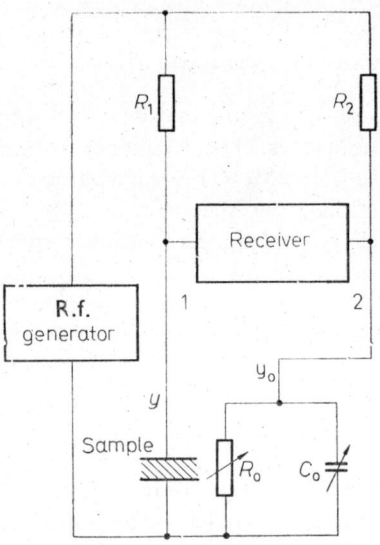

Fig. 2.51 Scheme of an admittance comparator

R_0 and C_0 it is possible to balance the system to zero amplitude at the receiver. In this case the admittance of the sample arm is

$$Y = G + iB = \frac{R_1}{R_2} Y_0 \qquad 2.51$$

where Y_0 is the complex admittance of the reference arm

$$Y_0 = \frac{1}{R_0} - i2\pi \nu C_0 \qquad 2.52$$

EXPERIMENTAL METHODS

Thus, by reading the resistance and capacitance of the variable reference elements upon adjusting zero level at the receiver the admittance components G and B can readily be obtained. Values of ε' and ε'' can be calculated from the measured values of G and B as follows

$$\varepsilon'' = \frac{G}{2\pi vC}$$

$$\varepsilon' = \frac{B - 2\pi vC_0}{C}$$

2.53

where C_0 is the inactive capacity of the sample, i.e. the stray capacities and the capacity of the cable leading to the sample holder, C is the active capacity of the sample and v is the operating frequency.

The principle of admittance comparison can be used to design automatic dielectric spectrometers capable of recording the variation of G and B as a function of the frequency. Automatic admittance comparators use variable capacity elements (vacuum tubes or varactor diodes) in order to balance the system electronically. The scheme is given in Fig. 2.52. The principle of operation is based on the fact that the phase of the signal appearing at the receiver of the comparator shown in

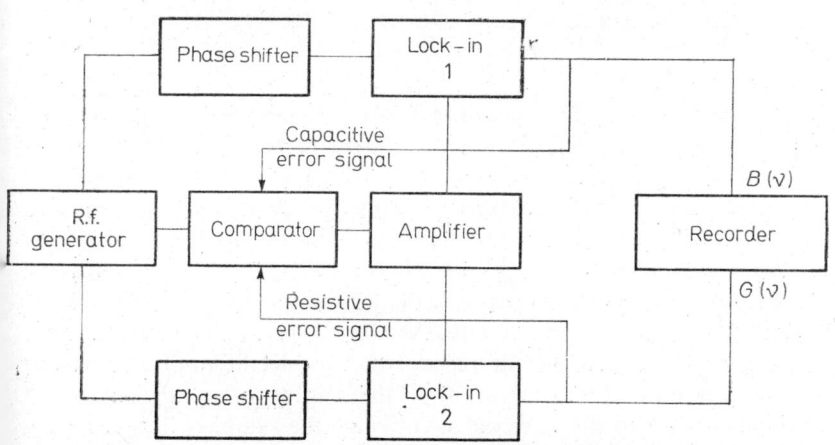

Fig. 2.52 Block scheme of an automatic admittance comparator

Fig. 2.52 depends on what kind of unbalance exists, resistive or capacitive. The unbalance signal is fed simultaneously to two lock-in detectors with reference signals shifted by 90° from each other. In this way one of the lock-in detectors will be sensitive to resistive unbalances, the other to capacitive ones. The d.c. signal of the lock-in detectors can thus be used as an error signal for balancing the comparator. If the phases of the lock-in detectors are correctly set the system will always keep balance even if the driving frequency is changing. The output signals of the lock-in detectors are proportional to the amount of unbalance, i.e. to the variation of G and B as a function of the frequency. These signals can be recorded directly.

Since dielectric absorption bands are rather wide, the frequency of the r.f. generator should be swept through a wide range. It is absolutely necessary that the relative phases of the lock-in detectors should not change during the course of the measurement. The signal amplifier should also be phase-correct over the whole frequency range. Automatic admittance comparators can be used at intermediate frequencies between 1 kc/s and 100 Mc/s. It is not possible to cover the whole range by a single apparatus; about three of them are needed, one for the low frequency band (1–100 kc/s), one for the radio frequency band (0.1–50 Mc/s) and one for ultra-high frequencies (50–100 Mc/s).

2.5.4 MEASUREMENT AT MICROWAVES

The difficulties in measuring dielectric absorption bands are incveased at microwave frequencies. As shown in Section 2.1 microwave bands are much narrower than radio frequency bands covered by a single technique. Such absorption bands can practically be measured only point by point, with the disadvantages mentioned above.

Another difficulty arising in evaluating ε' and ε'' from microwave measurements is that dimensions of the samples are of the same order as the operating wavelength. As a result of the multiple reflexion of the wave between the interfaces of the sample, the amplitude attenuation will not be related simply to the absorption coefficient α and the phase shift of the wave passed through the sample will be dependent on sample thickness. Owing to these difficulties there seems to be no possibility of displaying dielectric absorption spectra in the microwave region by direct recording, as done at intermediate frequencies. Nevertheless, it is always possible to work at a fixed frequency by measuring ε' and ε'' as a function of the temperature.

This will provide valuable information about changes in the physical structure of the material.

A large number of methods have been developed for measuring ε' and ε'' in the microwave region. Only the basic principles will be mentioned here.

Resonance methods. Cavity resonators can be used for measuring complex permittivities, especially for low loss materials. Upon introducing the sample the resonance frequency of the cavity will be shifted as a result of change in the wavelength inside the dielectric material. The change in the resonant frequency is thus directly connected with the dispersive part ε' of the complex permittivity. Since the sample introduces losses in the cavity, its Q-factor will be decreased correspondingly to an increase of the cavity resonance line width. The calculation of ε' and ε'' from the measured cavity resonance curve is extremely easy in gaseous samples or vapours, for the cavity is completely filled with the sample and thus no multiple reflexion can take place. An illustrative example of the change of a cavity resonance curve upon introducing gaseous samples is shown in Fig. 2.53. The empty cavity exhibits a rather sharp resonance line at about 18,000 Mc/s (1.6 cm wavelength). The half-width of the line is measured by means of frequency markers obtained by mixing the main frequency with the frequency of a local oscillator modulated by 2.5 Mc/s. The frequency of the microwave generator is swept through the cavity resonance and the line is displayed on a cathode-ray oscilloscope (CRO). This swept frequency is mixed with a fixed-frequency microwave generator to produce intermediate frequencies varying from zero beat to several Mc/s as the frequency is swept. This changing intermediate frequency is mixed again with a variable frequency generator in order to obtain sharp beat signals at the oscilloscope. In the case shown in Fig. 2.53 the variable frequency generator was set to 5 Mc/s; each harmonic of this frequency produces a zero beat with the changing intermediate frequency. The splitting of the marker signals is caused by the 200 kc/s selective amplifier used after r.f. mixing. Changing the frequency of the second microwave generator would shift the markers without changing their relative distance. Changing the frequency of the variable generator would change the distance between marker lines. In

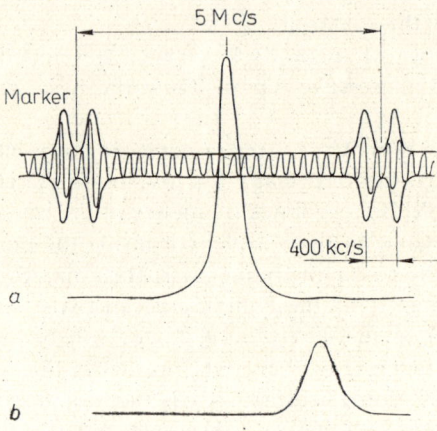

Fig. 2.53 Shift of a microwave cavity resonance curve upon introducing dielectric materials. Sample; water vapour. Frequency 16 Gc/s

this way frequency differences can be easily and accurately measured.

As shown in Fig. 2.53b the cavity resonance line is shifted appreciably upon introducing the sample and gets broader.

According to this principle there are many possible technical ways of measuring ε' and ε'' by using microwave cavities. This method is useful for gases and vapours. It can also be used for low loss liquids or solids, although the evaluation of ε' and ε'' from the experimentally measured frequency shifts and Q-changes is more complicated. For high loss liquids or solids this method cannot be used, because the cavity resonance curve gets too broad and its position cannot be located accurately.

Transmission line methods. For measuring high loss materials simple transmission line methods can be used. The system consists of a microwave generator, a standing wave indicator and a sample holder with a variable shorting plunger. At lower frequencies (1–3 Gc/s) coaxial waveguides are used with cylindrical samples. At higher frequencies (3–50 Gc/s) hollow rectangular or circular standard waveguides are used, with disc-shaped samples. Evaluation of ε' and ε'' from the measured data is not very simple. The error is dependent on the sample thickness, which can only be chosen correctly when the approximate values of ε' and ε'' are already known. Thus, generally, the measurements have to be repeated at different sample thickness.

2.5.5 DIELECTRIC SPECTROSCOPY AT VERY LOW FREQUENCIES

The method of investigating very slow dielectric polarization was first introduced by Hamon[2.22] in 1952. The method involves analysis of transient currents flowing across the specimen as a response to voltage steps. The principle of the measurement is shown in Fig. 2.54. The sample is placed between condenser plates and the current response to voltage steps is recorded. Voltage steps can be regarded as superpositions of frequencies from zero up to high order components distributed continuously. Thus the measured current response curves can be transformed to frequency responses by using the method of Fourier transformation. The current response curves are found to be of the following form

$$I(t) = \beta t^{-m} \qquad 2.54$$

where β and m are experimental constants, t is the time, I is the current flowing across the specimen under the action of a voltage drop.

It can be shown by Fourier transformation that the $I(t)$ response function for unit voltage drop and unit cell capacity can be expressed as follows

$$I(t) = \frac{1}{\pi} \int_0^\infty \varepsilon(t) \exp(2\pi i v t) \, dv \qquad 2.55$$

where ε is the complex permittivity of the sample.

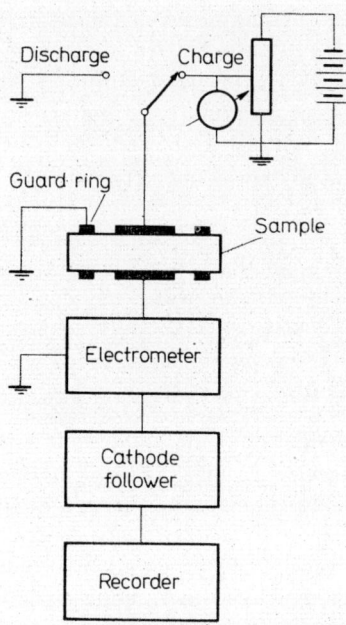

Fig. 2.54 Apparatus for measuring slow dielectric polarization. After Hamon[2.22]

The imaginary part of the complex permittivity can be expressed as a function of the frequency by the following transformation formulae:

$$\varepsilon''(v) = \frac{t}{\rho(t)} 1.8 \times 10^{13} \qquad 2.56$$

$$v = \frac{0.63}{2\pi t}$$

where $\rho(t)$ is the apparent resistivity of the sample which can be calculated directly from $I(t)$. Equations 2.56 hold only if the

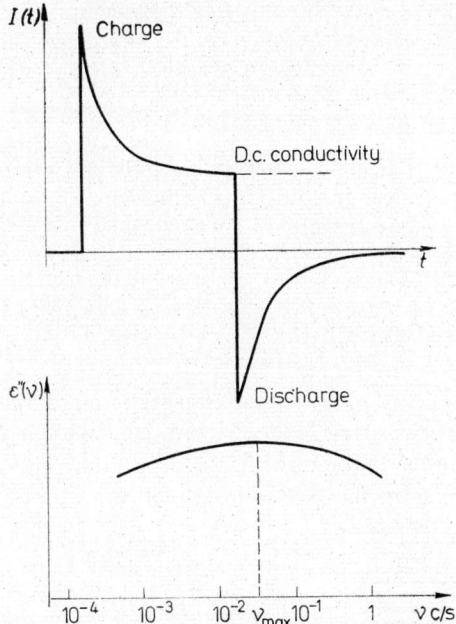

Fig. 2.55 Typical charge–discharge curves in poly(methyl methacrylate)

exponent m in Equation 2.54 is between 0.3 and 1.2.

Using Equations 2.56 the experimental $I(t)$ or $\rho(t)$ curve can be transformed into $\varepsilon''(v)$, i.e. into the dielectric absorption spectrum at the frequency band defined by the voltage step. The transformation can only be used in the region where Equation 2.56 holds. If it does not, other transformation formulae should be used (cf. Williams[2.23]).

A typical charge and discharge curve is shown in Fig. 2.55. By switching the voltage on first the current rises sharply, corresponding to the sudden change in the electric field, then decays down to an equilibrium current corresponding to the d.c. conductivity of the material. The time needed for reaching equilibrium varies from tenths of a second to hours. Upon removing the voltage and short-circuiting the sample with the electronic ammeter the discharge curve can be recorded.

Using Equations 2.56 the $\varepsilon''(v)$ spectrum can be obtained. The sample is poly(methyl methacrylate). As can be seen from the transformed $\varepsilon''(v)$ curve (Fig. 2.55) there is a distinct absorption band at very low frequencies in this polymer indicating highly hindered orientation of large polar groups.

2.5.6 MEASUREMENT OF D.C. CONDUCTIVITIES

The continuous transfer of electric charges in matter can be measured by the straightforward method of measuring the current flow across the specimen under the action of a voltage. The conductivity of the sample is defined by the following equation

$$\sigma = \frac{d}{A} \cdot \frac{I}{V} \qquad 2.57$$

where σ is the conductivity in $ohm^{-1}\ cm^{-1}$, d is the thickness of the sample, in cm, I is the current, V is the applied voltage in volts, A is the cross-sectional area of the sample.

Conductivities are generally constant only in a limited region of electric field strengths inside the sample. By increasing the field strength (voltage) in many cases the current is raised linearly (ohmic region). In some materials the mobile charge carriers are used up after a voltage limit; saturation is reached (Fig. 2.56). In other cases the currents get higher at high electric field strengths than those predicted by Ohm's law. This over-ohmic region of conductivity is caused by the so-called 'hot' charge carriers. With further increasing the field strength, the sample is burned through and its conductivity is raised to very high values.

From the data of current field strength measurements preliminary information can be derived about the nature of charge carriers.

Temperature dependence of d.c. conductivities. Electric conductivities are greatly influenced by physical and chemical impurities in the material. This is why bulk conductivity values seldom represent the properties of the pure material. From the temperature dependence

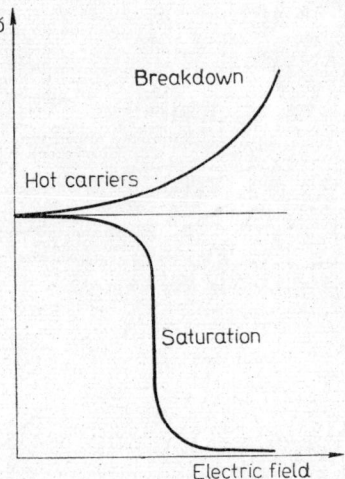

Fig. 2.56 Dependence of d.c. conductivity on the applied electric field strength

of σ some hints can be obtained about the conductivity mechanism.

Typical plots of $\ln \sigma$ against $1/T$ are shown in Fig. 2.57. Curve A represents metallic conductivities, i.e. the increase in temperature results in a decrease in the σ-values. Region B represents simple insulating materials or intrinsic semiconductors with a conductivity changing exponentially as a function of the temperature (see Equation 1.19 in Chapter 1). The values of σ in these materials are in the order of 10^{-10} to 10^{-20} ohm^{-1} cm^{-1}, which are very low compared with those of metals 10^{-1} to 10^{3} ohm^{-1} cm^{-1}. The activation energies according to

Equation 1.19 are expressed as follows

$$E_a = \frac{\ln(\sigma_2/\sigma_1)}{\dfrac{1{,}000}{T_2} - \dfrac{1{,}000}{T_1}} \; 8.5 \times 10^{-2} \text{ eV} \qquad 2.58$$

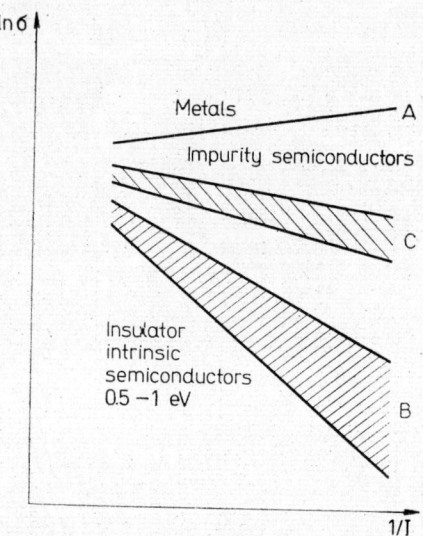

Fig. 2.57 The temperature-dependence regions of electric conductivities

where σ_1, σ_2 are the conductivities measured at T_1, T_2 °K. Values of E_2 are in the order of 0.1–2 eV. Most of the organic compounds fall in this range. Finally group C represents the impurity semiconductors having small activation energies and relatively high conductivities. Most of the inorganic semiconductors and some organic ones are of this type. Their activation energies are usually small, < 0.1 eV.

Irreversible changes. In stable materials Arrhenius plots like those of Fig. 2.57 should be exactly the same by repetitive heating and cooling cycles. In these cases activation energies of the conductivity can be regarded as a substantial factor describing the electrical properties of the compound. Changes in the activation energy upon reheating and cooling indicate permanent physical or chemical changes.

Electrode effects. Charges can be injected into the sample from the electrodes. Thus the measured values of d.c. conductivity are usually

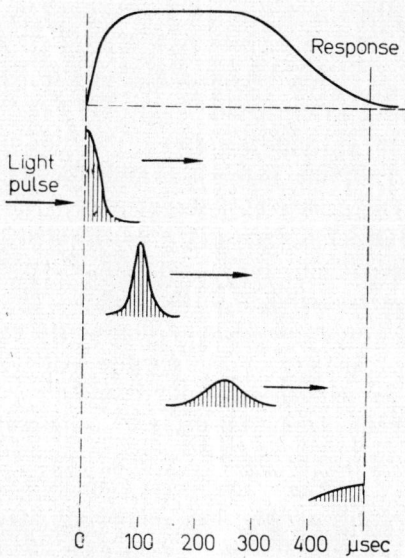

Fig. 2.58 The principle of the pulsed photoconductivity method for measuring charge carrier mobilities in organic materials. After Kepler[2.25]

slightly dependent on the electrode material. This is a serious factor, which makes d.c. conductivity data very difficult to analyze from the point of view of structure determination. Although it is evident that d.c. conductivities should be strongly dependent on physical and chemical structure, quantitative results are very hard to obtain.

Measurement of charge-carrier mobilities. The most interesting materials for structure determination are the insulating or semiconducting organic compounds. From straightforward d.c. conductivity measurements only a little qualitative information can be derived. More promising are the methods of photoconductivity, where the process of activation of the charge carriers can be separated from the process of carrier propagation. As an example of this the *pulse photoconductivity* method is described in some detail. The main features of this method are the following.[2.25]

1. Activation of charge carriers near the surface of the material by irradiation with an ultraviolet light pulse.
2. Recording the current responses to the short irradiating pulses. From the length of the pulses the mobilities of the carriers can be calculated. The principle of the method is shown in Fig. 2.58. The short u.v. pulse (1 μsec) creates a bunch of carriers near the surface of the material. Under the action of the electric field the positive carriers (holes) are combined at the negative electrode, while the negative carriers would move across the sample to reach the opposite electrode (+). The drift of the negative carrier bunch represents a current pulse, which can be amplified and recorded by a cathode ray oscilloscope. The length of this pulse is determined by the time needed for the carriers to reach the opposite electrode, i.e. to travel through distance d. The whole area under the current response curve is proportional to the total number of charges.

By inverting the polarity of the electrodes the mobility and concentration of the holes (positive carriers) can be determined.

Using the pulsed photoconductivity method, various important experimental facts can be obtained about the formation and motion of charge carriers in solids. The formation of carriers can be investigated by varying the wavelength of the light pulse. From the temperature dependence of the carrier mobilities some information can be derived about the structure of the solid. It is also possible to measure the dependence of the carrier concentrations upon the light intensity.

These measurements are interesting for solid state physicists rather than chemists at the moment. However, the basic problems of charge transfer in solid and liquid systems are connected with the chemical and physical structure. In the last few years several attempts have been made to relate chemical and physical structures to conductivity data. In organic single crystals, as for example in anthracene, the mechanism of conductivity is rather well studied. In other systems, for example in polymers, interpretation of the experimental results is difficult.

A separate field is the study of ionic conductors in the liquid and solid state. This problem belongs mainly to electrochemistry and thus will not be discussed here.

REFERENCES

2.1 Gordy, W., Smith, W. V. and Trambarulo, R. F., *Microwave Spectroscopy*, J. Wiley, New York (1953).
2.2 Kisliuk, P. and Townes, C. H., Molecular Microwave Spectra Tables, *J. Res. Nat. Bur. Stand.* **44**, 611 (1950).
2.3 Gordon, J. P., Zeiger, H. J. and Townes, C. H., *Phys. Rev.* **95**, 282 (1954).
2.4 Feher, G., *Bell System Techn. J.* **36**, 449 (1957).
2.5 *Atlas Spektrov EPR* (ed. V. V. Voevodsky), Izd. Akad. Nauk. S.S.S.R, Moscow (1962).
2.6 Allen, L. C., Gladney H. M. and Glarum, S. H., *J. Chem. Phys.* **40**, 3135 (1964).
 Glarum, S. H., *Rev. Sci. Instr.* **36**, 771 (1965).
 Mohos, B., *Lecture held at the ESR-NMR Summer School*, Jablonna, Poland (1965).
2.7 Dixon, W. T. and Norman, R. O. C., *J. Chem. Soc.* 3119 (1963).
2.8 Freeman, R. and Whiffen, D. H., *Mol. Phys.* **4**, 321 (1961).
2.9 Anderson W. A. and Freeman, R., *J. Chem. Phys.* **37**, 85 (1962).
2.10 *NMR Applications*, published by Japan Electron Optics Laboratory, **19** (1965).
2.11 Hahn, E. L., *Phys. Rev.* **80**, 580 (1950).
2.12 Carver, T. R. and Slichter, C. P., *Phys. Rev.* **92**, 212 (1953).
2.13 Overhauser, A. W., *Phys. Rev.* **92**, 411 (1953).
2.14 Feher, G., *Phys. Rev.* **103**, 500 (1956). **105**, 1122 (1957).
 Hyde, J. S., *J. Chem. Phys.* **43**, 1806 (1965).
2.15 Anderson, W. A., *J. Chem. Phys.* **37**, 1373 (1962).
 Freeman, R. and Anderson, W. A., *J. Chem. Phys.* **42**, 1119 (1965).
2.16 Gordon, S. L. and Baldeschwieler, B. D., *J. Chem. Phys.* **43**, 76 (1965).
2.17 Unland, M. L., Weiss, V. and Flygare, W. H., *J. Chem. Phys.* **42**, 2138 (1965).
 Cox, A. P., Flynn, G. W. and Wilson, E. B., *J. Chem. Phys.* **42**, 3094 (1965).
2.18 Battaglia, A., Gozzini, A. and Polacco, E., *Bull. Colloque Ampère* **9**, 171 (1960).

2.19 Böttcher, *Theory of Electric Polarisation*, Elsevier, New York (1952).
2.20 Fröhlich, F., *Theory of Dielectrics*, Oxford Press (1949).
2.21 Moriamez, C., Moriamez, M. and Arnault, R., *Bull. Colloque Ampère* **47** (1961).
2.22 Hamon, B. V., *Proc. Nat. Engrs. Monographs*, **27**, 99 (1952).
2.23 Williams, G., *Trans. Faraday Soc.* **58**, 1041 (1962).
2.24 Robinson, F. N. H., *J. Sci. Instr.* **42**, 653 (1965).
2.25 Kepler, R. G., *Organic Semiconductors* (ed. J. J. Brophy and J. W. Buttrey), Macmillan Co., New York, 1 (1962).
2.26 Drain, D. E., *Proc. Phys. Soc.* **62A**, 301 (1949).
2.27 Abragam, A., *The Principles of Nuclear Magnetism*, Oxford Univ. Press (1961).

3
Free Radicals

One of the most important achievements of radio frequency and microwave spectroscopy is the possibility of investigating free radical structures and reactions. As shown in Chapter 1, free radicals can be studied directly by measuring electron spin resonance of their unpaired electrons. Free radicals, i.e. compounds having unpaired valence electrons, are formed in the following ways.

1. *Formation of radicals during the course of reactions.* Radical reactions substantially involve fission of chemical bonds and formation of transient radical products. Some types of reactions are initiated by radicals, in others, intermediate radical products are formed and more or less stabilized during the course of the reaction, in others still the reaction may lead to formation of stable radical products. Examples in the next sections and those given in Chapter 5 show that ESR provides a unique way of determining the concentration and structure of stable or transient radicals.

2. *Formation of radicals by irradiation.* Chemical bonds may be broken by irradiating the compound with rays of sufficiently high energy. In some cases radicals can be produced by illumination with natural light, as in some materials involved in biological photosynthesis. Ultraviolet irradiation results in radical formation in many cases and high energy irradiation with gamma-rays, accelerated electrons, α-particles, protons or neutrons always produce radicals. The study of these radicals is essential in photochemistry and in radiation chemistry. Examples of this are given in Chapter 5.

3. *Radicals formed by mechanical treatment.* Chemical bonds can be broken by mechanical treatment, such as milling, machining

pressing etc. Radicals formed in this way can be stabilized at low temperatures and measured by ESR. These radicals may initiate secondary reactions (mechano-chemical initiation) leading to the transformation of the products during the course of processing. An example of the ESR study of radicals produced by mechanical treatment is given in Section 3.3.

4. *Radicals produced by gaseous discharges.* Solid samples treated by gaseous discharge may react with the ionized or atomized products of the discharge resulting in intermediate radicals trapped in the solid. By using the ESR technique it is possible to observe the primary radical products present in the discharge and also the radicals trapped in the solid.

In the subsequent sections general ways of studying radical structures in the solid, liquid and gaseous phase will be discussed.

3.1 DETERMINATION OF RADICAL CONCENTRATIONS

As shown in Section 1.2 the intensity of the ESR spectrum lines is proportional to the total number of unpaired electrons present in the sample. ESR spectra of radicals can usually be distinguished from those of paramagnetic ions by the difference in the g-values. The unpaired electrons in the paramagnetic ions are localized in an inner electronic shell and consequently have greater spin–orbit coupling than loosely coupled radical electrons. Greater spin–orbit couplings would shift the ESR lines from the free electron spin value ($g = 2.00229$) appreciably. Usually there are no paramagnetic ions at all in the system investigated, and thus the ESR spectrum appearing during reaction or under irradiation can be directly attributed to radical products. In irradiation of complex organic or inorganic crystals besides radicals, paramagnetic colour centres (e.g. trapped electrons) are formed as well. The g-values of the colour centre lines are very close to the free spin value; the hyperfine splittings caused by the neighbouring nuclei are also similar to those observed in radicals.

If the total width of the spectrum $\overline{\Delta H}$ is much smaller than the operating resonance field H_0

$$\overline{\Delta H} \ll H_0$$

or in frequencies: 3.1

$$\Delta v \ll v_0$$

the total concentration of radicals present in the sample can be expressed as follows

$$[R] = C \frac{T}{H_1^2 v_0^2} \int_{-\Delta \overline{H}_{1/2}}^{+\overline{\Delta H}_{1/2}} f(H - H_0)\, d(H - H_0) \qquad 3.2$$

where C is a technical constant depending on the spectrometer sensitivity; T is the absolute temperature, °K; H_1 is the amplitude of the microwave magnetic field at the position of the sample in gauss, v_0 is the operating frequency; $f(H - H_0)$ is the line-shape function (Gaussian or Lorentzian, see Section 2.2). The integration must cover all the lines coming from the radical no matter how they are split by hyperfine interaction.

Condition 3.1 is fulfilled in all practical cases. The total widths of radical spectra $\overline{\Delta H}$ are in the order of 100 gauss, the resonant fields are around 3,000 gauss by using conventional X-band spectrometers. Thus

$$\frac{\overline{\Delta H}}{H_0} \approx 0.03$$

Equation 3.2 permits only determination of relative concentrations with respect to a standard material containing known concentrations of unpaired spins. If the sample and standard spectra are recorded on the same spectrometer setting, at the same temperature, the relative concentrations are given by

$$\frac{[R]_{\text{sample}}}{[R]_{\text{standard}}} = \frac{I_{\text{sample}}}{I_{\text{standard}}} \qquad 3.3$$

where I stands for

$$\int_{H_0-\Delta H_{1/2}}^{H_0+\Delta H_{1/2}} f(\mathbf{H})\,d\mathbf{H}$$

in Equation 3.2. This integral can simply be obtained by electronic or graphic double integration of the recorded derivative spectrum. For accurate determination of the integrals it is necessary that the recorded curve be the true derivative of the absorption line $f(\mathbf{H})$. This can only be fulfilled if

$$\mathbf{H}_M \ll (\Delta\mathbf{H})_{1/2} \qquad 3.4$$

where $(\Delta\mathbf{H})_{1/2}$ is the half-width to maximum amplitude of the spectrum lines, $\mathbf{H}_M$ is the field modulation amplitude.

As shown in Section 2.2, condition 3.4 can only be fulfilled by serious reduction of the signal level. Thus at low radical concentrations the error in the relative value $I_{\text{sample}}/I_{\text{standard}}$ will be great either because condition 3.4 is not fulfilled, or because the signal-to-noise ratio is decreased by using too small field modulation amplitudes. In such cases the first moment of $f(\mathbf{H} - \mathbf{H}_0)$

$$\langle M \rangle = \int_{-\infty}^{+\infty} (\mathbf{H} - \mathbf{H}_0) f(\mathbf{H} - \mathbf{H}_0)\,d(\mathbf{H} - \mathbf{H}_0)$$

can be used instead of I for describing the total number of spins. By using the same spectrometer setting

$$\frac{[R]_{\text{sample}}}{[R]_{\text{standard}}} = \frac{\langle M \rangle_{\text{sample}}}{\langle M \rangle_{\text{standard}}} \qquad 3.5$$

The first moments can also be determined graphically or electronically from the recorded derivative spectrum (see Section 2.3).

3.1.1 CHOICE OF STANDARD MATERIALS

The error in radical concentration measurements depends greatly on the choice of standard materials. In cases when determination of absolute concentrations is important it is advisable to have a set of standard materials with different spin concentrations and line widths in order to be able to choose the most suitable ones for the given radical system. The main problems in selecting standard materials are summarized briefly as follows.

1. The spin concentration of the standard must be stable at the temperature of the measurement.
2. The spin concentration of the standard sample must be measured independently. From this point of view the paramagnetic salts are preferable to radicals, since their spin concentrations are known. If stable free radicals or charcoals are used for standards the absolute concentrations are to be measured independently.
3. The line width of the standard sample should not differ appreciably from that of the sample. By using different linewidths and identical field modulation amplitudes for recording the sample and standard spectra the error introduced by the shape distortion will be different resulting in greater difference in the measured ratios of $I_{sample}/I_{standard}$ or $\langle M \rangle_{sample}/\langle M \rangle_{standard}$.
4. The dielectric and static magnetic behaviour of the standard and those of the sample should not differ very much. Great difference in the losses would affect the technical factor C in Equation 3.2 by changing the Q-value of the resonator and also the microwave field amplitude at the position of the sample H_1. The true microwave field acting inside the samples is determined not only by the microwave power and Q-value of the cavity, but by the dielectric permittivity of the sample as well. Therefore it is not advisable to measure radical concentrations in aqueous solutions against a standard dissolved in a non-polar solvent, or against a low loss solid standard.

As can be seen from the points stated above determination of absolute radical concentrations by ESR is a rather delicate problem. The estimated error is about $\pm 30\%$. Fortunately in most cases chemists are interested rather in changes of radical concentrations or merely in radical structures. For measuring radical concentration changes, any standard material with constant spin concentration can be used. In these cases the standard spectra are recorded simultaneously with the unknown one. The method of doing this will be described somewhat later. Changes in radical concentrations can be measured with an accuracy of about 5%.

3.1.2 EXAMPLES OF STANDARD MATERIALS

Data of some standard materials used in ESR spectroscopy are given below.

α,α-Diphenyl-β-picrylhydrazyl (DPPH). This material is widely used in ESR spectroscopy for checking technical parameters of the spectrometer, as standard for g-shift measurements and spin concentration measurements. The structure of the compound is the following

The radical is prepared from the corresponding hydrazine; it can be obtained commercially. In the solid state it is a polycrystalline powder of dark green colour. Its molecular weight is 394 and contains 1.5×10^{21} radicals/cm^3, stable up to 75 °C in the absence of strong light. Upon prolonged standing in air the concentration is slightly decreased.

In the solid state DPPH exhibits a sharp, slightly asymmetric ESR line with a g-value of 2.0037 ± 0.0002. The g-value is slightly anisotropic. The line width is 2.8–3.0 gauss for the polycrystalline sample, depending on the method of preparation. In single crystals the lines are sharper (1.8 gauss); the widths are dependent on the crystal orientation. The line shape in single crystals is Lorentzian, in crystalline powders asymmetric, and at higher frequencies a doublet is observed due to g-anisotropy (Fig. 3.1). In solutions in benzene, toluene or tetrahydrofuran a hyperfine splitting is observed from the two nitrogens ($I = 1$) and in deoxygenated solvents from the ring protons too.[3.1]

The total width of the spectrum is 130 gauss, the splitting from ^{14}N nuclei is 22 gauss. The spectra are shown in Fig. 3.1c and in Fig. 3.2.

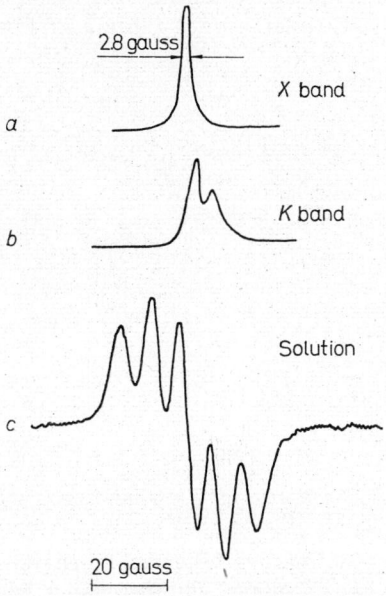

Fig. 3.1 ESR spectra of α,α-diphenyl-β-picrylhydrazyl.
a — Polycrystalline solid, X-band; *b* — K-band; *c* — In benzene solution

As follows from these data DPPH is not very suitable as a standard for concentration measurements. It is, however, very useful for rapid checks of spectrometer sensitivity, resolution and other technical parameters.

Peroxylamine disulphonate. This material, also known as nitrosyl disulphonate, is very valuable because it produces stable radicals in aqueous solutions. The chemical structure is the following

$$\begin{array}{c}{}^{-}O_3S\\{}^{-}O_3S\end{array}\!\!\!>\!\dot{N}O^{-}$$

Interaction of the unpaired electron with the ^{14}N nucleus results in a very well resolved triplet with a splitting of 13.00 ± 0.07 gauss.

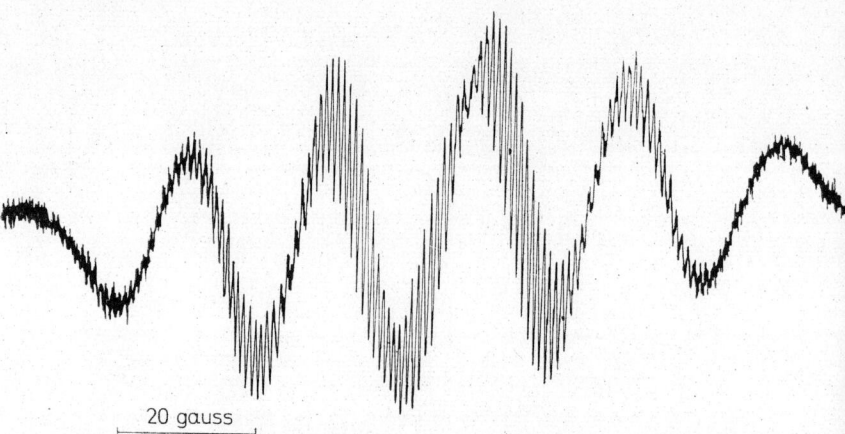

Fig. 3.2 ESR spectrum of α,α-diphenyl-β-picrylhydrazyl in deoxygenated toluene solution. Courtesy of B. Mohos, Budapest

The g-value corresponding to the centre line is 2.0057 ± 0.0001. The line widths are 0.26 ± 0.02 gauss.[3.2] The spectrum is shown in Fig. 3.3. The absolute concentration of the radicals can be determined by chemical ways (e.g., iodine titration).

p,p'-Disulpho-α-diphenyl-β-picrylhydrazyl (SDPPH). In the solid state this stable radical exhibits a single line at $g = 2.003 \pm 0.001$; the line width is 3.2 gauss. Like peroxylamine disulphonate it produces stable radicals in aqueous solution.[3.3] The hyperfine structure consists of 5 lines like that of DPPH corresponding to the interaction with the two ^{14}N nuclei. The chemical structure of the compound is the following

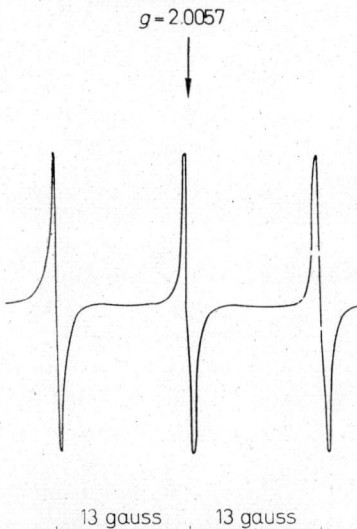

It can be used as a standard by studying radicals in aqueous solutions.

Charred dextrose. Charred dextrose is prepared from anhydrous dextrose by heating it at 200 °C under nitrogen atmosphere for several days. Then the temperature is increased up to 300 °C and the sample is slowly cooled down to room temperature. After powdering, the

Fig. 3.3 ESR spectrum of peroxylamine disulphonate in aqueous solution. After Blois[3.2]

temperature is raised again up to 560 °C and the material is charred. After degassing, the material is melted in vacuum and sealed.

The sample exhibits a very sharp (0.6 gauss) spectrum line at $g = 2.0023 \pm 0.0003$. The main advantage of this material is that the g-value and the line width remain unchanged over a very wide temperature range from 1.4 °K to 508 °K. The spin concentration is 5×10^{20} spins/cm^3; it remains stable up to 560 °C.

The temperature stability makes this material very useful when radical concentrations are to be measured as a function of temperature.

Cupric sulphate $CuSO_4.5H_2O$. Salts of the iron group elements can be used directly as standard materials. The only difficulty is that owing to the large spin–orbit coupling the g-values are very anisotropic. The g-values of $CuSO_4.5H_2O$ single crystals are 2.267, 2.236 and 2.086 for the three principal crystal orientations. Since these values are considerably higher than those of the radicals, the standard lines appear at much lower fields than signal lines. This makes it possible to record signal and standard lines in a single wider field sweep.

The line widths are of the order of 100 gauss. Therefore cupric sulphate is a suitable standard material for solid radicals. In crystalline powders the g-value is 2.22; the line is highly asymmetric showing fine splittings as those shown in Fig. 2.17.

$CuSO_4.5H_2O$ cannot be used at higher temperatures for the loss of crystalline water results in a serious change of the spectrum line.

Mn^{++} doped magnesium oxide. The dipole–dipole interaction among paramagnetic ions can be reduced very effectively by dissolving them in diamagnetic solid matrices. In the case of Mn^{++} doped MgO, the interaction of the unpaired electron with the Mn nuclei ($I = 5/2$) is very well resolved. The spectrum of a powdered sample is shown in Fig. 3.4. The number of lines is 6, corresponding to the 5/2 spin (for explanation see Fig. 1.3). The concentration of Mn^{++} ions is in the order of 10^{-5}/mole. Standard samples can be prepared from very pure MgO mixed with MnS in the concentration required and then fused at 1,200 °C.

It is fortunate that radical lines appear around $g = 2.0037$ in between two Mn^{++} lines as shown by the arrow in the spectrum of

Fig. 3.4. Thus, spectra with total widths less than 50 gauss can easily be recorded simultaneously with the $Mn^{++}-MgO$ standard without the need of additional techniques.

Ruby. Ruby is aluminium oxide (Al_2O_3) doped with Cr^{+++} ions, red ruby contains about 0.1 % chromium ions; the spin concentration

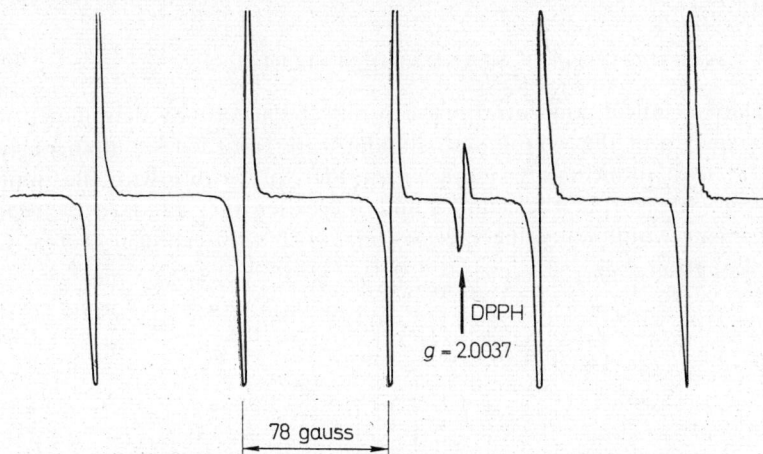

Fig. 3.4 ESR spectrum of Mn^{++} doped magnesium oxide containing a small amount of α,α-diphenyl-β-picrylhydrazyl (DPPH)

in pink ruby is about 0.01 %. The spin of the Cr^{+++} ions is 3/2; the possible orientations, $+3/2$, $+1/2$, $-3/2$ and $-1/2$, are partially quenched by the strong crystalline fields. The external magnetic field would increase this splitting, resulting in four electron spin levels. Thus, ruby has ESR lines with g-values strongly dependent on the crystal orientation.

Since the Cr^{+++} concentrations of the ruby samples are not known, each standard crystal should be measured against another standard. Ruby is therefore useful in measuring relative concentrations rather than absolute ones. The ruby line can easily be shifted out of the range of the radical spectrum to be measured and thus the standard and signal can be recorded in one sweep. The advantage of ruby over $CuSO_4.5H_2O$ is that it can be used over a fairly large temperature range. The line widths available in the operating region are about 50–100 gauss, depending on the crystal orientation. Thus ruby can be used as a standard for measuring radical concentration changes in the solid state, where widths are of the same order.

3.1.3 POSITIONING OF STANDARD MATERIALS

Relative radical concentrations can only be accurately determined if the spectra of the sample and standard are recorded simultaneously or at least immediately one after another, preferably with the same sweep. Measuring the sample and thereafter the standard by replacement would introduce serious errors in the technical factor C in Equation 3.2.

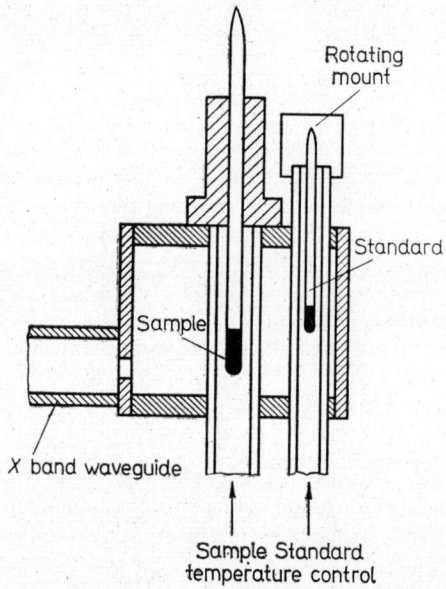

Fig. 3.5 Positioning of a standard material in the ESR cavity resonator

The simplest solution is to use single crystal standards having highly anisotropic g-values and to shift the standard lines out of the range of the spectrum to be measured. A possible position of such crystals is shown in Fig. 3.5. The quartz tube where the crystal

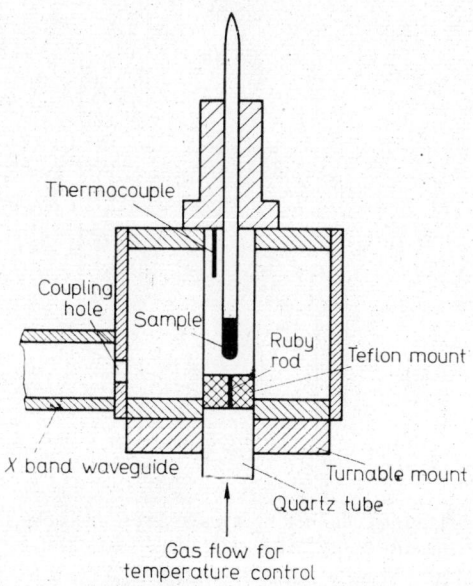

Fig. 3.6 Positioning of a ruby standard in the ESR cavity resonator

($CuSO_4.5H_2O$) is placed can be rotated in order to adjust the required shift from the signal to be measured. The temperature of the standard crystal is kept constant by nitrogen or air flow in the usual way. Using this system the temperature of the sample may differ from that of the standard. This method is useful when changes in radical concentration are to be measured at fixed temperatures (recombination of radicals or reactions). It is not useful for measuring the effect of temperature dependence on spectra.

In Fig. 3.6 a way of positioning ruby standards is shown. The crystal is mounted in the quartz tube used for temperature control. Upon rotating the quartz tube the ruby line can be shifted; by shifting the quartz tube along the axis of the cavity the intensity of the stand-

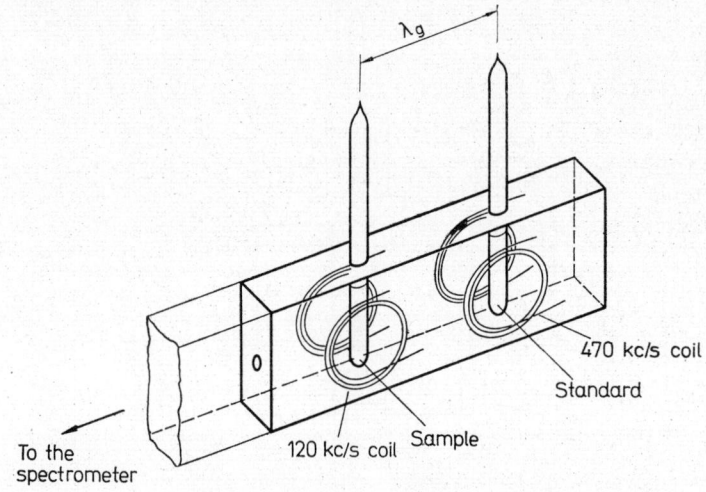

Fig. 3.7 Double channel ESR cavity for measuring radical concentrations. After Köhnlein and Müller[3.4]

ard line can be changed. In this case the temperature of the standard is the same as that of the sample.

Standard lines exhibiting spectra at the same field range as the radicals to be measured can be recorded simultaneously by using the double channel system illustrated in Fig. 3.7. A higher mode cavity is used with two field modulating coils excited by two separate frequencies (e.g. 120 kc/s and 470 kc/s). The signals obtained at the detector of the ESR spectrometer are separated by selective amplifiers and lock-in detectors, and, thus, the standard and signal spectra can be recorded simultaneously. Since both samples are placed in the same resonator, variations of dielectric loss of the sample would affect the standard signal as well by decreasing the Q-value of the resonator. An advantage over the previously described methods is that it is not necessary to extend the scanning ranges and thus the time required for displaying a spectrum at a given sensitivity level can be shortened. A slight disadvantage is that the sensitivity is somewhat reduced by the reduction of the ratio of the effective sample volume to the total volume of the cavity (see the discussion in Section 2.2).

Using the double channel method the ratio signal/standard concentrations can be calculated even if the gain of the channels is not equal. Thus a single standard can be used at very wide signal levels. To compensate differences in the channels each measurement should be repeated by exchanging the samples. The unknown concentration is then given by [3.4]

$$\frac{[R]_x}{[R]_0} = \left(\frac{I_{x1}}{I_{02}} \frac{I_{x2}}{I_{01}}\right)^{1/2} \qquad 3.6$$

where $[R]_x$ is the unknown concentration, $[R]_0$ is the known spin concentration of the sample, I_{x1}, I_{x2} are the values obtained by double integration of the spectrum corresponding to the unknown radical measured at positions 1 and 2, respectively, I_{01}, I_{02} are the corresponding values for the standard signals. The gains of channel 1 and channel 2 are unchanged by exchanging the sample with the standard.

3.2 DETERMINATION OF RADICAL STRUCTURES

Structures of free radicals can be determined very efficiently by observing the hyperfine interaction of the unpaired electron with the magnetic nuclei of the molecule. The unpaired electron of the free bond is usually spread over a large part of the molecule. The probability density function of the unpaired electron called *spin density* is the most important factor in characterizing radical structures. Chemists denote radicals by a point at the atom where the valence is supposed to be free. In reality, the electron corresponding to this valence is delocalized over a large area. Thus the question, where is the unpaired electron, means determination of its spin density over the molecule. Delocalization of the unpaired electrons is well illustrated by the hyperfine splitting of α,α-diphenyl-picrylhydrazyl (DPPH) mentioned in the previous section. At low resolution the 5-line spectrum shown in Fig. 3.1c indicates that the unpaired electron is delocalized to both nitrogens (for explanation see Fig. 3.11). Further analysis shows that the spin density is the same at each nitrogen. (For calculation of spin densities see the discussion later.)

At higher resolution, in deoxygenated solution in tetrahydrofuran or toluene the hyperfine coupling with the α,α-ring protons is also shown (Fig. 3.2). This means that the unpaired electron is spread over the rings and interacts with the ring protons. This is why the complex spectrum of Fig. 3.2 is obtained.

Determination of radical structures thus means determination of the spin densities in the given molecule. This can be done by a thorough analysis of the measured hyperfine spectra. The theory of hyperfine interaction is well developed. It is possible to calculate spin densities from the measured hyperfine splitting in most cases. For interpretation of complex spectra such as in Fig. 3.2, electronic computers are used.

However, instead of calculating the exact spin densities, in research work very valuable qualitative information can be derived from the ESR spectra by using simple considerations based on general theory. Spectra given in the subsequent sections will be interpreted in this way although it is possible to derive much more information from them by more detailed analysis.

3.2.1 HYPERFINE INTERACTION

Hyperfine interaction is a coupling between the spin system of the unpaired electrons and that of the surrounding magnetic nuclei. As shown in Section 1.2 coupling with a nucleus of spin I would split each electron spin energy level into $2I + 1$ sub-levels. The hyperfine splittings of ESR spectra arise from magnetic dipole transitions between levels with the same nuclear spin state, i.e.

$$\Delta m_I = 0$$

where $m_I = -I, -I + 1 \ldots I + 1$ is the quantum number of transitions between the nuclear spin levels. An example of hyperfine splitting was given in Fig. 1.3 for manganous ions with $I = 5/2$. The corresponding hyperfine spectrum is illustrated in Fig. 3.4 for Mn^{++} doped MgO. The hyperfine spectrum consists of six equidistant lines of equal amplitude.

It is clear from the simplified arguments given in Section 1.2 that from the number of the hyperfine lines one can tell exactly what nuclei interact with the unpaired electron of the radical. Interaction with a single proton results in two lines with equal amplitude. ^{14}N nuclei ($I = 1$) would cause a splitting into three lines, as in the case of peroxylamine disulphonate shown in Fig. 3.3.

The next important experimental parameter is the distance between the hyperfine lines, the hyperfine splitting A_I, the distance in gauss between the hyperfine lines split by a nucleus of spin I. The value of A_I is determined by the nuclear moment of the nucleus and by the strength of the hyperfine coupling. As shown in Fig. 1.3, in atomic hydrogen the doublet is separated by $A_I = 507$ gauss.

In peroxylamine disulphonate the separation of the lines is $A_I = 13$ gauss, in Mn^{++} doped MgO $A_I = 70$ gauss.

Contact interaction. As mentioned in Chapter 1 there is a quantum-mechanical interaction between the unpaired electron and the nucleus. It is only different from zero in those electronic states where the spin density is not zero at the position of the nucleus. The splitting is

$$A_I \propto g_I \mu_i g_s \mu_s \rho(I) \qquad 3.7$$

where $\rho(I)$ is the electron spin density at the nucleus, g_I, g_s are the nuclear and electronic splitting factors, μ_i, μ_s the nuclear and Bohr magnetons, respectively. For illustration of electron densities the simple case of atomic hydrogen is shown in Fig. 3.8. In the ground state the only electron of the hydrogen atom has a density distribution of spherical symmetry (*s*-state). Molecular σ-states have similar density distributions. In these cases the density is very high at the position of the nucleus (proton) and thus the contact interaction is high. This is why such high hyperfine splitting is observed in gaseous atomic hydrogen in the ground state. The first excited p-state exhibits a different density distribution as shown in Fig. 3.8. Molecular π-states have similar distributions with zero densities at the nucleus. In these cases no contact hyperfine interaction should be observed.

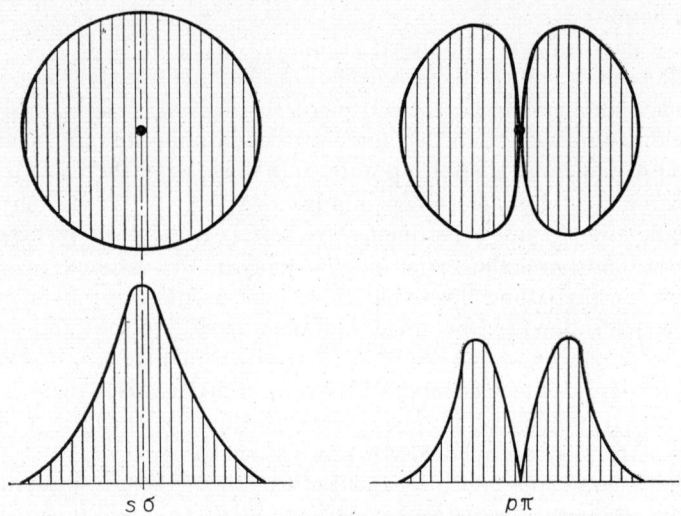

Fig. 3.8 Schematic representation of electron densities in atoms or molecules

According to the discussion of Abragam and Pryce[3.5] the interaction with a nucleus of spin I would split the ESR spectrum lines in the following way

$$\mathbf{H} = \frac{h\nu_0}{g_s\mu_s} + A_I m_I + \frac{1}{2}\left(\frac{\nu_0}{g_s\mu_s}\right)^2 A_I^2 \left[I(I+1) - m_I^2\right] \qquad 3.8$$

where g_s is the electronic g-factor; μ_s is the Bohr magneton, A_I is the hyperfine splitting constant caused by nucleus of spin I in gauss, $\mathbf{H}$ is the magnetic field, ν_0 is the operating frequency, $m_I = -I$, $-I + 1 \ldots + I$.

The second term in Equation 3.8 is the first-order hyperfine splitting constant A_1 measured in frequency or rather magnetic field strength units. The third term is called *second-order hyperfine splitting;* it can only be observed in favourable cases at high resolution. The first term is the interaction of the polarizing magnetic field with the electron moments. Its order of magnitude is 3,000 gauss at $v_0 = 9$ Gc/s, the second HFS term is about 30 gauss (90 Mc/s), the third (second-order) term is in the order of 0.1 gauss (0.3 Mc/s).

Intermixing of electronic states; configurational interaction. According to the simplified picture given above about contact interaction, there should be no hyperfine splitting at all in molecules with π-state unpaired electrons such as in aromatic compounds in solution.

In fact, in all aromatic compounds a splitting of about 23 gauss is observed. Even in atomic hydrogen the splitting of 507 gauss is too much for that calculated from the s-electron density.

The reason of these discrepancies is that atoms and molecules are never in pure states; there are always perturbations which intermix all the possible states to result in an electron density distribution having $s(\sigma)$ and $p(\pi)$ character as well. The ratio of intermixing of excited states depends on the symmetry of the molecule. On the basis of the ESR data using the methods of quantum chemistry it is possible, however difficult, to calculate the spin densities for different chemical groups.

Despite the difficulties caused by the configurational interaction the spin density $\rho(I)$ at a nucleus I can be approximated by the following simple formula (McConnell)[3.6]

$$\rho(I) = \frac{A_I}{Q_{CH}} \qquad 3.9$$

where A_1 is the measured hyperfine splitting in gauss; Q_{CH} is a constant for C–H bonds having a value of 23 to 28, depending on the type of the molecule.

The total width of the spectrum is

$$\overline{\Delta H} = Q_{CH} \sum_{i=1}^{N} \rho_i(I) \qquad 3.10$$

where N is the number of nuclei in contact with the radical electron density.

Thus, the total width of the spectrum should be approximately the same for radicals of the same structure. As will be shown in the next section, this holds fairly well for aromatic molecules with $Q \approx 28$ gauss. At least the spin densities calculated from Equations 3.9 and 3.10 give fairly good qualitative information on how the unpaired electron is delocalized over the molecule.

Dipole–dipole hyperfine interaction. Besides the isotropic interaction in solids or in highly viscous media an orientation-dependent interaction exists between the electron and nuclear spin systems. This is the classical interaction between two magnetic dipoles expressed by Equation 2.9 in Chapter 2. In single crystals, the splitting of the spectral lines caused by this interaction depends on the orientation of the crystals with respect to the polarizing magnetic field. The splitting in gauss is expressed in terms of the anisotropic splitting constant B_1; instead of using Equation 3.8, the spectrum lines are described as follows

$$\mathbf{H} = \frac{h\nu_0}{g_s\mu_s} - [A_I + B_I(3\cos^2\vartheta - 1)]m_I \qquad 3.11$$

The second-order term of Equation 3.8 is omitted for simplicity; ϑ is the angle between a crystalline principal axis and the polarizing magnetic field.

In fact, in crystals of arbitrary symmetry B_I is a tensor. Usually its principal values are given in the direction of the principal axes of the crystal. By rotating the crystal around an axis and measuring the variation of the hyperfine splitting it is possible to separate the isotropic part A_I, which does not depend on the orientation from the anisotropic part B_I which is changed as $3\cos^2\vartheta - 1$.

In polycrystalline or amorphous media all possible crystal orientations are present. Thus $3\cos^2\vartheta - 1$ is averaged to result in a broadening of the spectral lines.

3.2.2 HYPERFINE INTERACTION WITH MANY NUCLEI

Since the radical electron is usually delocalized over the whole molecule or at least a large part of it, the unpaired electron comes into contact-interaction with many nuclei. From the number and intensity distribution of the lines of the measured spectrum one can tell how many nuclei interact with the radical electron. The simplest case is interaction with n equivalent nuclei, i.e. when the spin density of the unpaired electron is distributed uniformly over the area of the nuclei. Another case is when N groups of nuclei containing n equivalent elements interact with the electrons. The most complicated cases are those when non-equivalent nuclei with different spins are present in the molecule. Such complicated spectra usually cannot be interpreted directly; they can be simulated by the use of electronic computers (spectrum accumulator: see Section 2.2).

Interaction with n equivalent nuclei of spin 1/2. This is the most common case in organic radicals where chiefly carbon and hydrogen atoms are present. The natural ^{12}C atom has no magnetic moment, the ^{13}C with $I = 1/2$ is present in an abundance of 1.2% which is usually too small to be detected. Thus in hydrocarbons, hyperfine splitting is limited to protons with $I = 1/2$. A simple example of interaction with equivalent protons is the 1,4-benzo-semiquinone ion radical

This radical is formed as an intermediate product during the course of reduction of quinones or oxidation of hydroquinones. The reaction is discussed in Chapter 5. The spectrum consists of 5 lines with an intensity ratio of $1 : 4 : 6 : 4 : 1$.

The spectrum can easily be interpreted by supposing that the radical electron is spread uniformly over the ring, i.e. the spin densities are equal at the 4 ring protons. Thus each proton will cause an equal splitting of the energy level corresponding to the scheme illustrated in Fig. 3.9. As shown there, each additional proton would split the energy levels by the same a_I (splitting constant in energy units), result-

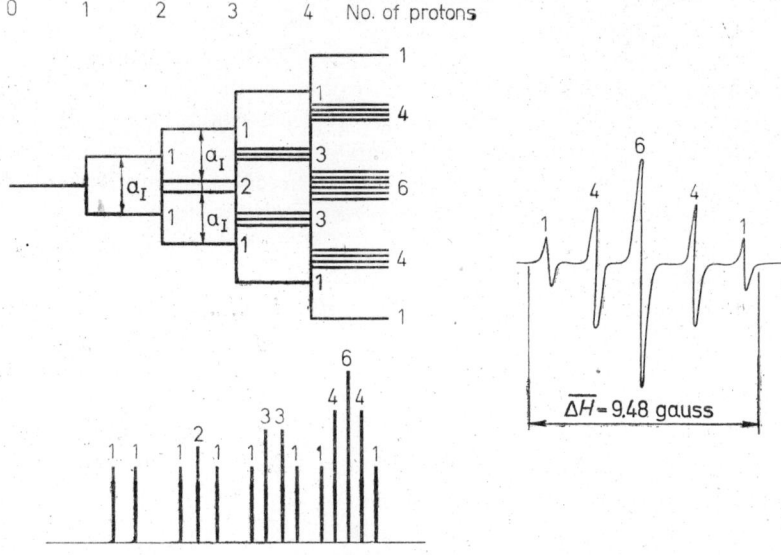

Fig. 3.9 ESR hyperfine splitting from 1–4 equivalent protons. The spectrum of the 1,4-benzosemiquinone radical ion

ing in $n + 1$ levels from n protons with intensity ratios given by the binomial coefficients. In the case of 1,4-benzo-semiquinone, $n = 4$, the total number of lines is 5, the binomial intensity ratio is 1 : 4 : 6 : 4 : 1. The binomial coefficients are given in the Appendix, Table 6.2.

In the case of equivalent nuclei of spin 1/2 hyperfine splitting spectra can easily be interpreted. The conditions for equal spin densities are

1. the hyperfine lines should be equidistant, defining a single splitting A_1;

2. the intensity ratios should be equal to the binomial coefficients of Table 6.2.

A rough estimation of the spin densities can be made by using Equations 3.9 and 3.10. As shown in Fig. 3.9 the total width of the 1,4-benzo-semiquinone spectrum is $\overline{\Delta H} = 9.48$ gauss.[3.7] The splitting is $A_p = 4.28$ gauss. The density at the ring protons is roughly

$$\rho \text{ (protons)} = \frac{\overline{\Delta H}}{Q_{CH}} = \frac{9.48}{28} = 0.34$$

Therefore about 34% of the spin density is distributed uniformly at the ring protons; the remaining 66% must be on the oxygens:

```
          0.33
           Ȯ
           |
   0.087  ╱ ╲  0.087
         ‖ — ‖
   0.087  ╲ ╱  0.087
           |
           O
          0.33
```

The numbers represent the spin densities at the corresponding place. The radical electron is thus somewhat localized to the oxygen atoms. The situation can be visualized as follows

```
           O
      H    |    H
       ╲  ╱ ╲  ╱
        ╲╱ — ╲╱
        ╱╲   ╱╲
       ╱  ╲ ╱  ╲
      H    |    H
           O
```

The radical electron cloud represented by the ellipse inside the ring is attracted by the oxygen atoms. This results in the measured spin density distributions.

Interaction with groups of equivalent 1/2 spin nuclei. Using the energy splitting scheme of Fig. 3.9 it is possible to construct theoretical spectra of hyperfine interaction with N groups of 1/2 spin nuclei with n spins in each groups. The total number of the lines is $(n + 1)^N$. There are N different splitting constants. Some of the lines might coincide. Interpretation of such spectra can usually be made only by using electronic computers because of the great number of the lines. The spectrum can be described as follows

$$\mathbf{H}(i,j) = \mathbf{H}(0) + \sum_{i=-n}^{+n} \sum_{j=-N}^{+N} m_i A_j \qquad 3.12$$

where the first summation corresponds to the n equivalent nuclei with magnetic quantum number m, the second to the N groups with different splitting constants A_j.

As an example the high resolution ESR spectrum of the tetracene radical ion is considered on the basis of the discussion of J. S. Hyde and H. W. Brown.[3.8] Dissolving tetracene in concentrated sulphuric acid, rather stable positive radical ions are formed. The chemical structure is the following

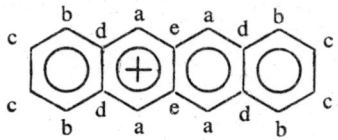

According to the symmetry of this molecule there are three groups of ring protons ($N = 3$), each containing 4 equivalent protons ($n = 4$). The groups of protons are denoted by a, b, c, the carbon atoms where no protons are present by d, e. The total number of the lines should be $(4 + 1)^3 = 125$.

By measuring at very high resolution (5 Mgauss) in dilute solution 85 lines have been found. On the basis of Equation 3.12, assuming that some lines coincide, the following hyperfine splittings have been calculated

$A_a = 5.06$ gauss $(i = 4)$

$A_b = 1.69$ gauss $(i = 4)$

$A_c = 1.03$ gauss $(i = 4)$

total width $\overline{\Delta H} = 31.12$ gauss

The spin densities calculated from Equation 3.9 using a Q_{CH}-value of 28 gauss are the following

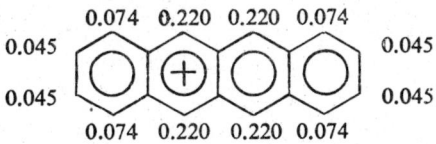

Thus the radical electron is partially localized on the central rings. The spin density distribution is symmetric.

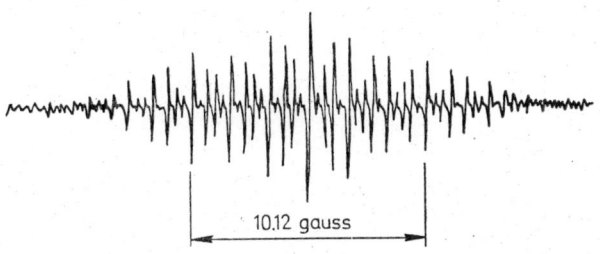

Fig. 3.10 The ESR spectrum of the tetracene radical ion. After Hyde and Brown[3.8]

The tetracene spectrum shown in Fig. 3.10 is used as a standard to check spectrometer resolution.

Interaction with nuclei of arbitrary spin. If the unpaired electron is in contact interaction with nuclei having spins other than 1/2, the energy levels will be split into $2I + 1$ sub-levels by each nucleus. The energy level diagram can be built up systematically according to the principle given above. An example is shown in Fig. 3.11 for interaction with two nuclei of spin $I = 1$.

The first scheme and spectrum correspond to equal interaction with two nitrogen nuclei. This is the case of the stable radical α,α-

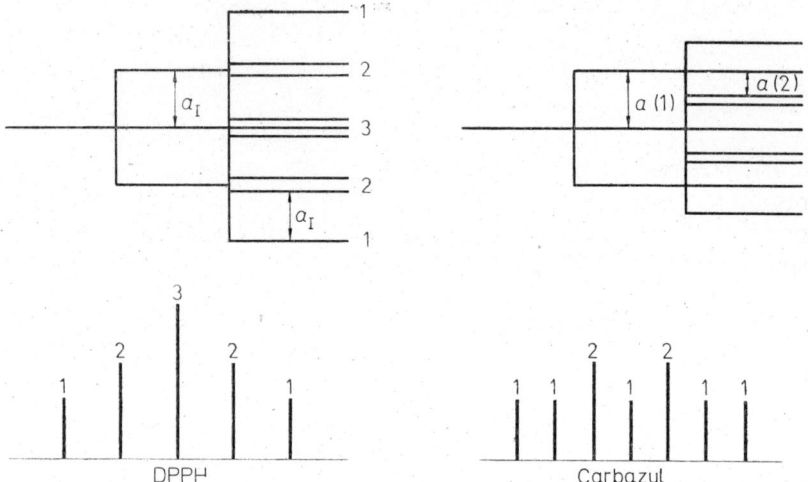

Fig. 3.11 Hyperfine interaction in α,α-diphenyl-β-picrylhydrazyl and in N-aminocarbazyl

diphenyl-picrylhydrazyl, discussed in the previous section, as a standard material

where the radical is equally distributed between two nitrogen atoms. The corresponding spectrum is shown in Fig. 3.2; it consists of 5 lines with $1 : 2 : 3 : 2 : 1$ intensity ratios.

The second scheme is that of N-picryl-9-amino carbazyl[3,9]

Its spectrum consists of 7 lines as shown in Fig. 3.11. At higher resolution these lines are further split by interaction with the ring protons. The 7-line spectrum can be interpreted by assuming that the spin densities are not equal:

$$\frac{a(\alpha N)}{a(\beta N)} = \frac{1}{2}$$

In this case only two of the possible 9 energy levels overlap, resulting in the 7-line spectrum observed (see Fig. 3.11).

Further splitting of the spectrum of DPPH taken at high resolution (Fig. 3.2) indicates that the spin density at the ring protons is not zero. If all the 10-ring protons in the two α-rings were equivalent, each line of the 7-line spectrum should be further split into 11 lines with the binomial intensity distribution. The resolution is not high enough for determination of the spin densities at the ring protons, but the fact that it is different from zero is important information about the radical structure.

Hyperconjugation. If a double bond or a system of double bonds is followed by a single bond, as for example in methyl-substituted aromatic radicals, the double bond character may be partially transferred to the single bond. In the ESR spectrum of methyl-substituted 1,4-benzo-semiquinone radicals the spin density is thus found to be the same at the methyl protons as at the ring protons.

The ESR spectrum of the 2-methyl-1,4-benzo-semiquinone radical ion

$$\begin{array}{c} \dot{O} \\ H \diagup\!\!\!\diagdown CH_3 \\ | - | \\ H \diagdown\!\!\!\diagup H \\ O \end{array}$$

consists of seven lines with an intensity ratio of 1 : 6 : 15 : 20 : 15 : 6 : 1 corresponding to the interaction with six equivalent protons. Similarly

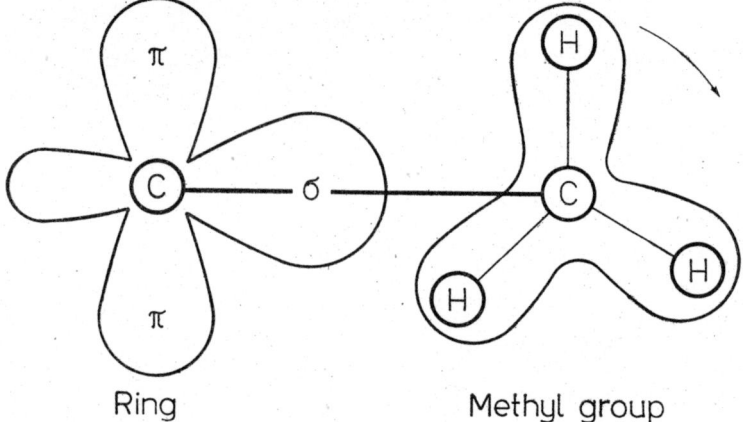

Fig. 3.12 Electron density distribution in a methyl group coupled to a ring carbon atom

all the protons are equivalent in the tetramethyl-1,4-benzo-semiquinone radical ion

$$\begin{array}{c} \dot{O} \\ H_3C \diagup\!\!\!\diagdown CH_3 \\ | - | \\ H_3C \diagdown\!\!\!\diagup CH_3 \\ O \end{array}$$

Here the spectrum consists of 13 lines with the binomial intensity ratio.

The electron configuration leading to the partial transfer of the conjugation to the C—C bond is shown in Fig. 3.12. The π-electron state of the ring is hybridized by configurational interaction. So the electron density is a mixture of s- and p-type distributions. The methyl group connected to the ring is rotating around the C—C axis, and so the methyl protons are equivalent. The experimental fact that they are also equivalent with the ring protons can only be explained by the transfer of conjugation to the C—C bond. The situation can be visualized in the following, somewhat unorthodox, way

indicating that the radical electron is distributed uniformly among the CH_3 groups.

It is interesting that the radical electron density can penetrate even two C—C bonds. In di-t-butyl-1,4-benzo-semiquinone

besides the splitting from the two ring protons the 18 t-butyl proton splittings can be observed. The spin densities at the butyl protons are, of course, much smaller than those at the ring protons. The splitting caused by the butyl protons is about 0.05 gauss.

Negative spin densities. By definition the spin density function can be positive or negative depending on the orientation of the spin with respect to the polarizing magnetic field. The integral of the density over the whole molecule is

$$\int \rho_I \, dV = \pm 1$$

According to the assumption of McConnell,[3,10] the interaction of the unpaired electron with the paired ones makes possible the formation of

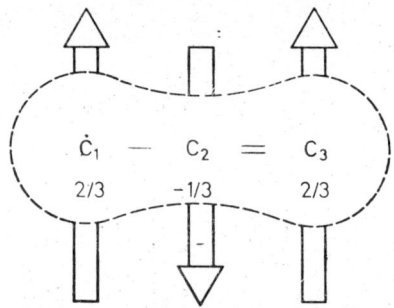

Fig. 3.13 Illustration of the negative spin density in allyl radicals

electron spin states opposite to the polarization of the whole system. If the electron as a whole is in 'up' (positive) direction with respect to the polarizing field H_0, some part of its density may be polarized 'down' (negative). One can visualize the unpaired electron as a cloud magnetized in some places 'upward' (positive spin density) and in other places 'downward' (negative spin density).

The situation is illustrated in Fig. 3.13 for allyl radicals. The spin density at the two side carbon atoms is $+0.64$; at the centre atom there is a negative spin density -0.24. The total splitting measured experimentally is 36 gauss, considerably larger than Q_{CH} (23–28 gauss).

The total splitting is thus expressed, instead of by Equation 3.10, as follows

$$\overline{\Delta H} = Q_{CH} \sum_{i=1}^{N} |\rho(i)| \qquad 3.13$$

where N is the number of magnetic nuclei in the system. The hyperfine splitting does not depend on the sign of the spin density. Therefore negative spin densities cannot be observed directly by electron spin resonance. If the total shift is greater than the Q-factor of the bond (in the present case the $C-H$ bond)

$$\overline{\Delta H} > Q_{CH}$$

somewhere there exist negative spin densities. The position can be derived by theoretical arguments.

Interaction with ^{13}C nuclei. Although the natural abundance of ^{13}C is only 1.2%, in some cases it is possible to observe the hyperfine interaction of the unpaired electron with the 1/2 spin ^{13}C nuclei. In some cases ^{13}C-labelled samples are used. Using this technique it is possible to measure spin densities at those places in the molecule, where no hydrogen or other magnetic nuclei are present. In the case of the 1,4-benzo-semiquinone radical ion ^{13}C splitting makes it possible to measure the densities in positions 1 and 4 near the oxygen atoms:

while from the proton splitting only spin densities at carbons 2, 3, 5 and 6 can be determined. Das and Fraenkel succeeded in detecting ^{13}C satellites in natural abundance.[3.11] The ^{13}C satellites of the central line of the proton hyperfine spectrum (5 lines) are shown in Fig. 3.14.

The spin density at positions 1 and 2 can be calculated by considering the ^{13}C carbonyl group:

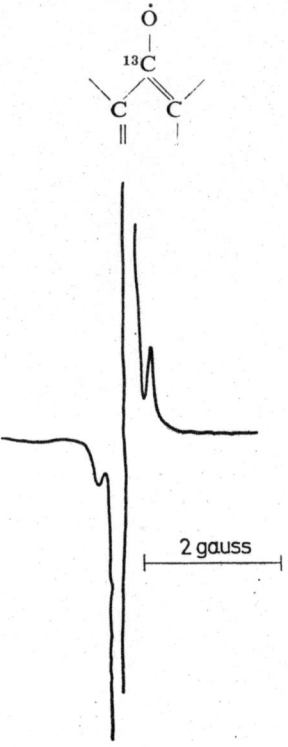

Fig. 3.14 Satellites caused by hyperfine interaction with ^{13}C nuclei in 1,4-benzosemiquinone. After Das and Fraenkel[3,11]

According to the theory developed by Fraenkel[3,12] the splitting is given by

$$\Delta H_{CO}^{C} = 16.2\rho_{C_1} - 13.9(\rho_{C_2} + \rho_{C_6}) + Q_{CO}^{C}\rho_{C} + Q_{OC}^{C}\rho_{O} \quad 3.14$$

FREE RADICALS

Here $\rho_{C_1}, \rho_{C_2}, \rho_{C_6}$ are the π-electron spin densities at carbon atoms 1, 2 and 6 respectively, Q_{CO}^C and Q_{OC}^C are the parameters describing the spin density distribution on the carbonyl groups, called sigma–pi parameters.

In Equation 3.14 the densities ρ_{C_2} and ρ_{C_6} can be calculated by the proton splittings:

$$\Delta H_i^H = Q_{CH}^H \rho_i^H$$

$$\Delta H_C^C = 35.6\, \rho_{C_1} - 13.9\, (\rho_{C_2} + \rho_{C_6})$$

3.15

Using Equations 3.14 and 3.15 together with the law of conservation of spin density

$$\Sigma \rho_i = 1 \qquad 3.16$$

it is possible to calculate the spin densities at all sites from the measured ^{13}C and proton splittings.

The sigma–pi parameters Q_{CO}^C and Q_{OC}^C have been determined by a semi-empirical method on the basis of ^{13}C splitting data of different carbonyl groups. The numerical values are the following

$$Q_{CO}^C = 17.7 \text{ gauss}$$
$$Q_{OC}^C = -27.1 \text{ gauss}$$

Using these values the ^{13}C splitting at carbon 1 in 1,4-benzo-semi-quinone is given as follows

```
              0.1450
                Ȯ
                |   0.1796
       0.0877  ╱──╲  0.0877
               ║  ║
       0.0877  ╲──╱  0.0877
                |   0.1797
                O
              0.1450
```

Thus the structure of this radical is exactly known.

Such exact pictures of the distribution of the radical electron can only be given for a few simple radicals for the time being. In more complex cases only less accurate information can be derived from the ESR hyperfine spectra.

3.3 RADICALS IN THE SOLID STATE

Resolution of ESR spectra are seriously limited in the solid state by anisotropic dipole–dipole interactions and by the effects of g-anisotropy. As discussed in Section 2.2 the ESR line widths are determined by the interactions among electronic spins and by the anisotropic interaction of the electron spins with the magnetic nuclei of the system.

According to Equation 2.9, dipole–dipole interaction between electron spins depends on their relative orientation and on the distance between them. In solids the orientation effects are not averaged out and thus the local field will be slightly different at each radical as a result of the others. Assuming an average distance of 1 Å between radicals, dipole–dipole interaction would broaden the lines by $\Delta H_{1/2} =$ 20–30 gauss in cases of random orientation of spins (amorphous or polycrystalline media). According to the arguments of Section 3.2 the total hyperfine splitting for C—H groups is 23–28 gauss. Therefore in *amorphous or polycrystalline radicals proton hyperfine splitting usually cannot be observed.*

Dipole–dipole interaction between electron spins can be reduced by increasing the average distance between the paramagnetic centres. In solid solution of ionic or radical paramagnetic centres in diamagnetic media, well resolved hyperfine splittings can usually be observed. This is the case in Mn^{++} doped MgO (Fig. 3.4) where the average distance between Mn^{++} ions is quite large in the diamagnetic MgO, and thus the line width is effectively reduced. The line of $MnSO_4$ or $MnCl_2$ is about 300 gauss while in the diluted system widths of 3–5 gauss are observed depending on the Mn^{++} ion concentration.

In stable radicals in the solid state, line widths are usually small because of the strong exchange interaction between electron spins.

For the stable standard radicals mentioned in Section 3.1 (diphenylpicrylhydrazyl, e.g.) line widths of 3 to 10 gauss are observed in the undiluted solid state. However, hyperfine structure is destroyed by the strong exchange interaction.

Fortunately, in most cases the radical concentration in the solid state is low enough to resolve larger hyperfine splittings. The usual range of line widths is 10–100 gauss. Although this corresponds to a resolution of about three powers of ten lower than that of liquids, some very important information about structures can be gained.

Since in the solid state the dipole–dipole interaction among electron spins, the g-value and the anisotropic part of the hyperfine interaction are dependent on the orientation, it is always advisable to work with single crystals whenever it is possible. In this way the orientation broadening effects are eliminated.

3.3.1 TRAPPING

Radicals which are usually very reactive in the gaseous and the liquid state are stabilized, i.e. trapped, in the solid phase by their reduced mobilities. At sufficiently low temperatures all radicals, even hydrogen atoms, can be trapped. Examples of this are given in Chapter 5. At elevated temperatures some degrees of freedom are liberated in the solid resulting in an increased mobility of the radicals. At sufficiently high temperatures even in the solid state radicals become extremely reactive and combine with each other or react with surrounding molecules. The effect of the physical structure upon recombination of radicals is illustrated in Fig. 3.15 for radicals trapped at low temperatures in irradiated phenol measured by Lebedev, Michailov and Buben.[3.13] Radical concentrations are plotted against time at different temperatures. As shown, recombination is stopped at a certain level at each temperature indicating that only a part of the trapped radicals have been released. Similar stepwise radical recombination curves have been observed at elevated temperatures in polymers.[3.14] In Fig. 3.16 recombination curves for polychlorotrifluoro-ethylene irradiated in air are shown at different temperatures. The trapped radicals have the

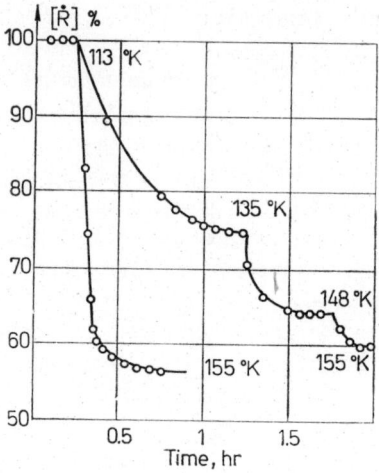

Fig. 3.15 Stepwise recombination of radicals in irradiated phenol at low temperatures. After Lebedev, Michailov and Buben[3.13]

following structure

$$\sim \underset{\underset{F}{|}}{C} - \underset{\underset{F}{|}}{\overset{\overset{\overset{\cdot}{O}}{|}}{\underset{|}{C}}} - \underset{\underset{F}{|}}{\overset{\overset{Cl}{|}}{C}} \sim$$

Upon increasing the temperature and simultaneously recording the ESR spectrum against a ruby standard (see Section 3.1) the kinetic curve of recombination can easily be obtained. The recombination steps are clearly seen. Prolonged standing at elevated temperature results in no significant change in the radical concentration after equilibrium has been reached.

The study of trapping conditions is extremely important in solid state reactions, in photolysis and radiolysis of solids. Some examples of this are given in Chapter 5.

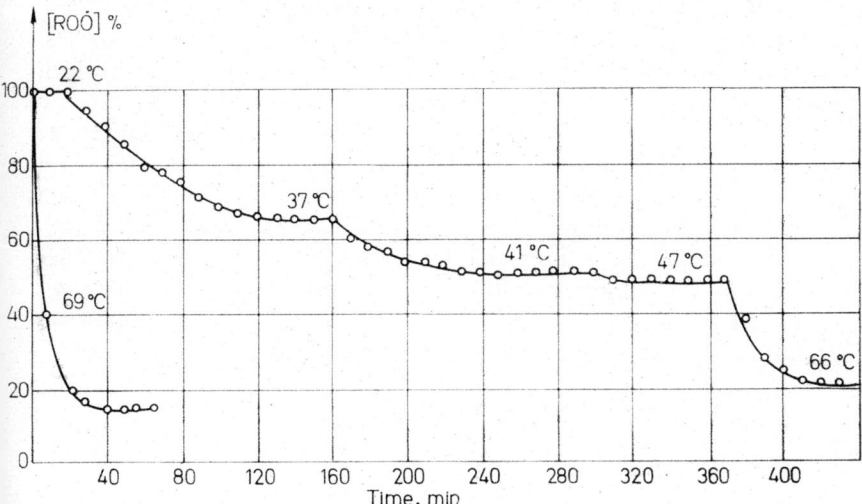

Fig. 3.16 Stepwise recombination of radicals in irradiated poly(chlorotrifluoro-ethylene) at high temperatures[3,14]

3.3.2 RADICALS IN SINGLE CRYSTALS

In single crystals besides contact (isotropic) hyperfine interaction, dipole–dipole (anisotropic) interaction also exists. Therefore a part of the observed hyperfine splitting A will not be dependent on crystal orientation, another part B will. The hyperfine spectrum is described by Equation 3.11 for interaction with a single nucleus. If N groups containing n equivalent nuclei are present:

$$\mathbf{H}(i,j) = \mathbf{H}(0) + \sum_{i=-n}^{+n} \sum_{j=-N}^{+N} m_i | A_j + B_j (3\cos^2 \vartheta - 1) | \qquad 3.17$$

where summation i refers to the n equivalent nuclei with a splitting of A_j, B_j, summation j refers to the N groups of equivalent nuclei, ϑ is the angle between the static magnetic quantum number of the nucleus.

Generally the anisotropic splitting constant B is a tensor with three principal values along the three symmetry axes of the crystal. The variation of the splitting is usually plotted against angle ϑ, with which the sample is rotated about the principal axis.

Ammonium tartrate. The anisotropy of the hyperfine splitting in single crystals is illustrated by the example of irradiated ammonium tartrate.[3,15] Single crystals of about $8 \times 4 \times 2$ mm were grown from aqueous solutions and irradiated with gamma rays. The radical concentration was determined by comparing the signal intensities with known quantities of DPPH. Irradiation with a total dose of 1 Mrad results in a trapped radical concentration of about 10^{18} spins/ml.

The probable radical structure is the following

$$\begin{array}{l} O{=}C{-}O^- \quad NH_4^+ \\ | \\ HO{-}C{-}H \\ |\nearrow \\ HO{-}C^{\bullet} \\ | \\ O{=}C{-}O^- \quad NH_4^+ \end{array} \qquad A = 2.1 \text{ gauss}$$

Hyperfine interaction with the $\beta-H$ atom results in an isotropic splitting (doublet) with $A = 2.1$ gauss. Apparently there is only a small interaction with the OH groups. In deuterated samples the same splitting is observed:

$$\begin{array}{l} O{=}C{-}O^- \quad NH_4^+ \\ | \\ DO{-}C{-}H \\ |\nearrow \\ DO{-}C^{\bullet} \\ | \\ O{=}C{-}O^- \quad NH_4^+ \end{array} \qquad A = 2.1 \text{ gauss}$$

However, the line widths are smaller in the deuterated compound, indicating that some anisotropic coupling with the OH protons is present. Some of the spectra are shown in Fig. 3.17. Upon rotating the crystal around the principal axes shown in the figure the angular variation of the splitting can be determined and from that the principal values of B calculated. The result is $B = 0.7$; 1.0; and 4.6 gauss.

The extreme conditions shown in Fig. 3.17 are $H \parallel a$ with $3\cos^2\vartheta - 1 = 0$, where the doublet collapses. $H \parallel c$ $3\cos^2\vartheta - 1 = 1$, the splitting is maximum (4.6 gauss).

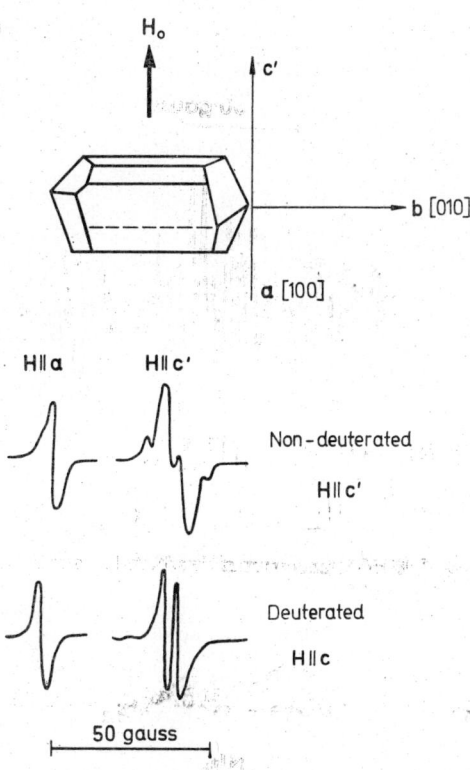

Fig. 3.17 ESR spectra of an irradiated single crystal of ammonium tartrate at different orientations. After Rao and Anderson[3.15]

ε-*Caprolactam.* The resolution on single crystals is sometimes rather good. In irradiated ε-caprolactam single crystals a 24-line structure is found[3.16] as a result of the interaction with a nitrogen and three hydrogen nuclei. The radical structure is the following

$$\begin{array}{c} \text{NH} \\ \text{CO} \quad \text{ĊH} \\ (CH_2)_4 \end{array}$$

The exact steric structure of the molecule is not known.

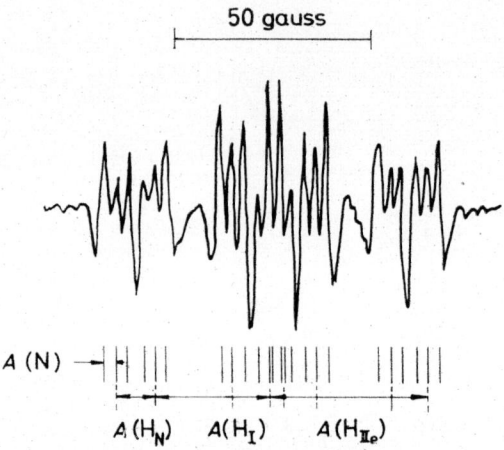

Fig. 3.18 ESR spectrum of irradiated ε-caprolactam. After Kashiwagi and Kurita[3.16]

From the analysis of the spectrum shown in Fig. 3.18 the following couplings between the radical electron and nuclei are revealed

Each coupling is characterized by three principal values corresponding to the three principal axes of the monoclinic crystal. The numbers correspond to the coupling constants in gauss in that orientation where the splitting is the highest. The existence of these couplings was also checked by observing the spectra of partially deuterated crystals.

dl-Aspartic acid. In hydrocarbon radicals hyperfine coupling constants are found to be isotropic for β-hydrogens and anisotropic for α-positions. This is illustrated by irradiated aspartic acid

$$\text{HOOC}-\underset{\underset{H}{|}}{\overset{\overset{H}{|}}{C}}-\underset{\underset{H}{|}}{\overset{\overset{NH_2}{|}}{C}}-\text{COOH}$$

Upon X- or γ-irradiation the following radical is formed[3.17]

$$\text{HOOC}-\underset{\underset{H^{(3)}}{|}}{\overset{\overset{H^{(2)}}{|}}{C}}\xrightarrow{\overset{A_2}{}\overset{A_1}{}}\underset{\underset{H^{(1)}}{\overset{A_3}{\nearrow}\overset{B_1}{\searrow}}}{\overset{\cdot}{C}}-\text{COOH}$$

The hyperfine splitting can be explained by considering two isotropic coupling constants A_2 and A_3 for the β-protons and a single anisotropic one for the α-proton B_1. The COOH hydrogens are not coupled to the radical; by deuteration

$$\text{DOOC}-\underset{\underset{H}{|}}{\overset{\overset{H}{|}}{C}}-\underset{\underset{H}{|}}{\overset{\cdot}{C}}-\text{COOD}$$

the same splittings are observed. Therefore $A_1 = 0$.

Radicals in oriented polymers. Anisotropic hyperfine splittings can be observed in polymers oriented by stretching. The splittings are different when measured with the polarizing magnetic field H_0 parallel or perpendicular to the direction of the stretching. As an example, radicals in irradiated stretched polypropylene will be discussed in some detail on the basis of investigations made by Fischer and Hellwege.[3.18]

In stretched atactic polypropylene irradiated *in vacuo* at room temperature, the following allyl-type radicals are formed

$$\sim \underset{\underset{H}{|}}{\overset{\overset{H}{|}}{C}} - \underset{\underset{CH_3}{|}}{\overset{\overset{H}{|}}{C_2}} - \underset{\underset{H}{|}}{C_1} = \underset{\underset{CH_3}{|}}{C_0} - \underset{\underset{H}{|}}{\dot{C}_1} - \underset{\underset{CH_3}{|}}{\overset{\overset{H}{|}}{C_2}} \sim$$

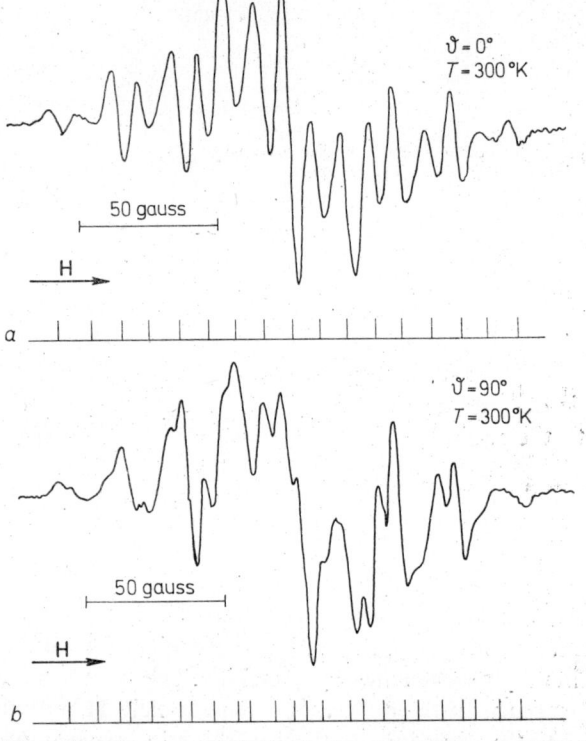

Fig. 3.19 ESR spectra of stretched polypropylene at different orientations. The polymer was irradiated *in vacuo*. After Fischer and Hellwege[3.18]

If the polarizing magnetic field is parallel to the direction of the stretching a 17-line spectrum is observed with a splitting of 17 gauss. Upon rotating the sample a 24-line spectrum appears with two splitting constants. At 90° orientation with respect to H_0 the splittings are maximum: 4.5 and 2.3 gauss (Fig. 3.19b)

Analysis of the spectra taken at different orientations shows that the following nuclear spin systems interact with the radical electron.

1. The β-methyl protons at carbon 0. These are equivalent because of the rotation of the group. The interaction with these protons is isotropic with a coupling constant A_0.
2. The 2 equivalent β-protons at carbons 2. Since they are in β-position the interaction is isotropic with a splitting constant A_2.
3. The α-protons at carbons 1. The interaction is anisotropic with a maximum coupling constant B_1. There is an isotropic part A_1.

According to Equation 3.17 the spectrum is described by

$$H = \frac{h\nu_0}{g\mu_s} + [A_1 + B_1(3\cos^2\vartheta - 1)]\sum_{i=-1}^{+1} m_i +$$

$$+ A_0 \sum_{i=-3}^{+3} m_i + A_2 \sum_{i=-2}^{+2} m_i \qquad 3.18$$

The total number of lines is $(2 + 1)(3 + 1)(2 + 1) = 36$, the number of anisotropic lines which should not change upon changing orientation is 12. The observed spectra can be explained by partial coincidence of the lines. The isotropic coupling constants are the following

The radical is evidently distributed along carbons 1, 2 and 0 and the double bond is partially transferred by hyperconjugation.

The anisotropic component of the coupling with the two α-protons is 0.75 gauss.

The orientation dependence of the coupling in some cases enables one to choose between two possible structures. In polypropylene irradiated and measured at *low temperature* the following structures are possible

$$\sim\underset{H}{\overset{H}{C}}-\underset{CH_3}{\overset{H}{C}}-\underset{H}{\overset{H}{C}}-\underset{H}{\overset{\cdot}{C}}-\underset{H}{\overset{H}{C}}-\underset{CH_3}{\overset{H}{C}}\sim \qquad A$$

$$\sim\underset{H}{\overset{H}{C}}-\underset{CH_3}{\overset{H}{C}}-\underset{H}{\overset{H}{C}}-\underset{CH_3}{\overset{\cdot}{C}}-\underset{H}{\overset{H}{C}}-\underset{CH_3}{\overset{H}{C}}\sim \qquad B$$

Structure A has an α-hydrogen near the radical which should cause anisotropic splitting. In structure B only β-protons are present and correspondingly the splitting should be isotropic. In fact no anisotropy has been observed in the samples irradiated at low temperature; the spectrum consists of 8 equivalent lines with a splitting of 22.5 gauss. Thus structure A is excluded.

3.3.3 RADICALS IN POLYCRYSTALLINE AND AMORPHOUS MEDIA

Although the resolution of ESR spectra in polycrystalline and amorphous media is very poor, some important information about the delocalization of the radical electron can be derived from analyzing the line shapes. Unresolved or poorly resolved hyperfine structures have been analyzed by Lebedev and his co-workers using an electronic computing machine. A series of typical spectra have been published.[3.19] The curves are calculated for typical interaction types with various line widths. Line widths are represented by the parameter

$$\beta = \frac{\Delta H_0}{A} \qquad 3.19$$

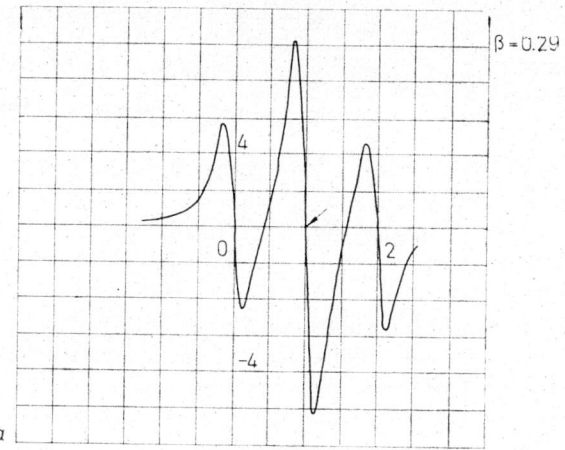

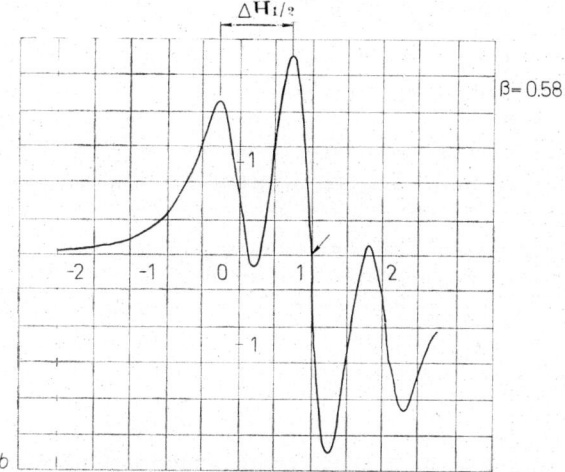

Fig. 3.20 Theoretical hyperfine splitting due to 2 protons at line width parameters 0.29 and 0.58, respectively. Lorentzian line. After Lebedev et al.[3-19]

where A is the hyperfine splitting constant in gauss, ΔH_0 is the width of the derivative line measured between maximum and minimum points. An example of the effect of increasing line width on the shape of the spectrum is shown in Fig. 3.20. The simple case of interaction with two protons is considered. In high resolution three lines appear with a splitting A. This case is approximated by the first curve where $\beta = 0.29$; the splitting constant A is unity.

In this well-resolved spectrum the splitting constant A, line width ΔH_0 and intensity ratios I_1/I_2 can be directly observed. Spectrum b corresponds to a β-parameter of 0.72, i.e. the line width is not much smaller than the hyperfine splitting constant. The distance between the zero intersections of the spectrum is no longer the splitting constant. In such cases it is more convenient to measure the distance between maxima $\Delta H_{1/2}$ as indicated in the figure. The true splitting constant can be calculated from

$$A = \frac{\Delta H_{1/2} - \Delta H_0}{k - 1} \qquad 3.20$$

where ΔH_0 is the line width, k is the number of hyperfine components.

In this way splitting constants can be determined with an accuracy of 15% even in poorly resolved spectra. Even in those cases, when hyperfine splitting is not resolved at all, comparison between theoretical and experimental line shapes makes it possible to determine the width factor β. Since line widths can usually be varied by changing the temperature of the sample, a fair approximate value of the splitting constant can be obtained even from completely unresolved lines.

An example of theoretical spectra as a function of parameter β is given in Figs 3.20, 3.21 and 3.22. Equal interaction with two nuclei of spin 1/2 is calculated with Lorentzian line shapes. In Fig. 3.20 a fairly well-resolved spectrum is shown with a binomial intensity ratio of 1 : 2 : 1; the splitting is unity. Although the β-parameter is small, the

Fig. 3.21 Theoretical hyperfine splitting due to 2 protons at line width parameters 0.83 and 1.16, respectively. Lorentzian line. After Lebedev et al.[3-19]

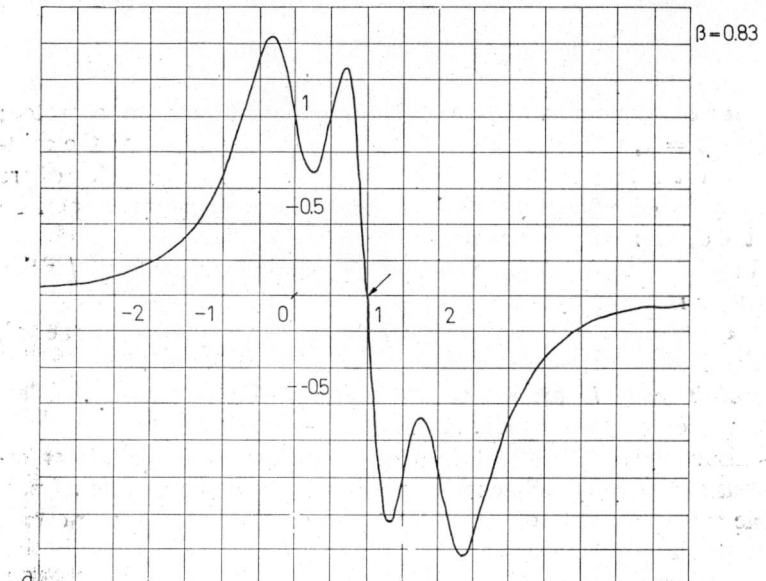
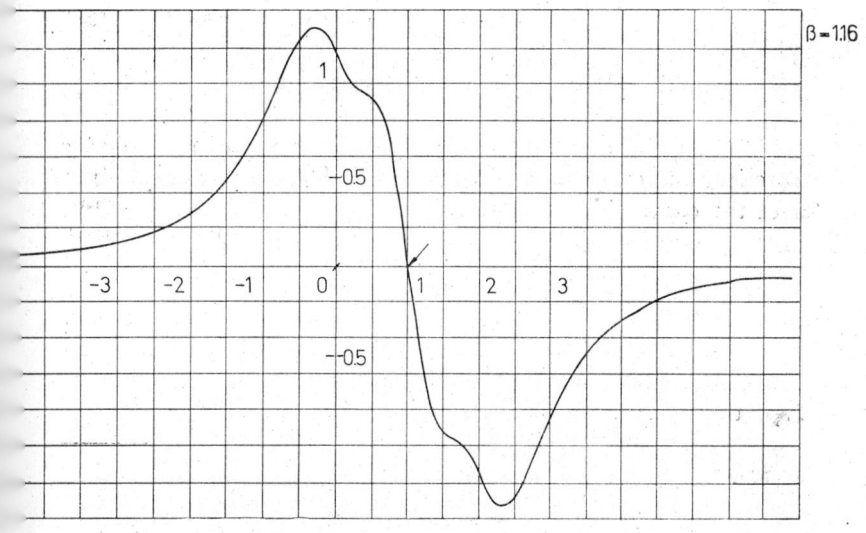

zero intersections do not provide the exact splitting constant A; it must be calculated by using Equation 3.20. Increasing the line widths makes the resolution poorer. At $\beta = 0.83$ the components are just separated; above $\beta = 1$, where the individual line widths are equal to the splitting constant, no splitting is observed at all. However, line shapes are strictly correlated with β which can thus be determined by comparison of the experimental line with the theoretical ones.

In practice approximate values of the splitting constants and/or line widths are in many cases known. Thus analysis of unresolved line shapes permits one to say how many nuclei interact with the radical, which is one of the most important problems to decide.

g-Anisotropy in amorphous and polycrystalline substances. In less delocalized radicals, as e.g. $X-\dot{S}$ and $O-\dot{O}$, there is an appreciable *g*-anisotropy as a result of the increased spin-orbit coupling. In amorphous and polycrystalline media the *g*-values are averaged to all possible orientations. The result is an asymmetric spectral line or even a fine structure such as shown in Fig. 2.17.

In systems of cylindrical symmetry the *g*-value is expressed as follows

$$g^2 = g_\perp^2 \sin^2 \vartheta + g_\parallel^2 \cos^2 \vartheta$$

where $g_\parallel$ is the *g*-value measured in orientation parallel to the polarizing field, $g_\perp$ is that measured in perpendicular orientation. The line shape for random orientations depends on the values of $g_\parallel$ and $g_\perp$. Theoretical curves are shown in Fig. 3.23.[3.20] The asymmetry parameter of the curves is

$$\delta = \frac{\mathbf{H}_\parallel - \mathbf{H}_\perp}{\Delta \mathbf{H}_{1/2}} \qquad 3.21$$

where $\mathbf{H}_\parallel$ and $\mathbf{H}_\perp$ are the resonance fields corresponding to *g*-values $g_\parallel$ and $g_\perp$, respectively, $\Delta \mathbf{H}_{1/2}$ is the half-width to full amplitude of the spectral line.

Fig. 3.22 Inhomogeneously broadened Lorentzian line formed by hyperfine interaction with 2 protons. Line width parameters 1.45 and 2.89 for curves *a* and *b*, respectively. After Lebedev et al.[3.19]

β = 1.45
1
0.5
−0.5
−1
−3
−2
−1
0
1
2
3

β = 2.89
0.4
−0.4
−5
0
5

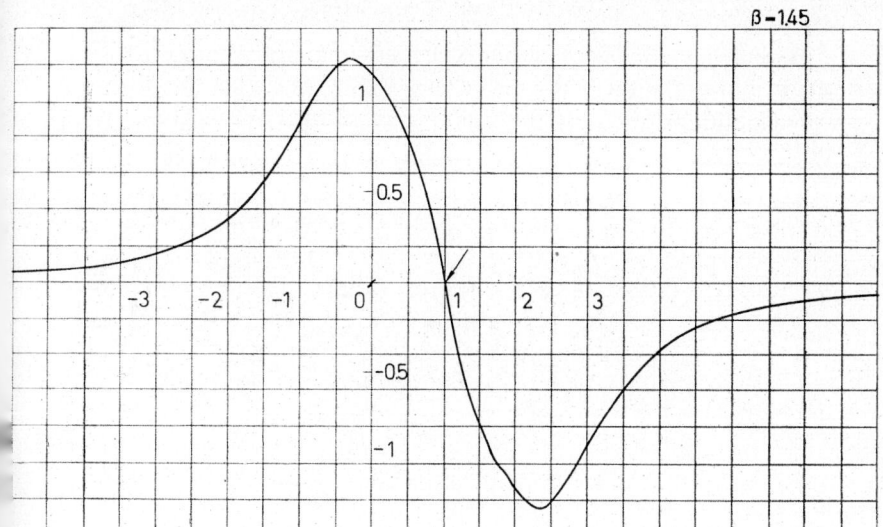

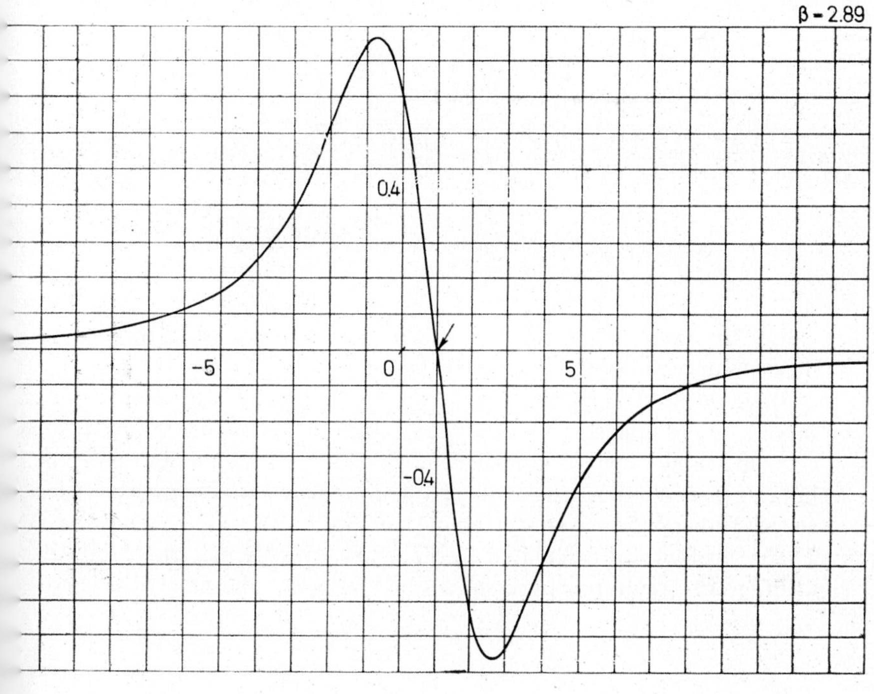

As shown in the figure the line is almost symmetrical for $\delta = 1$, where the g-splitting is the same as the line width. At higher g-anisotropy, (smaller line widths) the line splits to result in fine structure, as e.g. in the case of $\delta = 4$–12. From the observed line shapes approximate

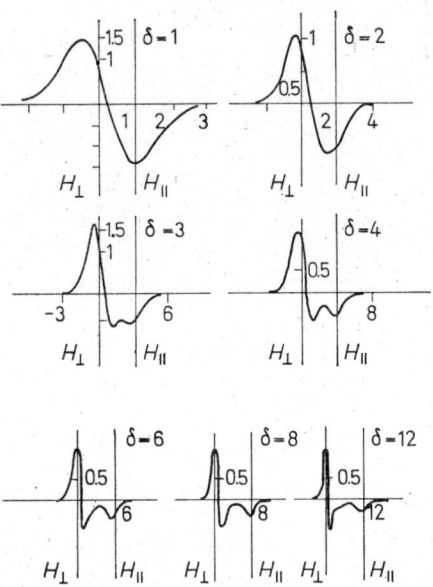

Fig. 3.23 Theoretical line shapes of randomly oriented radicals. After Lebedev[3.20]

values of ΔH_0, $g_\perp$, $g_\parallel$ can be determined by comparing the experimental curves with the calculated ones.

Since the splitting caused by g-anisotropy depends on the operating field, it is possible to get higher resolution at higher fields and at the corresponding higher frequencies. For this it is advisable to work at the highest possible frequencies. Investigation of g-anisotropy has been successfully applied to study radical orientation in *stretched polymers*. Theoretical curves calculated by Lebedev for polymer chain radicals at different orientations are shown in Fig. 3.24. The first row corresponds to oriented samples with different orientations of the magnetic field H_0, molecular symmetry axis M and radical symmetry axis R.[3.21] As shown in the figure a single, almost symmetrical line should be found for radicals oriented parallel to the molecular axis and

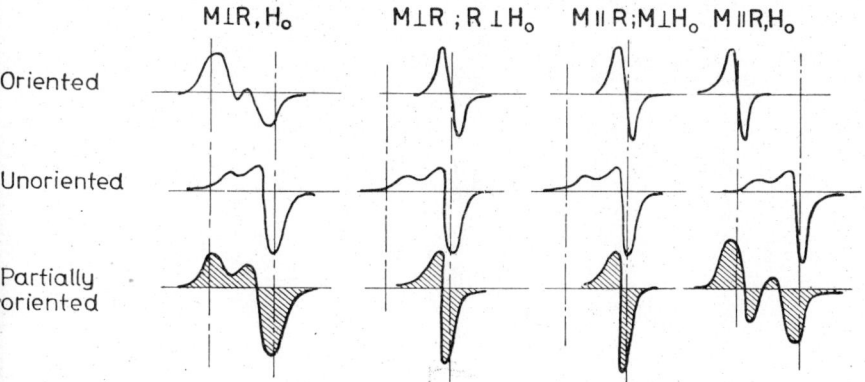

Fig. 3.24 Comparison of line shapes of oriented and partially oriented radicals. After Lebedev[3.21]

parallel to the polarizing magnetic field $R \parallel M, H_0$. A fine structure appears for $M \perp R, H_0$. The second row corresponds to unoriented radicals. The line shape is the same for each orientation, but the line is shifted. The third row corresponds to a real case, where randomly oriented radicals are present together with oriented ones. The spectral line shapes are different for each orientation. Usually an asymmetric singlet is obtained such as for $M \perp R; R \perp H_0$ and $M \parallel R$, $M \perp H_0$. At the other two orientations a fine structure appears.

The anisotropy of g-values is illustrated by the example of irradiated polypropylene[3.18] (Fig. 3.25). Similar curves are obtained in irradiated poly-(tetrafluoro-ethylene).[3.21] The unoriented (unstretched) polymers exhibit characteristic asymmetric lines upon irradiating the polymers in air. The radicals are of ROȮ type with unpaired electrons localized on the oxygen. The stretched polymers exhibit different spectra at different orientation of H_0 with respect to the direction of the stretching

M. In parallel orientations single, asymmetric lines are observed similar to theoretical curves of $M \perp R$, H_0 and $M \perp R$; $M \perp H_0$ in the mixed case. This is easy to interpret, for stretching only results in a partial orientation of the radicals; others remain randomly oriented.

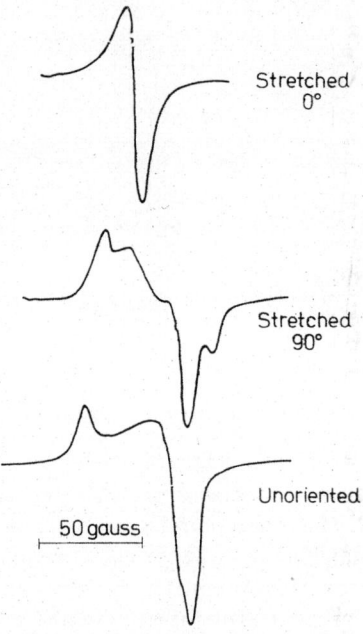

Fig. 3.25 Orientation dependence of the peroxy radicals in irradiated polypropylene. After Fischer and Hellwege[3.18]

Upon rotating the sample through $90°$ line shapes similar to cases $M \perp R$, H_0 and $M \parallel R$, H_0 are observed.

Irradiated sodium citrate pentahydrate. For illustrating a complex study of radicals in single crystals, the radicals formed by irradiating

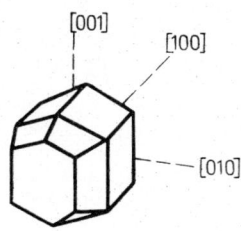

single crystals is discussed in some detail by D. B. Russell.[3.22] The crystal can be grown from saturated aqueous solution of sodium citrate. The crystal is orthorhombic, the principal axes are shown in Fig. 3.26. Upon irradiating the crystal at room temperature with ^{60}Co

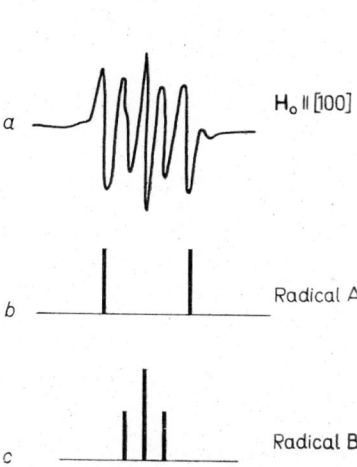

Fig. 3.26 ESR spectrum of an irradiated single crystal of sodium citrate. After Russell[3.22]

gamma rays it turns yellow. The ESR spectrum taken at $H_0 \parallel$ [100] is shown in Fig. 3.26a. Upon increasing the microwave power the intensity of the side lines of the spectrum is increased more strongly than that of the inner lines. The temperature dependence of the lines is also different, and thus it has been concluded that two radicals are present in the system, one giving a doublet with a splitting of 91 Mc/s, the other a triplet with 1 : 2 : 1 intensity ratio split by 16 Mc/s. The corresponding spectra are shown in Fig. 3.26b and c. Radical B giving the triplet is the following

$$
B \quad
\begin{array}{c}
H \\
| \\
H-C-COONa \\
| \\
\dot{O}-C-COONa \\
| \\
H-C-COONa \\
| \\
H
\end{array}
$$

The interaction with the two protons results in a triplet of 1 : 2 : 1 in $H_0 \parallel$ [100], [001] direction and a doublet of 1 : 1 in [010] direction. The principal values are B (16; 21.5; 15). At intermediate orientations the radical produces four lines.

The other radical A exhibiting a doublet at [100] position is the following

$$
A \quad
\begin{array}{c}
H-\dot{C}-COONa \\
| \\
HO-C-COONa \\
| \\
H_2C-COONa
\end{array}
$$

The splitting is due to the single proton. The dependence of the splitting constant on the orientation of the crystal is shown in Fig. 3.27. The observed splittings are plotted against the angle of rotation around [011] and [010] axes. The cosine-law of Equation 3.11 holds exactly. The principal values of the hyperfine splitting tensor are (91.1; 54.2; 36.7).

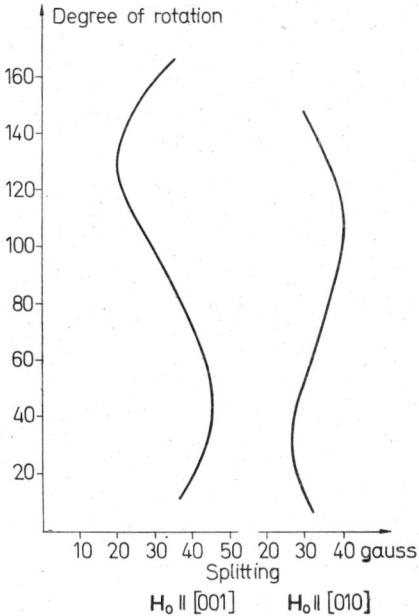

Fig. 3.27 Dependence of the hyperfine splitting on the crystal orientation. Irradiated sodium citrate. After Russell[3.22]

3.3.4 ELECTRON-NUCLEAR DOUBLE RESONANCE

As discussed in Chapter 2 electron-nuclear double resonance permits one to observe ESR hyperfine splittings in those cases when originally they are hidden by the large dipole–dipole broadening. The reason is that the ENDOR line widths are determined by the *nuclear* relaxation times, which are several orders of magnitude longer than electronic relaxation times. Thus it is possible to increase the resolution of ESR spectra appreciably by using the ENDOR technique without loss in sensitivity.

The earlier ENDOR experiments were performed at very low temperatures, usually at the temperature of liquid helium (4.2 °K).

Recently ENDOR spectra have been observed at liquid nitrogen temperatures, and using very high r.f. fields, even at room temperatures. This is important in the study of free radicals, because many crystal structures change upon cooling down to liquid nitrogen temperatures

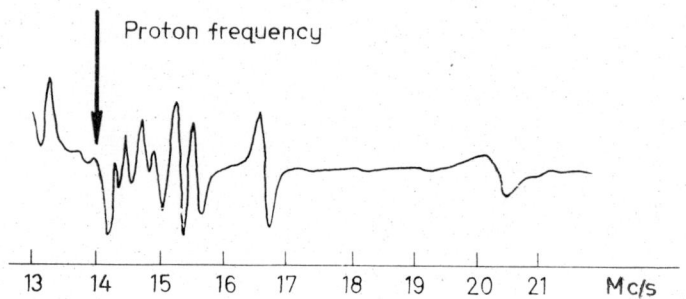

Fig. 3.28 Electron-nuclear double resonance spectrum of irradiated adipic acid. After Kwiram[3.23]

with a corresponding change in the hyperfine splitting constants. The pulse technique developed by Hyde, mentioned in Section 2.4, makes it possible to observe ENDOR lines near the free proton frequency, where spectra taken by ordinary technique would be hidden by a broad (2 Mc/s) depolarization signal.

As an example of this technique the ENDOR signal taken at $-90\ °C$ in an irradiated single crystal of adipic acid is shown in Fig. 3.28 after A. Kwiram.[3.23] The irradiated (X-rays) single crystal is placed in an ESR cavity and the main field is set to give the largest ESR signal. The second, high-power r.f. field is swept through the range of 13–21 Mc/s and the spectrum is recorded. The radical structure is the following

$$\underset{HO}{\overset{O}{\diagdown}}C-\underset{H_\alpha}{\overset{}{C}}-\underset{H_\beta}{\overset{H_\beta}{C}}-\underset{H_\gamma}{\overset{H_\gamma}{C}}-\underset{H_\delta}{\overset{H_\delta}{C}}-\underset{O}{\overset{OH}{\diagup}}C$$

The interaction of the radical electron with the α- and β-protons can be observed by ordinary ESR technique. The splitting from H_γ and H_δ

and that of the carboxyl groups cannot be resolved because of the inhomogeneous broadening. As shown in Fig. 3.28 the lines of H_γ and H_δ are very well resolved in the pulsed ENDOR spectrum. The individual line widths are 300 kc/s at room temperature, and 120 kc/s at -90 °C where the spectrum of Fig. 3.28 has been recorded. The lines can be identified by partial deuteration of the compound.

3.3.5 RADICALS FORMED BY MECHANICAL TREATMENT

Machining solid materials essentially involves breaking of chemical bonds. The radicals formed in this way can be trapped at low temperatures at sufficiently high concentrations for ESR study. For obtaining the high concentration of radicals required for the ESR measurements the mechanical treatment must be made at low temperatures, usually at the temperature of liquid nitrogen. A simple arrangement for doing this is shown in Fig. 3.29. The material is placed in a test tube which is

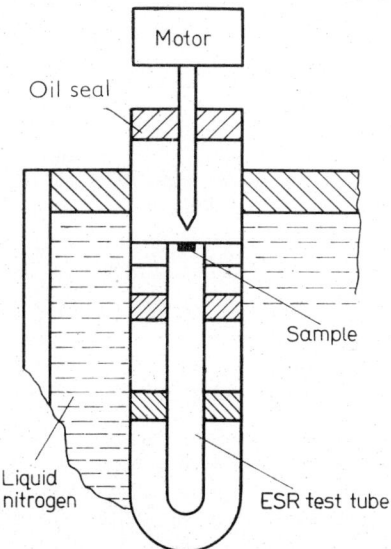

Fig. 3.29 Arrangement for mechanical destruction of polymers at low temperatures. After Lázár and Szőcs[3.24]

evacuated and placed in the dewar vessel containing liquid nitrogen. The sample is drilled *in vacuo*, the drill input being sealed by an oil system. The fragments of the material are collected in an ordinary quartz test tube used in ESR measurements. The radicals trapped in

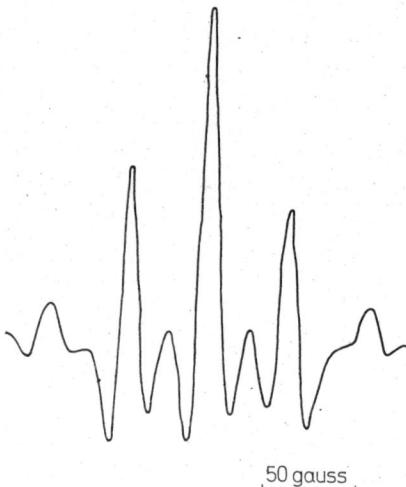

Fig. 3.30 Second derivative ESR spectrum of radicals formed by mechanical destruction of poly(methyl methacrylate) at 77 °K. After Butyagin[3,25]

this way can be measured rather easily at low temperatures by using a conventional ESR spectrometer.[3,25]

Recently radicals obtained by mechanical destruction of various polymers have been investigated. Some of the results are summarized in Table 3.1 after Butyagin.[3,25] Different kinds of destruction have been studied: grinding, drilling the samples separately or *in situ* in the cavity of the ESR spectrometer. All experiments were performed at the temperature of liquid nitrogen.

In Fig. 3.30 the ESR spectrum of poly(methyl methacrylate) radicals produced by mechanical destruction are shown after the measure-

Table 3.1

RADICALS FORMED BY MECHANICAL DESTRUCTION OF POLYMERS
(AFTER BUTYAGIN[3.25])

Polymers		No. of HFS lines	Splitting gauss
Polyethylene	$\sim CH_2-\dot{C}H-CH_2\sim$	6	20
Polystyrene	$\sim CH_2-\dot{C}H-$	3	24
Poly(methyl acrylate)	$\sim CH_2-\dot{C}H(COOCH_3)$	3	29
Poly(vinyl acetate)	$\sim CH_2-\dot{C}H(OCOCH_3)$	3	17
Poly(vinyl alcohol)	$\sim CH_2-\dot{C}H(OH)$	3	
Poly(methyl methacrylate)	$\sim CH_2-\dot{C}(CH_3)COOCH_3$	9	27
Polyisobutylene	$\sim CH_2-\dot{C}(CH_3)_2$	15	29
Polymethylstyrene	$\sim CH_2-\dot{C}(CH_3)C_6H_5$	5	21

ments of Butyagin.[3.25] The spectrum has been obtained just after the mechanical treatment at 77 °K. As in the case of ultraviolet irradiation (see Chapter 5) this spectrum is interpreted as corresponding to different radical species present simultaneously. Upon storing the sample at elevated temperatures the less stable species decay. Further reaction of radicals produced by mechanical destruction will be discussed in Chapter 5.

3.4 RADICALS IN THE LIQUID STATE

In the liquid state the reactivity of radicals is mainly determined by their chemical structure. The physical surroundings have much less effect than in the solid state.

Simple radicals containing single bonds are found to be highly reactive in solution. In such systems a dynamic equilibrium concentration of the radicals can be measured by ESR provided that the stationary concentration is high enough. In some cases non-stationary processes can also be studied. Examples of this are given in Chapter 5.

Radicals having conjugated double bonds are found to be relatively stable in solution. Examples of this have already been given in Sections 3.1 and 3.2 (diphenyl-picrylhydrazyl, e.g.). The stability of such systems is due to the delocalization (resonance) of the radical electron. This is why most ESR data are available for aromatic systems, which are easy to measure even in dilute solutions.

Transient radicals are much more difficult to measure by ESR, because the concentration is in most cases too low. In these cases the continuous flow system described in Chapter 2 should be used.

The lifetime of the individual radicals has no effect on the ESR spectra provided it is long compared with the reciprocal operating frequency 10^{-10} sec. If the average lifetime of the radicals approaches this limit, the ESR lines would be seriously broadened and immersed in the noise level.

The stationary concentration of the radicals present in the sample is often not high enough for ESR measurement. This is a serious restriction; many intermediate radical products cannot be detected in the liquid state because of the lack of sensitivity. From this respect it is better to work in the solid state at the lowest possible temperature, where sensitivity is improved.

On the other hand, the resolution is tremendously improved in the liquid state by averaging out the dipole–dipole broadening effects, and therefore much more information can be gained about the radical structure. Thus if very small radical concentrations (below 10^{14} spins/ml) are to be measured it is better to work at low temperatures in the solid state. If there are enough radicals in the system it is better to work in solutions. Examples on investigating intermediate radical products are given in Chapter 5. Some examples of the 'freezing-in' technique are also quoted in order to illustrate how reactions can be stopped and intermediate radicals trapped at very low temperatures.

In this section stable radicals or intermediate radicals in high stationary concentrations in solutions are considered. Examples on non-stationary intermediates are given in Chapter 5.

3.4.1 SOLVENT EFFECTS

The hyperfine splitting and resolution of ESR spectra have already been discussed in Section 2.2 and Section 3.2. It has been shown that in favourable cases, as for example in the tetracene-radical ion dissolved in sulphuric acid, a resolution of 0.03 gauss is achieved. At such a high resolution the effect of the solvent molecules on the line widths and splittings must also be considered.

Collisions with the solvent molecules. The most straightforward transfer of energy from the solvent molecules to radicals is due to collisions. The effect of collisions on electron spin states is measured by the change of the spin–lattice relaxation time of the radical electrons. As shown in Section 1.2 the spin–lattice relaxation time is a measure of how the spin system is connected with the surrounding 'lattice'. Evidently it must be affected by collisions. The average time between collisions, called correlation time τ_c, is expressed according to Bloembergen, Purcell and Pound[3.26] as follows

$$\tau_c = \frac{4\pi r_0}{3kT} \eta \qquad 3.22$$

where r_0 is the effective radius of the ion or molecule where the radical electron is located, η is the viscosity of the medium, kT is the thermal energy.

The spin–lattice relaxation time of the radical electron is approximately

$$T_1 \propto \frac{1 + 4\pi v_0^2 \tau_c^2}{\tau_c} \qquad 3.23$$

where v_0 is the operating frequency.

T_1 can be measured independently by the saturation method or more accurately by the spin–echo technique. The temperature dependence of T_1 is usually found as

$$T_1 \propto \exp\left(\frac{E_0'}{kT}\right) \qquad 3.24$$

For the correlation time by theoretical arguments a similar relationship holds

$$\tau_c = \tau_c^0 \exp\left(\frac{E_D}{kT}\right) \qquad 3.25$$

where E_D is the activation energy of the radical diffusion.

Using equations 3.23, 3.24 and 3.25 by a more detailed analysis, fair approximate values of τ_c can be obtained by measuring the temperature dependence of the spin–lattice relaxation time T_1. Some typical values are the following

$$T_1 = 10^{-5} \text{ to } 10^{-6} \text{ sec} \qquad T = 30\,°C$$
$$\tau_c = 10^{-9} \text{ sec} \qquad r_0 = 7\,\text{Å}$$

For many practical cases $\omega^2\tau_c^2 \gg 1$, and thus Equation 3.23 is reduced to

$$T_1 \propto \tau_c$$

The above statements hold fairly well for aromatic ion-radicals in dilute solutions.

Electron transfer. The radical electron may be transferred to a neighbouring molecule M to produce a new radical $\dot{M}$. The reaction scheme is the following

$$\dot{R} + M \rightleftarrows \dot{M} + R \qquad\qquad 3.26$$

where $\dot{R}$ is the radical, R is the non-radical product left after the electron has been transferred to a solvent molecule M. Such new radical species have been observed in tetracene by Hyde and Brown.[3.8] The signal corresponding to $\dot{M}$ is a broad single line indicating that the exchange reaction is fast, i.e. the effective lifetime of radical $\dot{M}$ is short. This weak broad line cannot be observed directly with the strong tetracene spectrum (Fig. 3.10). It can only be detected by saturating the sharp tetracene lines at high microwave power level. At saturation the 85-line spectrum of Fig. 3.10 vanishes and the broad background line appears, for it is not saturated at this power level. The g-value of the $\dot{M}$ line in the tetracene–H_2SO_4 system is 2.0019 ± 0.00006 in contrast with that of tetracene 2.00250 ± 0.00003.

Electric and magnetic local fields. The local field at a radical electron is affected by the diamagnetic shielding of the surrounding solvent molecules. Strongly polar molecules can also affect the electron spin state *via* the spin–orbit coupling just as in crystalline solids. For highly solvated radical ions in polar solutions this effect can be quite appreciable. At high resolution the shifts caused by the electric and magnetic fields of the solvent molecules can be observed. The following main effects are observed

1. The hyperfine splitting constants A_j are slightly dependent on the solvent molecules. The change in the splitting is not uniform for all nuclei; some splittings are less affected by the solvents than others.
2. The relaxation times will be different for each solvent and also for the couplings, resulting in a variation of the line widths.
3. The lines are shifted slightly as a result of the g-anisotropy by the local electric fields of the solvent ions or molecules.

As an example let us consider the hyperfine splitting in thio–indigo radicals[3,27]

Because of the symmetry of this molecule there are four groups of spins each containing two protons. The total number of the HFS lines is thus, according to Equation 3.11, $(2 + 1)^4 = 81$. The total width of the spectrum in solutions of acetone is 6.774 gauss, in ethanol 7.346 gauss. The hyperfine splitting constants in acetone are the following

in ethanol

Numbers denote the splitting constants in gauss caused by the corresponding proton. Line widths in thio-indigo are of the order of 0.06 gauss.

The effect of solvent upon line widths is illustrated in Fig. 3.31. The radical is the following

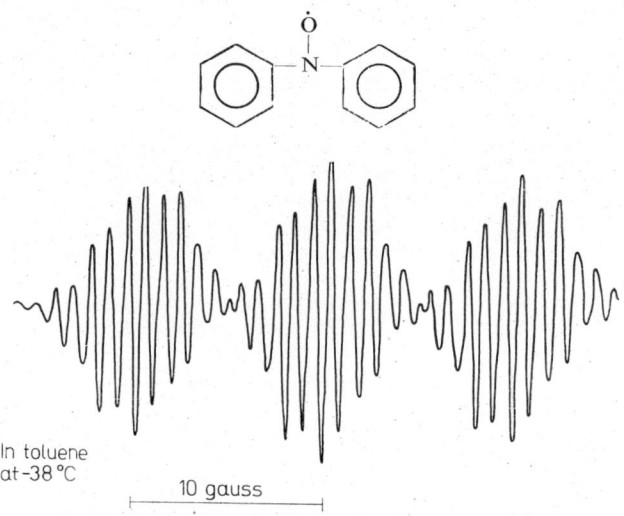

Fig. 3.31 ESR spectrum of diphenyl nitrosyl radicals in solution. Courtesy of Mohos and Tüdős[3.28]

The highest splitting is observed from the ^{14}N nucleus, resulting in a triplet with a splitting constant of 9.66 gauss. Each line of this triplet is further split by the ring protons. From the spectrum the following hyperfine coupling constants have been derived[3.28] with $A_N = 9.66$ gauss coupling with ^{14}N, $A_m = 0.86$ gauss coupling with the ring protons in *meta*-position, and $A_{o,p} = 4.86$ gauss with the *ortho*- and *para*-protons.

The line widths in toluene solutions are 0.3 gauss; the resolution is good. In solutions in ethanol, lines are broadened and the hyperfine structure from the ring protons is very poorly resolved. The splitting constants are also slightly changed.

An interesting example of the variation of electron relaxation times for different nuclear spin states is shown in Fig. 3.32. The radical is the following[3.29]

$$\begin{array}{c}
\dot{O} \\
| \\
H_3C\diagdown\quad\diagup N\diagdown\quad\diagup CH_3 \\
\quad C\qquad\qquad C \\
H_3C\diagup\ |\qquad\quad |\ \diagdown CH_3 \\
\quad H_2C\qquad CH_2 \\
\diagdown\quad\diagup \\
C \\
\| \\
O
\end{array}$$

The ^{14}N splitting is high. There is an additional splitting from the methyl groups although they are separated from the radical by two carbon atoms. The spectrum of Fig. 3.32 has been swept from right to left with decreasing fields. The line widths at the low field side are appreciably larger than those at the high field ^{14}N component, indicating that electronic relaxation times are shorter for ^{14}N $m_I = -1$ than for $m_I = +1$.

Oxygen effect. The ESR line widths in solutions have been found to depend strongly on the oxygen content of the sample. Samples saturated by air usually exhibit broad lines; high resolution can only be obtained by careful degassing. It can be done roughly by bubbling pure nitrogen through the samples. For accurate measurements the samples should be degassed by repetitive freezing–melting cycles *in vacuo*.

The line-broadening effect of oxygen is explained by its paramagnetism. The local field generated by oxygen molecules is fairly large.

Since oxygen molecules in the sample are randomly oriented, the local fields are averaged to produce broad lines.

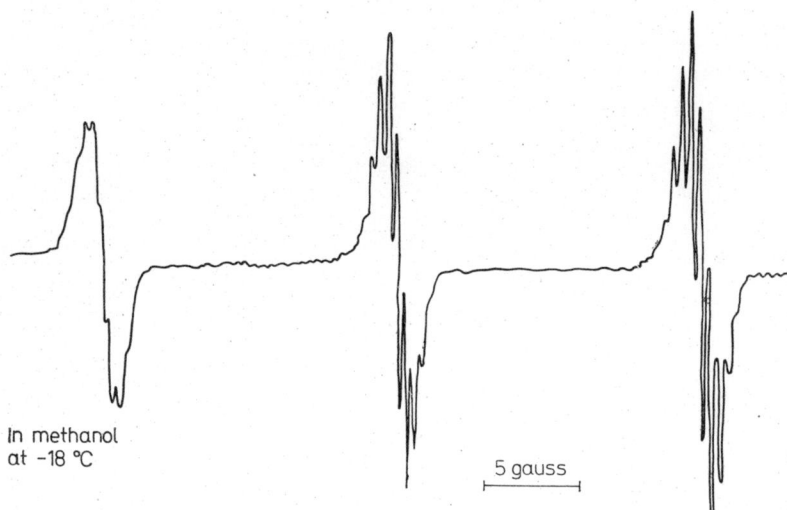

Fig. 3.32 The variation of line widths as a function of nuclear magnetic quantum number. Radical

Courtesy of Mohos and Tüdős [3.29]

Second-order hyperfine interaction. At very high resolution in the liquid state it is possible to observe second-order splittings according to the third term of Equation 3.7. As shown in Section 3.1, the second-order splitting is 0.1 gauss. The splitting has been observed in the pyracene radical ion:

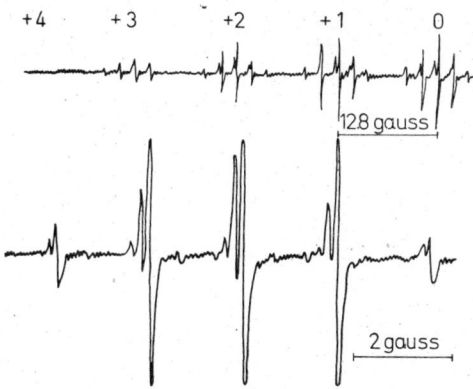

This radical ion is formed when pyracene is dissolved in concentrated sulphuric acid. There are two proton–spin systems in this radical, 8 α-

Fig. 3.33 Second-order hyperfine splitting of pyracene. After de Boer and Mackor[3.30]

protons and 4 β-protons. The total number of lines is 45; 9 groups of quintuplets. The splitting constants are $A_\alpha = 12.80$ gauss and $A_\beta = -2.00$ gauss for the α- and β-protons, respectively. At higher resolution each line is further split as shown in Fig. 3.33. Only a single quintuplet group is shown. The second-order splitting is clearly resolved.

3.4.2 TYPICAL RADICALS IN SOLUTION

From the large amount of experimental data on the structure of stable and transient radicals in solution, a few types are selected in order to illustrate the possibilities. Some additional examples are also given in Chapter 5.

Aromatic radical ions. Aromatic radical ions can be prepared in solution by oxidation or reduction of aromatic hydrocarbons. Oxidation can be made with $SbCl_5$ in CH_2Cl_2; BF_3 in CF_3COOH, $SbCl_3$, $AlCl_3$ or simply by dissolving the hydrocarbons in concentrated sulphuric acid. The resulting radical cations exhibit highly resolved hyperfine splittings. Radical anions can be prepared by reduction with alkali metals in inert solvents or by electrolysis.

The observed hyperfine splittings can be interpreted fairly well on the basis of molecular orbital theory. The spin densities can be calculated relatively easily. The simplest aromatic radical is the benzene negative ion with the following spin–spin coupling constants and spin densities[3.31]

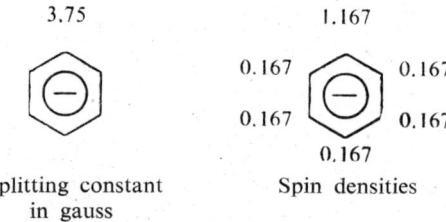

Splitting constant in gauss   Spin densities

The spin density is the same at each ring proton. Spin densities are calculated from the McConnell Equation 3.10 (see the discussion in Section 3.2). From a least squares analysis of the experimental data the following expressions have been derived for calculating spin densities in unsubstituted aromatic cations and anions, respectively[3.32]

$$A_{H_i}^- = (28.6 \pm 0.43)\, \rho_i$$

$$A_{H_i}^+ = (35.7 \pm 0.38)\, \rho_i$$

3.27

where $A_{H_i}^-$ and $A_{H_i}^+$ are the splitting constants of ring protons i in the negative and positive ions, respectively, ρ_i's are the corresponding spin densities.

Values of the splitting constant A_H obtained experimentally are plotted against calculated spin densities in Fig. 3.34.

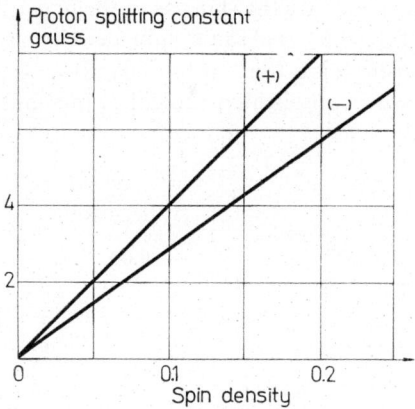

Fig. 3.34 Semi-empirical plots of spin density against splitting constants for positive and negative ion radicals. After Lewis and Singer[3.31]

As an example of positive and negative ions, splittings and spin densities of anthracene are given as follows[3.33]

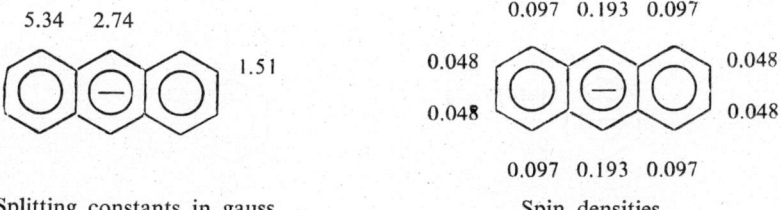

Splitting constants in gauss   Spin densities

the splitting constants in the positive ion are[3.31]

$$6.53 \quad 3.06$$

⬡⊕⬡ 1.38

Calculated spin densities are the same as in the negative ion. According to the chart of Fig. 3.34 the same spin density results in a higher splitting in the positive ion than in the negative.

A somewhat more complicated radical is phenanthrene[3.34]

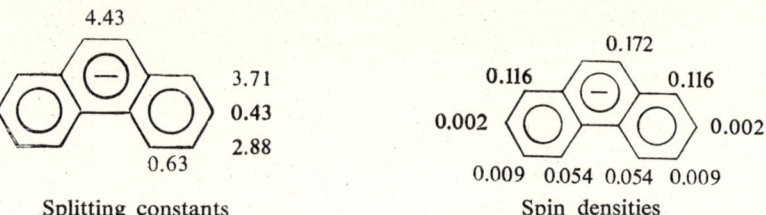

| Splitting constants in gauss | Spin densities |

Such radicals exhibit extremely complicated spectra. The best way of analyzing them is to introduce approximate coupling constants in an electronic computer and compare the computed spectra with the experimental ones. Equations 3.27 and the plot of Fig. 3.34 have been constructed by correlating a large amount of experimental results. For a detailed analysis see the paper of I. C. Lewis and L. S. Singer.[3.31]

Nitro-substituted radical anions. The group of aromatic radicals considered here can be prepared by electrolysis in dimethyl formamide solutions. The simplest product is nitrobenzene anion radical with the following spin densities

NO_2 $^{14}NO_2$ 9.70

⬡ 0.1417   ⬡ 3.36
   0.0451   1.07
0.170   4.03
Spin densities   Splittings in gauss

Further substitution results in a change in the spin densities. In 1,2-dinitrobenzene

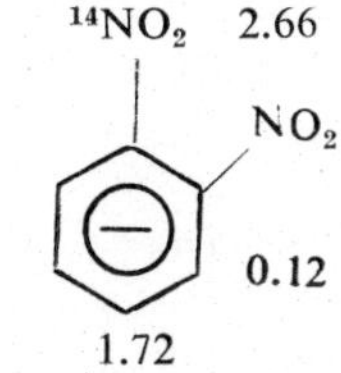

$^{14}NO_2$ 2.66, NO_2 0.12, 1.72

Splittings in gauss

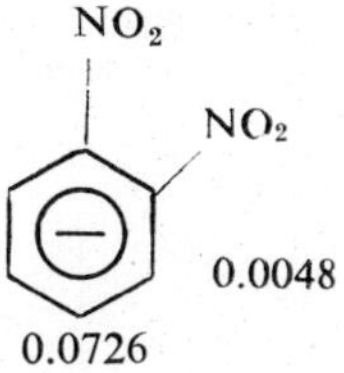

NO_2, NO_2 0.0048, 0.0726

Spin densities

The position of the substituting groups has also an effect. In 1,3-dinitrobenzene

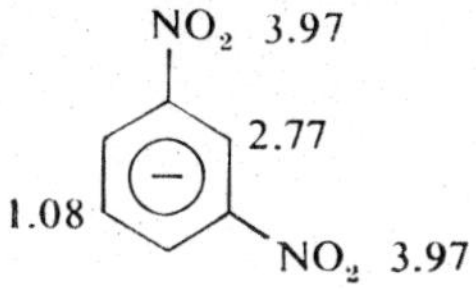

NO_2 3.97, 2.77, 1.08, NO_2 3.97

Splittings in gauss

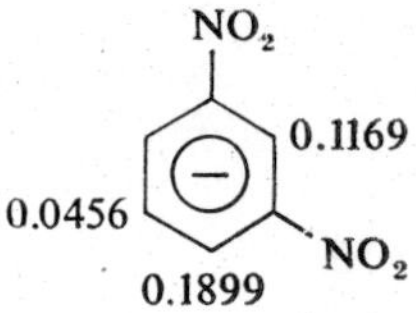

NO_2 0.1169, 0.0456, NO_2 0.1899

Spin densities

and in 1,4-dinitrobenzene.

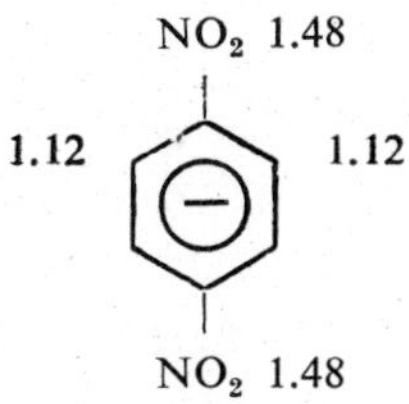

NO_2 1.48, 1.12, 1.12, NO_2 1.48

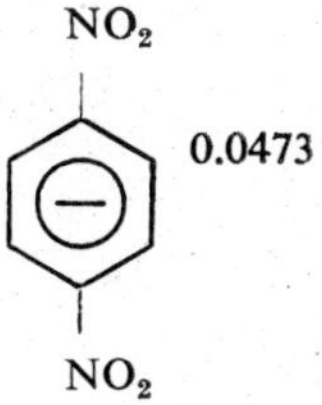

NO_2, 0.0473, NO_2

Respective data on some further derivatives are collected in Table 3.2. A detailed discussion can be found in the paper of Rieger and Fraenkel.[3,35]

The spin densities are calculated from the measured splitting constants using the McConnell equation

$$A_i^H = Q_{CH}\rho_i$$

Table 3.2

SPIN DENSITIES IN RADICAL ANIONS OF SOME NITROBENZENE DERIVATIVES (AFTER RIEGER AND FRAENKEL[3.35])

Radical-anion	Position	Spin density	Splitting constants gauss	
			in dimethyl formamide	in acetonitrile
Nitrobenzene	2	0.1417	3.36	3.39
	3	0.0451	1.09	1.07
	4	0.1700	4.03	3.97
	N	(0.2381)	9.70	10.32
1,2-dinitrobenzene	3	0.0048	0.11	0.42
	4	0.0726	1.72	1.63
	N	(0.1088)	2.66	3.22
1,3-dinitrobenzene	2	0.1169	2.77	3.11
	4	0.1899	4.50	4.19
	5	0.0456	1.08	1.08
	N	0.1042	3.97	4.68
1,4-dinitrobenzene	2	0.0473	1.12	1.12
	N	(0.1042)	1.48	1.74
4-nitrotoluene	1	0.1463	—	3.98
	2	0.0468	—	2.11
	3	0.1430	—	3.39
	N	0.2452	—	—
4-nitrobenzaldehyde	carbonyl C	0.0578	1.37	1.23
	2	0.0097	0.23	0.44
	3	0.1245	2.95	3.10
	5	0.0954	2.26	2.37
	6	0.0160	0.38	0.44
	N	—	5.11	5.83
4-nitroacetophenone	2	0.0219	0.52	0.66
	3	0.1236	2.93	2.95
	5	0.1122	2.66	2.95
	6	0.0219	0.52	0.66
	1	0.0283	0.77	0.66
	N	0.1716	5.86	7.02
3-nitroacetophenone	2	0.1295	3.07	
	4	0.1574	3.73	
	5	0.0460	1.09	
	6	0.1865	4.42	
	N	(0.2241)	8.97	
2,4-dinitrophenol dianion	3	0.2900	6.87	
	4	0.0797	1.89	
	N	0.1089	3.64	

where A_i^H is the isotropic splitting caused by proton i, ρ_i is the corresponding π-radical electron density, Q_{CH} is a constant for $C-H$ bonds; its value is usually taken as 23 gauss. For these systems the best fit with theoretical values has been found with $Q_{CH} = 23.7$

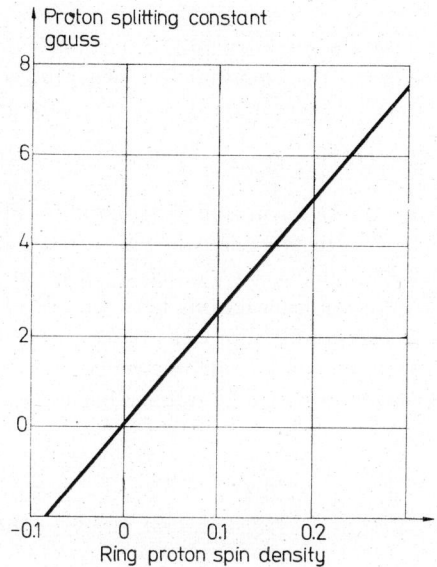

Fig. 3.35 Semi-empirical plot of spin densities at the ring protons of nitrobenzenes against hyperfine splitting constants. After Rieger and Fraenkel[3.35]

gauss. Thus the densities are only given at the ring protons, although the splitting from the ^{14}N nuclei is also observed. A plot of the experimental splitting constants against calculated spin densities is shown in Fig. 3.35. The mean deviation from the curve is in the order of $\pm 5\%$.

The observed splittings are considerably different if, for example, acetonitrile is used instead of dimethyl formamide. The splitting constants are given in the third and fourth column of Table 3.2. As shown, in acetonitrile the splitting constants become 5–10%

higher than in dimethyl formamide. In some cases abnormally large deviations are observed. In 1,2-dinitrobenzene at position (3) the deviation is $+270\%$. Generally the greatest solvent effects are found for the ^{14}N splitting constant, in the order of $+10\%$. This indicates that localized solvent complexes are most likely formed at the nitrogen atom.

The spin densities at the nitrogen and oxygen are much more difficult to evaluate. Extending the method of Karplus and Fraenkel[3.36] the nitrogen splitting constant can be expressed as follows

$$A_N = (S^N + Q^N_{NC} + 2Q^N_{NO})\rho_N + Q^N_{CN}\rho_{\dot{C}} + 2Q^N_{ON}\rho_{\dot{O}} \quad 3.28$$

where ρ_N, ρ_O, ρ_C are the electron spin densities at the nitrogen, oxygen and adjacent carbon atoms, respectively; S^N is the contribution of the nitrogen $1s$ electron density; Q^N_{ON}, Q^N_{CN}, Q^N_{NC} are the sigma–pi parameters describing the interactions between the ON, CN and NC bonds and the corresponding spin densities (see Section 3.2). In the case of aromatic radical ions of nitrobenzenes, from best fit with the experimental data the following values have been found for the sigma–pi parameters

$$S^N + Q^N_{NC} + 2Q^N_{NO} = 99.0 \pm 10.2 \text{ gauss}$$
$$Q^N_{NO} = 35.8 \pm 5.9 \text{ gauss} \quad 3.29$$
$$Q^N_{CN} \approx 0$$

In the nitro groups approximately

$$A_N = (23.1 \pm 1.4)\rho_N - (6.8 - 2.2)\rho_C \quad 3.30$$

where ρ_N and ρ_C, respectively, are the spin densities at the nitrogen and carbon adjacent to the nitro group.

3.4.3 ALIPHATIC RADICALS

As an example of aliphatic radicals in solution, $\dot{C}H_2COOR$-type radicals formed during the course of reaction of hydroxyl radicals with carboxylic acid esters are discussed in some detail.[3.37] For detecting the radicals the flow technique (see Chapter 2) is used. The reagents are mixed and pumped through the cavity resonator in order to establish stationary radical concentration there. The high resolution and sensitivity make it possible to detect hyperfine splitting from nuclei several bonds away from the place where hydrogen has been abstracted from the molecule.

Some observed splitting constants are the following

$$\overset{2.51}{\overset{\longleftrightarrow}{HCOO\overset{\cdot}{C}H_2}}$$
$$20.9$$

The splitting constants are given in gauss. The spectrum consists of a 1 : 2 : 1 triplet of doublets.

$$\overset{2.41\quad 24.5}{\overset{\longleftrightarrow\;\longleftrightarrow}{HCOO\overset{\cdot}{C}-CH_3}}$$
$$\underset{H}{|}\;19.6$$

In this radical 16 lines are observed. A quartet of 1 : 3 : 3 : 1 is due to the CH_3 group, which is doubled by the interaction with the α-proton. Each line is further doubled by the long-range interaction with the δ-COOH proton.

A more complex structure is the following

$$\overset{24.9\quad\quad 1.3}{\overset{\longleftrightarrow\;\longleftrightarrow}{H_3C-\overset{\cdot}{C}-COOCH_3}}$$
$$\underset{H}{|}\;20.3$$

where the splitting from the δ-methyl group is clearly resolved.

Generally in aliphatic radicals the coupling due to β-groups is between 20 and 24 gauss.

The spin densities can be calculated in good approximation from

$$A_\beta = 29.3\,\rho_\beta \qquad\qquad 3.31$$

Table 3.3

HYPERFINE SPLITTING IN SOME TRANSIENT ALIPHATIC RADICALS
(AFTER P. SMITH et al.[3.37])

$(CH_3)_2-\dot{C}-X$ group X	Splitting constant in gauss
H	24.7
C_2H_5	22.8
CH_3	22.7
CN	21.5
$COOCH_3$	21.3
COOH	21.3
OH	20.0

where A_β is the splitting of the β-group in gauss; ρ_β is the corresponding spin density. In Table 3.3 β-coupling constants for structures

$$H_3C-\dot{C}-X$$
$$|$$
$$CH_3$$

are given for different X-groups. As shown there the coupling constant is not very sensitive to the group X. The long-range coupling constants are much smaller than the α and β ones. The coupling can be explained by considering the hyperconjugation of the $C-O$ bond (see the discussion in Section 3.2).

$$R_1 \diagdown \qquad \diagup O$$
$$\quad \dot{C} \cdots C$$
$$R_2 \diagup \qquad \diagdown O-R_3$$

The spin densities at the oxygen atoms are not known. Since there is a coupling between the unpaired electron and group R_3, the spin density must be distributed all over the molecule.

3.5 GASEOUS RADICALS

Radicals in the gaseous phase can be studied by ESR and by observing Zeeman splittings in microwave molecular spectra. Rotation inversion energy levels of gases are split by a d.c. magnetic field. As

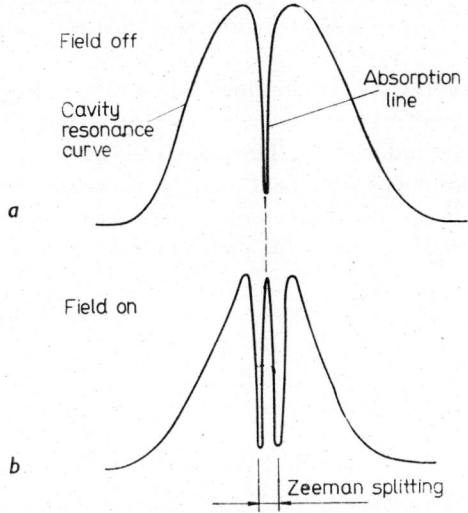

Fig. 3.36 Zeeman splitting of a microwave rotational spectrum line

mentioned in Chapter 1 the splitting of the energy levels is proportional to the magnetic moment of the molecule.

The technique for observing Zeeman effects in microwave molecular spectroscopy is very simple. Instead of an absorption cell a high Q-cavity resonator is used, placed between the pole-pieces of the electromagnet. The cavity is vacuum-tight in order to be able to work at reduced pressures. It is cooled by liquid nitrogen.

The rest of the system is conventional. If a molecular spectrum line is located in the range of the cavity resonance curve, it appears as shown in Fig. 3.36a. Upon switching on the magnetic field, the spectrum line will be split by the Zeeman effect as shown in Fig. 3.36*b*. It is also possible to sweep the frequency slowly through the cavity resonance line and display the Zeeman spectrum by the lock-in technique. By moving the cavity resonance curve to other rotational or inversion lines the Zeeman splittings of the corresponding energy levels can easily be determined. In favourable cases the hyperfine splitting caused by the interaction of the magnetic nuclei with the radical electron can also be observed.

A more straightforward technique is to induce direct transitions between the Zeeman energy levels. This can be done in two ways. Transitions can be induced by the interaction between the magnetic component of the microwave field and the magnetic moment of the molecule, as usual in ESR spectroscopy. In this case the intensity of the spectral lines is determined by the probability of the magnetic dipole transitions.

For gaseous radicals, however, it is possible to induce electric dipole transitions between the Zeeman levels split by the polarizing magnetic field. In these cases the molecular electric dipoles interact with the electric component of the microwave field. This technique is called *electric dipole ESR*.

The following main types of paramagnetic species can be investigated by the above-mentioned methods

1. Paramagnetic molecules such as oxygen and nitric oxide.
2. Free paramagnetic atoms produced by gaseous discharge.
3. Paramagnetic excited molecules or ions produced by gaseous discharges.
4. Gaseous free radicals formed during the course of reactions ($\dot{O}H$, $\dot{S}H$, $\dot{S}O$).
5. Transient paramagnetic molecules formed during gaseous reactions (CS).

For chemists, detection of intermediate radicals formed during gaseous reactions is the most important. Such measurements can be made by conventional ESR spectrometers equipped with gas-flow apparatus. Recently the electric dipole technique has gained interest because of its higher sensitivity.

3.5.1 ESR OF ATOMIC GASES

Atomized gases can be investigated by conventional ESR technique provided that the time elapsed from atomization to the measurement is short in comparison with the lifetime of the atomic products. Gases

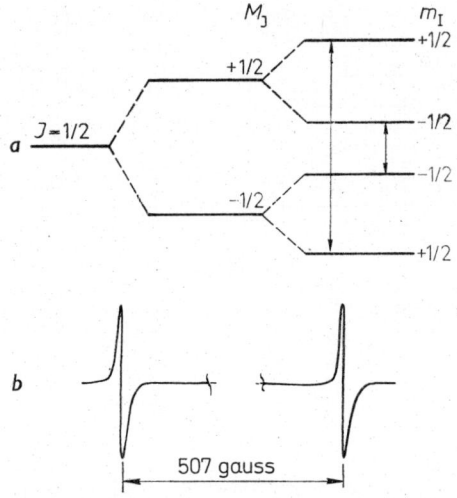

Fig. 3.37 Hyperfine splitting in atomic hydrogen

or vapours can be atomized either by microwave discharge or simply by thermal dissociation. A fast flow system is needed with positioning of the atomizer upstream and close to the cavity resonator.

The number of ESR spectral lines observed will depend on the rotational quantum number J and nuclear spin quantum number I. The total number of lines is $2J(2I + 1)$. For atomic hydrogen $J = 1/2$, $I = 1/2$ (proton), thus the total number of lines expected is 2. The energy level system is shown in Fig. 3.37a. The rotational ground state $J = 1/2$ splits into two Zeeman levels $m_J = \pm 1/2$. The hyperfine interaction of the electron with the proton further splits each level into two.

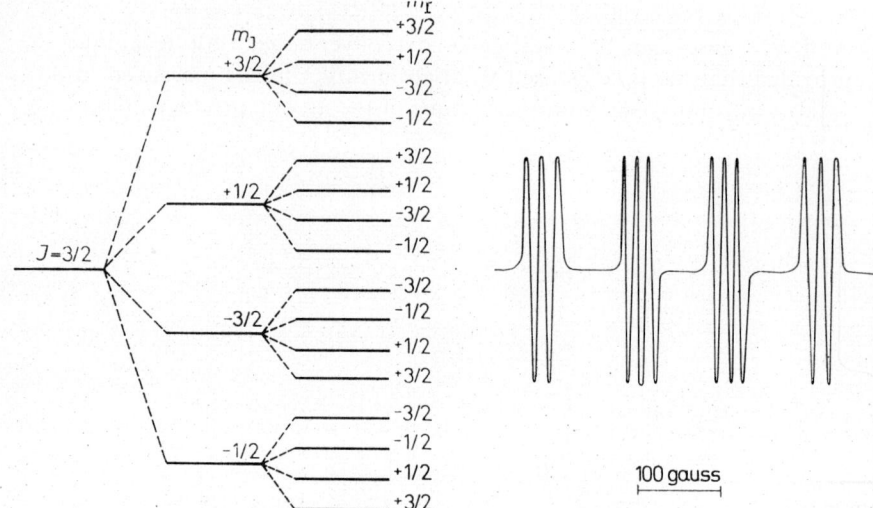

Fig. 3.38 Hyperfine splitting in atomic chlorine

For halogen atoms the rotational ground state is $J = 3/2$. The corresponding energy-level system for chlorine is shown in Fig. 3.38. Since ^{35}Cl has a nuclear spin $I = 3/2$, each rotational Zeeman level splits into four hyperfine levels as shown in the figure. The number of possible transitions is $2 \times 3/2\,(2 \times 3/2 + 1) = 12$. The corresponding spectrum is also shown. Four groups of triplets are observed according to the magnetic dipole selection rule $\Delta m_I = 0$. Generally, spectra of ^{35}Cl and ^{37}Cl are observed simultaneously; the lines of these isotopes are shifted slightly. Since the natural abundance of ^{35}Cl is 75.4% and of ^{37}Cl is 24.6%, the intesity ratios of the corresponding spectra are also different.[3.38]

Electric dipole ESR of diatomic transient radicals in the gaseous state $(\dot{O}H, \dot{S}H)$. Transitions between Zeeman energy levels of molecular rotation can be induced not only by microwave magnetic field but by microwave electric field as well. In states having large electric dipole moments the corresponding electric dipole ESR spectral lines are

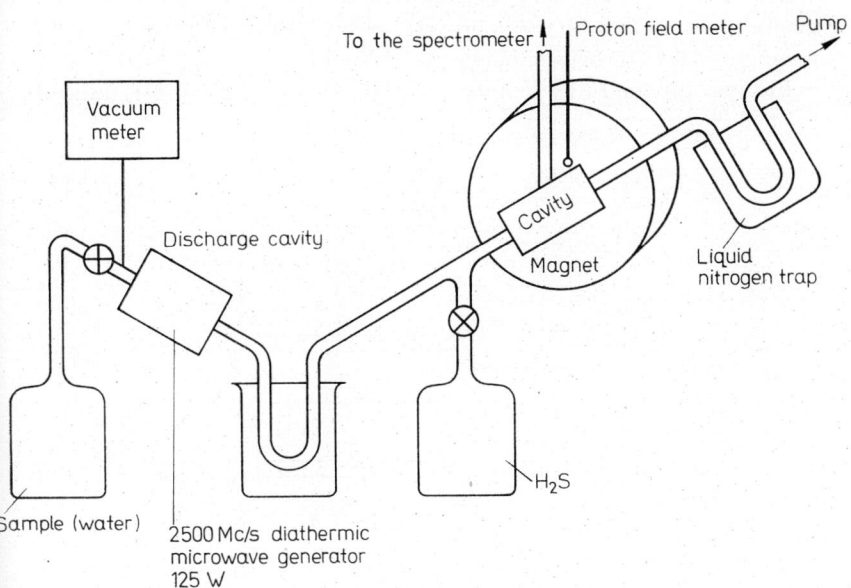

Fig. 3.39 Experimental scheme for observing electric dipole ESR spectra. After Radford[3.39]

considerably more intense than those taken in the usual way, by inducing magnetic dipole transition.

This technique is referred to as electric dipole ESR and is used currently for investigating transient gaseous radicals. For inducing electric dipole transitions between molecular Zeeman levels the gaseous sample must be placed in a microwave *electric field* perpendicular to the polarizing magnetic field H_0. The magnetic field is modulated in the usual way and the spectrum is displayed by sweeping H_0. For electric dipole ESR conventional spectrometers can be used by changing the cavity resonator only.

The gaseous reactants can be mixed immediately before entering the cavity. If necessary the components can be atomized or ionized by r.f. discharge. A simplified scheme is shown in Fig. 3.39. The sample is positioned so that both the microwave electric field and the microwave magnetic field are perpendicular to the main field.[3.41]

Electric dipole ESR is illustrated by the sample of ṠH radicals formed by reacting dissociated H_2O with H_2S.[3.39] Dissociation of water

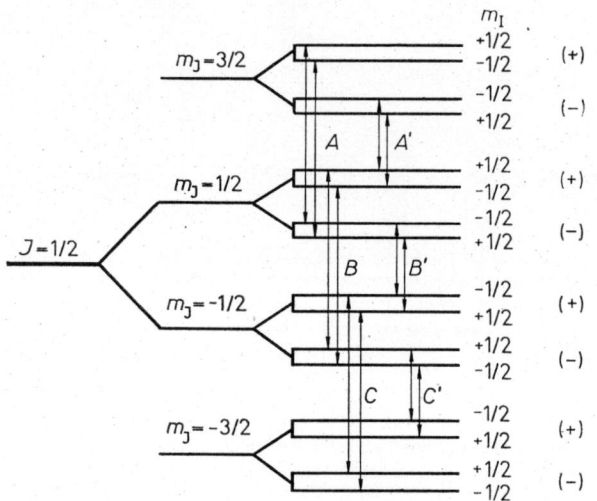

Fig. 3.40 Zeeman splitting of the energy levels of SH radicals. After McDonald[3.40]

vapour by r.f. discharge results in formation of Ȯ, and Ḣ radicals, which can be observed by magnetic dipole ESR. ȮH radicals formed can be measured by the electric dipole transition method. Magnetic and electric dipole ESR spectra can be observed simultaneously. After mixing H_2S with the dissociated stream, lines due to ṠO radicals appear. Adding H_2S in larger quantities causes electric dipole lines attributed to SH radicals to appear.

The simplified scheme of the Zeeman splitting of ṠH in the $J = 1/2$ and $J = 3/2$ rotational state is shown in Fig. 3.40.

Each rotational Zeeman level is doubled into the so-called Λ-levels. Each Λ-level is further doubled by the hyperfine interaction with the proton.

According to the scheme of Fig. 3.40 six groups of lines (doublets) should appear in the electric dipole ESR spectrum of ṠH. This has been found experimentally.[3.40] The spectrum is shown in Fig. 3.41. At X-band (9 Gc/s) the electric dipole transitions are around 7.500 gauss.

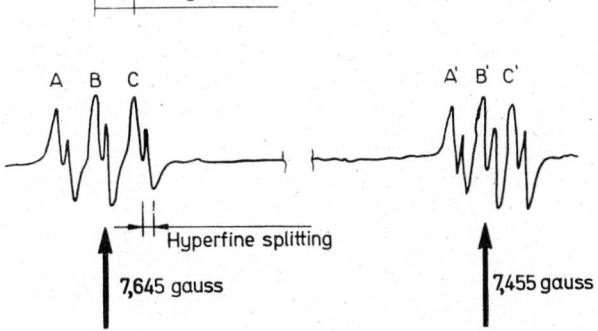

Fig. 3.41 Electric dipole ESR spectrum of gaseous SH radicals. After McDonald[3.40]

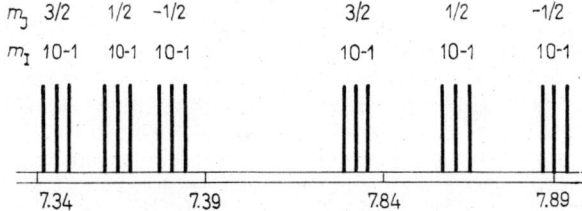

Fig. 3.42 The electric dipole ESR spectrum of gaseous OD radicals. After Radford[3.39]

The 6 groups can easily be identified and the hyperfine interaction is also resolved.

Similar spectra are observed for ȮH radicals produced by gaseous discharge in water vapour. In the deuterated radicals the hyperfine splitting caused by the deuteron nucleus is also observed.[3.39] The nuclear spin of the deuteron is $I = 1$, the Zeeman energy levels are thus split into $2I + 1 = 3$ lines according to $m_I = +1, 0, -1$. The corresponding spectrum of ^{16}ȮD radicals is shown in Fig. 3.42.

REFERENCES

3.1 Deguchi, Y., *J. Chem. Phys.* **32,** 1584 (1960).
3.2 Blois, M. S., *Free Radicals in Biological Systems.* Academic Press, New York (1961).
3.3 Putirskaya, G., Chernova, V. and Matus, I., *Acta Chim. Acad. Sci. Hung.* **21,** 289 (1959).
3.4 Kőhnlein, W. and Müller, A., *Z. Naturforsch.* **15b,** 138 (1960).
3.5 Abragam, A. and Pryce, M. H. L., *Proc. Roy. Soc.* A **205,** 135 (1951).
3.6 McConnell, H. M., *J. Chem. Phys.* **24,** 632, 764 (1956).
McConnell, H. M. and Chestnut, D. B., *J. Chem. Phys.* **28,** 107 (1958).
3.7 Venkataraman, B. and Fraenkel, G. K., *J. Amer. Chem. Soc.* **70,** 624 (1948).
3.8 Hyde, J. S. and Brown, H. W., *J. Chem. Phys.* **37,** 368 (1962).
3.9 Jarrett, H. S., *J. Chem. Phys.* **21,** 761 (1953).
3.10 McConnell, H. M. and Chestnut, D. B., *J. Chem. Phys.*, **27,** 721 (1957).
3.11 Das, M. R. and Fraenkel, G. K., *J. Chem. Phys.* **42,** 1350 (1965).
3.12 Karplus, M. and Fraenkel, G. K., *J. Chem. Phys.* **35,** 1312 (1961).
3.13 Lebedev, Ya. S., Michailov, G. P. and Buben, N. Ya., *Elementarnie Processi* (ed. Talrose), 184 (1965).
3.14 Dobó, J. and Hedvig, P., *Paper presented at the 2nd Symposium on Radiation Chemistry, Tihany, Hungary, 1966.* Published in 1967 by Akadémiai Kiadó, Budapest.
3.15 Rao, M. J. and Anderson, R. S., *J. Chem. Phys.* **42,** 2899 (1965).
3.16 Kashiwagi, M. and Kurita, Y., *J. Chem. Phys.* **40,** 1780 (1964).
3.17 Jaseja, T. S. and Anderson, R. S., *J. Chem. Phys.* **36,** 2727 (1962).
3.18 Fischer, H. and Hellwege, K. H., *J. Polymer Sci.* **56,** 33 (1962).
3.19 Lebedev, Ya. S., Chernikova, D. M. and Tikhomirova, N. N., *Atlas Spektrov EPR,* (ed. V. V. Voevodsky), Izd. Akad. Nauk. S.S.S.R., Moscow (1962). English edition, Consultants Bureau, New York (1963).
3.20 Lebedev, Ya. S., *Zhur. Strukt. Khim.* **3,** 151 (1962).
3.21 Lebedev, Ya. S., *Zhur. Strukt. Khim.* **3,** 21 (1962).
3.22 Russell, D. B., *J. Chem. Phys.* **43,** 1996 (1965).
3.23 Kwiram, A. L., *J. Chem. Phys.* **42,** 791 (1965).
3.24 Lázár, M. and Szőcs, F., private communication.
3.25 Butyagin, P. Yu., *Paper presented at the IUPAC Conference, Prague,* printed No. 236 (1965).
3.26 Bloembergen, N., Purcell, E. and Pound, R. V., *Phys. Rev.* **73,** 679 (1948).
3.27 Bruin, M., Bruin, F. and Heineken, F. W., *J. Chem. Phys.* **37,** 135 (1962).
3.28 Mohos, B. and Tüdős, F., (private communication).
3.29 Mohos, B. and Tüdős, F., (private communication).
3.30 de Boer, E. and Mackor, E. L., *Mol. Phys.* **5,** 493 (1962).
3.31 Lewis, I. C. and Singer, L. S., *J. Chem. Phys.* **43,** 2712 (1965).
3.32 McLahan, A. D., *Mol. Phys.* **3,** 233 (1960). See also ref. 3.31.

3.33 Aalbersberg, W. I., Hoijtink, G. J., Mackor, E. L. and Weijland, W. P., *J. Chem. Soc.* 3055 (1959).
3.34 Colpa, J. P. and Bolton, I. R., *Mol. Phys.* **6**, 273 (1963).
3.35 Rieger, P. H. and Fraenkel, G. K., *J. Chem. Phys.* **39**, 609 (1963).
3.36 Karplus, G. and Fraenkel K., *J. Chem. Phys.* **35**, 1312 (1961).
3.37 Smith, P., Pearson, J. T., Wood, P. B. and Smith, T., *J. Chem. Phys.* **43**, 1535 (1965).
3.38 Bertran-Lopez, V. and Robinson, H. G., *Phys. Rev.* **123**, 161 (1961).
Harvey, J. S. M., Kamper, R. A. and Lea, K. R., *Proc. Phys. Soc. London* **76**, 979 (1960).
3.39 Radford, H. E., *Phys. Rev.* **122**, 114 (1961).
3.40 McDonald, C. C. and Goll, R. J., *J. Phys. Chem.* **69**, 293 (1965).

4
Structure Determination and Analysis

4.1 GENERAL IDEAS

Radio frequency and microwave spectroscopy offer excellent ways for determining chemical structures. Microwave molecular spectroscopy has been extensively used for determining such molecular parameters as bond lengths, bond angles and molecular dipole moments. The data are collected in tables for practically all important molecules.

Recently high resolution nuclear resonance has gained much interest for analyzing complex molecular structures. As shown in Section 2.3 the resolution of NMR spectrometers is as high as a few parts in 10^9. Under such high resolution each molecule exhibits NMR spectra uniquely characteristic of its structure. From the number and position of the lines it is possible to determine the number, nature and steric position of the chemical groups in the molecule exactly and accurately. NMR is an absolute method; it is possible to determine structures from the measured spectra by theoretical ways. This is based on the fact that the magnitude and direction of local magnetic fields around the paramagnetic nuclei depend on the chemical structure. Paramagnetic nuclei act as tiny probes for measuring the distribution of local fields in the molecule. As shown in Sections 1.3 and 2.3 the following main experimental parameters are measured in NMR: the chemical shifts and spin–spin couplings. The problem of structure determination is how these experimental parameters are connected with the chemical structure. In the following sections examples will be given for the chemical shifts and spin–spin splitting of some important chemical groups.

4.1.1 DIAMAGNETIC SHIELDING

In order to relate NMR line shifts to structure we must investigate how local magnetic fields are built up inside the molecule. Let us consider first the simplest case of atomic hydrogen. As shown in Section 1.3, for proton resonance the system is placed in a high intensity magnetic field H_0. Under the action of this field a diamagnetic current will be induced in the electron cloud surrounding the proton. Accord-

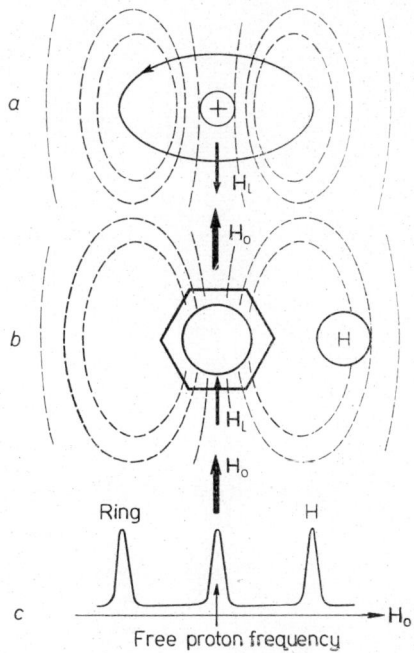

Fig. 4.1 Diamagnetic shielding of nuclei. *a* — Positive shielding in atomic hydrogen; *b* — Positive and negative shielding in an aromatic ring; *c* — Proton resonance line shifts

ing to the law of induction the local magnetic field generated by this current will be opposite to H_0 at the position of the proton, as shown in Fig. 4.1a. The resulting field at the proton will thus be somewhat smaller than H_0. The resonance frequency is

$$v_0 = \frac{g_I \mu_I (H_0 - H_l)}{h} \qquad 4.1$$

where g_I is the nuclear g-factor, μ_I is the nuclear magneton, h is Planck's constant, H_0 is the polarizing field, H_l is the local field generated by the diamagnetic current.

The frequency shift from the ideal line of the proton system is

$$\Delta v = \frac{g_I \mu_I H_l}{h} \qquad 4.2$$

Since NMR spectra are usually taken as a function of the magnetic field, but shifts are expressed in frequency (see Section 2.3 for explanation), the effect of diamagnetism of the electron cloud around the proton would shift the line towards higher fields. This type of shielding is called *positive diamagnetic shielding* (see Fig. 4.1c).

Another type of diamagnetic shielding is illustrated in Fig. 4.1b. It represents a benzene ring placed in a polarizing field H_0 perpendicular to the plane of the ring. The ring current induced by H_0 would generate a local field H_l which has the same direction at the ring protons as H_0. Inside the ring the direction of H_l is opposite to H_0. The resonance frequency of the ring protons is thus

$$v_0 = \frac{g_I \mu_I (H_0 + H_l)}{h} \qquad 4.3$$

which results in a shift of the line towards *low magnetic fields*, as shown in Fig. 4.1c. This type of shielding is called *negative diamagnetic shielding*.

As shown by this very simple consideration, the resonance line of protons coupled to an aromatic ring should always appear at lower fields than that of aliphatic protons and this is in fact observed in every case.

It is evident that the magnitude of the diamagnetic shielding and therefore the shift of the resonance lines depend on the structure of the electron cloud surrounding the nucleus. Since the electronic structure is directly connected with the chemical constitution, each chemical group should produce different local fields, i.e. different chemical shifts of NMR lines. The main range of the chemical shifts is illus-

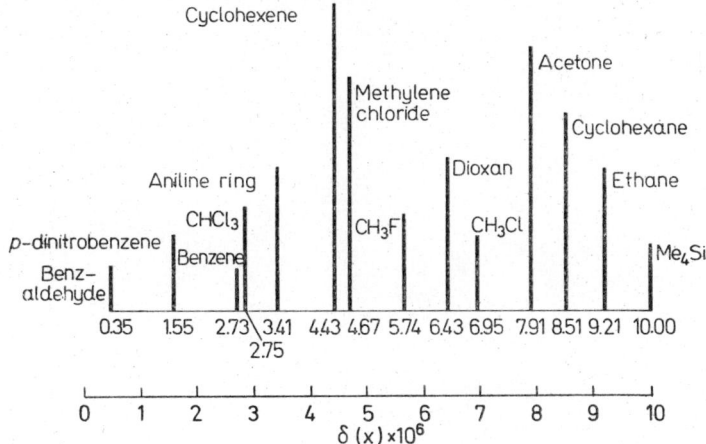

Fig. 4.2 Main range of proton resonance shifts

trated in Fig. 4.2, where the proton resonance spectral bands of some typical organic compounds are given. The total range is one part in 10^5 relative magnetic field, i.e. frequency units corresponding to absolute shifts at 100 Mc/s of 0–1000 c/s. The compound tetramethylsilane $(CH_3)_4Si$ has the highest positive shielding in this scale. It exhibits a narrow single proton-resonance line used generally as a reference signal in chemical shift measurements. The shift of $(CH_3)_4Si$ is conventionally taken equal to 10; and thus shifts are expressed in τ-units according to Equation 2.19.

As shown in Fig. 4.2 the proton resonance spectra of most organic molecules fall within the range of about 10 ppm. Some inorganic materials, especially metals, might have much higher chemical shifts, up to about 10,000 ppm. According to the resolution of 10^{-9} the scale of Fig. 4.2 can be extended by a factor of 1,000, making it possible to observe the multiplicity of the lines due to various functional groups and to spin–spin coupling.

4.1.2 SPIN–SPIN COUPLING

The nuclear spins in a molecule are not isolated, but coupled through molecular electrons. Thus the magnetic energy levels of one spin system

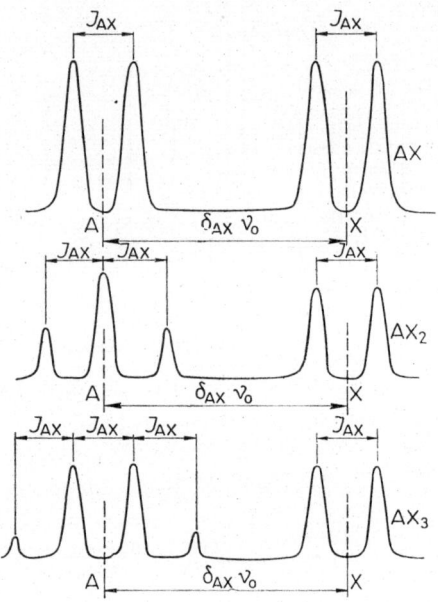

Fig. 4.3 Weak spin–spin interaction between protons

are influenced by the other with which it is coupled. Let us consider for example the splitting caused by protons of the methylene group, CH_2. The two methylene protons are equivalent, i.e. the local magnetic field is the same at each proton because of rapid rotation of the group. The possible spin orientations of the protons are the following

Proton 1 m_I	Proton 2 m_I	Relative population
+1/2	+1/2	1
+1/2	−1/2	2
−1/2	+1/2	
−1/2	−1/2	1

This corresponds to 3 different spin states for the assembly of 2 protons in the $-CH_2$ group with a relative population of 1 : 2 : 1. If this group interacts with a single proton, its line will be split into **three** with an intensity ratio of 1 : 2 : 1. This is illustrated in Fig. 5.6a where the $-OH$ signal in pure ethanol is split into three with an intensity ratio of 1 : 2 : 1 as a result of the spin–spin interaction with the methylene protons.

If a single proton interacts with a system containing 3 protons, (methyl groups) the possible spin states are the following

Proton 1 m_I	Proton 2 m_I	Proton 3 m_I	Relative population
+1/2	+1/2	+1/2	1
+1/2	+1/2	−1/2	
+1/2	−1/2	+1/2	3
−1/2	+1/2	+1/2	
−1/2	−1/2	+1/2	
−1/2	+1/2	−1/2	3
+1/2	−1/2	−1/2	
−1/2	−1/2	−1/2	1

Interaction of a proton with a methyl group would split the resonance line into four with an intensity ratio of 1 : 3 : 3 : 1 (see Fig. 4.3). Generally interaction with a spin system containing n equivalent nuclei of spin 1/2 would split the line of a single nucleus connected with the system into $n + 1$ lines with intensity ratios given by the binomial coefficients (see Table 6.2 in the appendix). The splittings between the lines are expressed in c/s and called spin–spin splitting constants J. Spin–spin splitting constants do not depend on the operating frequency. From the number of the lines and from their intensity ratios it is possible to tell how many nuclei there are in the group which causes the splitting. From the measured value of J one can tell how

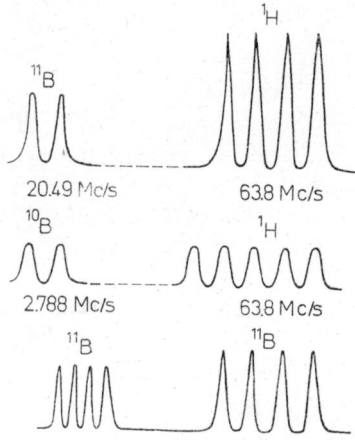

Fig. 4.4 Spin–spin splitting in ^{11}B–H groups

closely the two groups are coupled. Since spin–spin coupling is connected with the electronic and steric structure of the molecule, values of J provide valuable information in structure determination and analysis.

If nuclei of higher spins are coupled the situation is somewhat complicated but the problem can be solved exactly. In these cases the number of the lines in the absence of spin–spin coupling is $2I + 1$. The nucleus ^{11}B for example has a nuclear spin of $I = 3/2$, the corresponding number of lines is 4 (see Fig. 4.4). If this system interacts with protons or fluorine nuclei ($I = 1/2$) each line will split according to the number of protons, as shown in Fig. 4.3. Generally interaction with a system containing n spins of I will result in a splitting of each line into $2nI + 1$ components.

Figs 4.3 and 4.4 represent the cases when the spin–spin interaction is much weaker than the chemical shift between the spin systems. In this case spin–spin coupling constants can be obtained directly from the spectra. There is a conventional notation for such weak interactions:

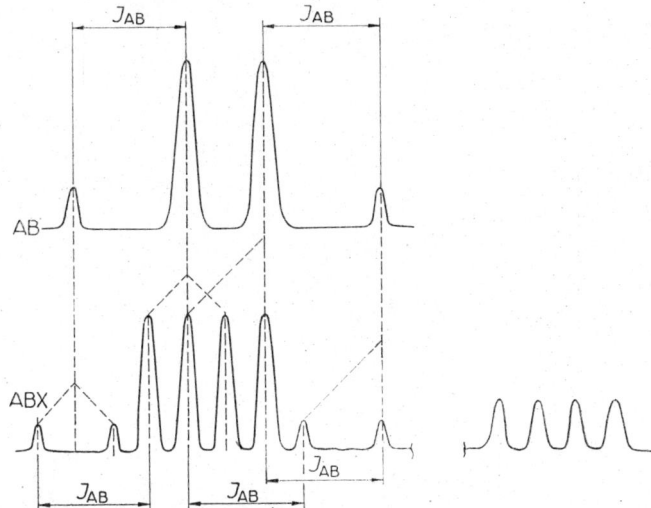

Fig. 4.5 Strong spin–spin interaction between groups of protons

the system observed directly is noted with A, the other system which is in weak interaction with it is X or Y. Thus AX denotes weak spin–spin interaction between system A and system X.

If spin–spin interaction is stronger than the chemical shift the spectrum becomes different. Strong spin interaction between two nuclei A and B results in a spectrum shown in Fig. 4.5 instead of that of Fig. 4.3.

The conventional notation describing the type of spin–spin interaction is the following

strong interaction $\qquad A_iB_jC_k$
$\delta_{AB} \leq J_{AB}$
weak interaction $\qquad A_iX_jY_k$
$\delta_{AX} \geq J_{AX}$

where the subscripts i, j, k denote the number of the spins in the corresponding system. Strong interaction between methyl group protons with methylene protons is thus described as A_2B_3. Weak interaction between methylene protons and hydroxyl protons is of type AX_2.

Evaluation of the chemical shifts and spin–spin coupling constants is somewhat complicated in cases of strong spin–spin splittings, especially in more complicated systems as A_2B_2, A_2B_3 and so on. As an example, in Fig. 4.5 an ABX-type spectrum is shown, i.e. strong interaction between systems A and B and weak interaction with system X. In these more complicated cases the spin–spin decoupling technique mentioned in Section 2.3 is used to make spectra simpler and thus separate chemical shifts from spin–spin couplings. With higher numbers of spins electronic computers are used for analyzing spectra.

4.1.3 ANISOTROPY OF LOCAL FIELDS

The local fields generated by induced diamagnetic currents depend on the relative orientation of the group with respect to the direction of the polarizing magnetic field. This anisotropy is, however, averaged out in gaseous and in liquid phase by the random motion of the molecules, provided that the intensity of the induced currents is not dependent on the orientation, i.e. in isotropic diamagnetic groups. However, in most cases the induced currents depend on the orientation with respect to the main field H_0.

The local field of a group formed by two atoms X and Y is considered at a distance r from the centre of the group and at an angle α from the axis. The effect of the diamagnetic current is approximated by a magnetic dipole

$$\mu = \chi H_0 \qquad 4.4$$

where χ is the diamagnetic susceptibility of the group.

In the case of axial symmetry two diamagnetic dipole moments can be considered corresponding to cases when H_0 is parallel to the symmetry axis or perpendicular to it

$$\mu_{\parallel} = \chi_{\parallel} H_0 \qquad 4.5$$
$$\mu_{\perp} = \chi_{\perp} H_0$$

In the liquid or gaseous state, where high resolution NMR spectra can be measured, all orientations are equally probable, and thus an average dipole moment is formed

$$\langle \mu \rangle = \mu_{\parallel} - \mu_{\perp} \qquad 4.6$$

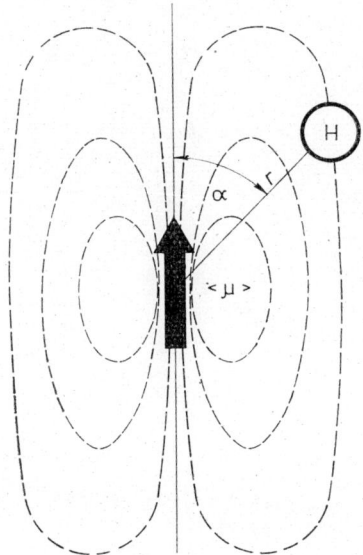

Fig. 4.6 The local field of an anisotropic group

The local field generated by this average moment is

$$\mathbf{H}_l = \frac{\langle \mu \rangle 3(\cos^2 \alpha - 1)}{3r^3} \mathbf{H}_0 \qquad 4.7$$

where r is the distance from the centre of the group, α is the angle with the symmetry axis, $\mathbf{H}_0$ is the polarizing field intensity (Fig. 4.6).

As can be seen from Equation 4.7 the sign of the local field is determined by the diamagnetic anisotropy of the group $X-Y$ and also by the steric factor $3\cos^2\alpha - 1$. In the region where $3\cos^2\alpha - 1 > 0$ and

$\langle \mu \rangle > 0$ the local field H_l is opposite to the main field H_0. This corresponds to positive shielding $\delta > 0$.

In the other region, where $3\cos^2 \alpha - 1 < 0$, the local field is in the same direction as the main field H_l, and thus the shielding is negative, $\delta < 0$.

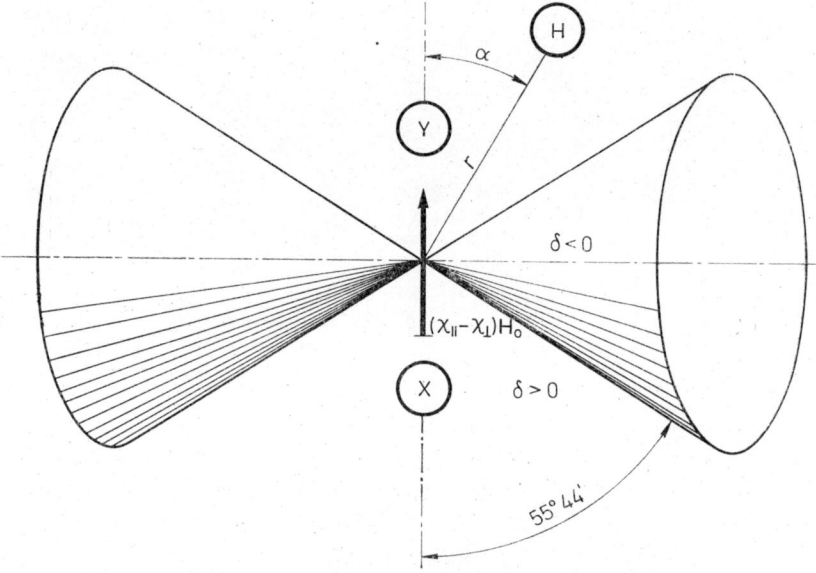

Fig. 4.7 Steric distribution of chemical shifts around a magnetically anisotropic group

The distribution of the local fields of an X—H group is shown in Fig. 4.7. It can be seen that the chemical shift of a proton located at an angle α from the axis of the group depends strongly on its steric position. The largest difference in the chemical shifts corresponds to positions parallel and perpendicular to the axis of the group ($\alpha = 0$, $\alpha = \pi/2$).

The diamagnetic anisotropy of groups is a property determined by the symmetry of the electron configuration. Thus in groups of cylindrical symmetry $\langle \mu \rangle$ is a given value for each group. The variation of the chemical shift of an additional atom or group is thus uniquely determined by the distance r and angle α. This is the main reason why high resolution NMR is such a powerful tool for the determination of stereochemical configurations. Examples of this will be given in Section 4.4.

4.2 DETERMINATION OF HYDROCARBON GROUPS

To illustrate how structure determinations can be made by measuring NMR chemical shifts and spin–spin couplings, groups containing C—H bonds are discussed in some detail. ^{12}C has no magnetic moment and thus the groups are to be determined by proton resonance only. The value of the chemical shift of a proton bound to carbon is mainly determined by the type of bonding. From the intensity of the spectrum and from the spin–spin splitting it is possible to determine the number of protons present in the group.

4.2.1 CHEMICAL SHIFTS IN HYDROCARBONS

Methyl groups. According to the rapid rotation of the group around the C-axis the three protons are equivalent, because the local field is averaged out. In the absence of spin–spin coupling, a methyl group would therefore exhibit a single proton resonance line with a chemical shift depending on the atoms or groups X to which it is bound:

$$\begin{array}{c} H \\ | \\ X-C-H \\ | \\ H \end{array}$$

According to a large number of experiments made on compounds having different X-groups it is concluded that the characteristic proton resonance lines of methyl groups fall within the range of $\tau = 5.5\text{--}9.8$. Similarly, the main shift ranges of methylene, methine, acetylene, and other groups have also been determined and collected.

Such chemical shift ranges for some important organic groups are given in Table 4.1 after Nukada and co-workers.[4.1] As shown there, it is possible to discriminate roughly between these groups, although some of the ranges overlap.

It is, however, rather easy to distinguish e.g. between methyl and methine groups in the range of $\tau = 5.5-8.5$, for the relative intensities of the lines are 3:1, corresponding to the ratio of protons in the groups. Further sub-division can be constructed by considering groups of the following type

$$-\underset{|}{\overset{|}{C}}-\underset{H}{\overset{H}{C}}-H$$

The corresponding chemical shift range is $\tau = 8.1-9.2$, depending on what other groups are connected to the second carbon atom. Some characteristic groups are shown in Table 4.2.

The resolution of NMR is high enough to follow this procedure to the next step

$$-\underset{|}{\overset{|}{C}}-\underset{|}{\overset{|}{C}}-\underset{H}{\overset{H}{C}}-H$$

The corresponding chemical shift range is $\tau = 8.9-9.2$. So the method is sensitive enough to feel the change of the local magnetic field at the methyl protons by substitutions separated by two $C-C$ bonds.

As shown in Tables 4.1 and 4.2 this successive approximation method permits the identification not only of main functional groups as CH_3, CH_2, CH etc., but their surroundings to 2 or 3 bond distances too. For identification of such groups only the chemical shifts must be determined exactly.

Tables of type 4.1 or 4.2 are available for most important groups in organic chemistry. In cases of special types of compounds a series of standard samples must be synthesized to find the chemical shift of unknown groups.

Table 4.1

PROTON SHIFTS IN C—H GROUPS

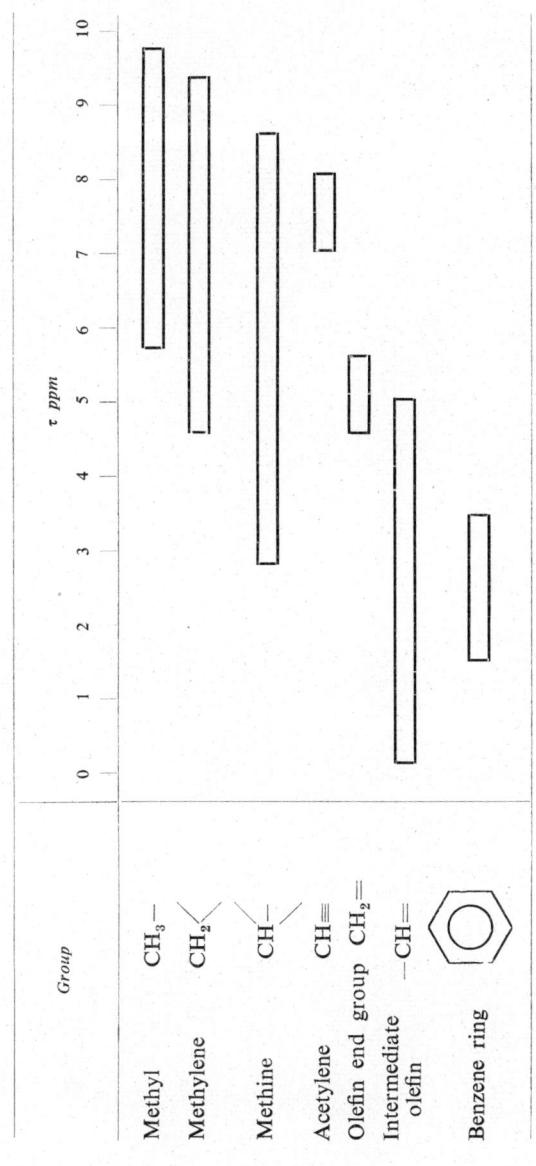

Table 4.2
Proton shifts in CH_3-C-X groups

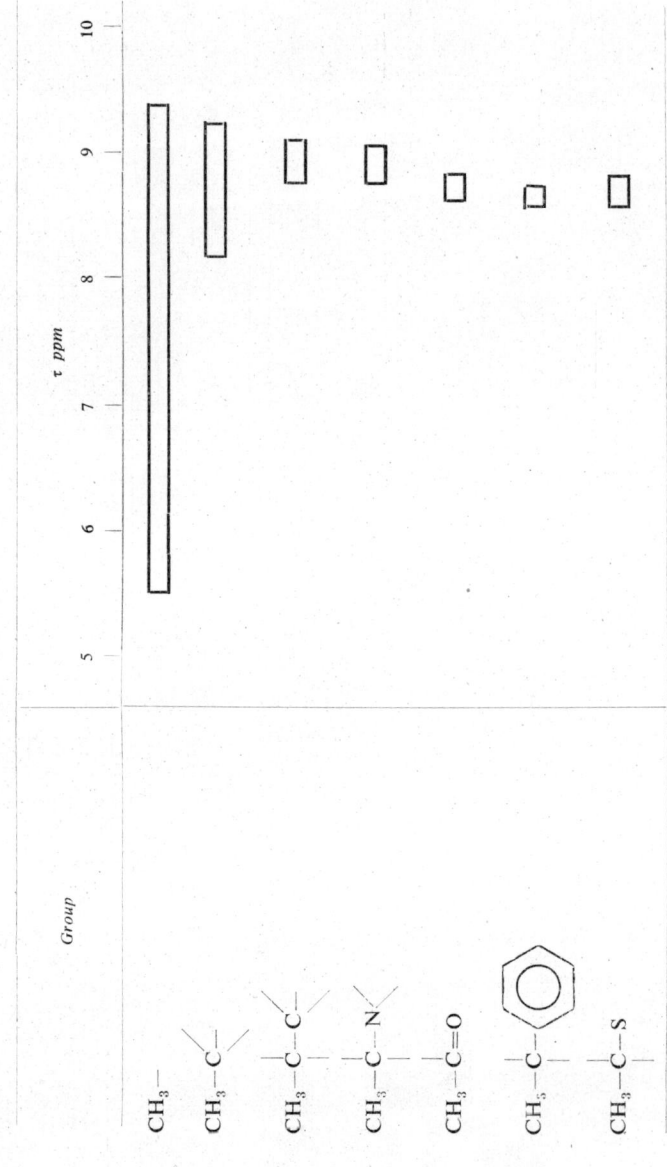

Determination of the surroundings of methyl groups is limited by resolution. The effects of substitution on the chemical shift ranges $\Delta\tau$ according to Table 4.2 are the following

first substitution $\Delta\tau \approx 4$ ppm
second $\Delta\tau \approx 1$ ppm
third $\Delta\tau \approx 0.3$ ppm
resolution 0.01 ppm

In most cases 2–3 bonds away from the methyl group there are some other magnetic nuclei in the compound. So the information gained from the methyl resonance can be supported by measuring the other resonant groups and also by observing the spin–spin couplings with these nuclei.

Methylene groups. Methylene groups have two bonds where atoms X and Y can be coupled

$$\begin{array}{c} H \\ | \\ X-C-Y \\ | \\ H \end{array}$$

According to Shoolery[4.2] there is a direct correlation between the electronegativity of groups X and Y and the chemical shift of the methylene protons. The shift can be calculated from the following equation

$$\tau(X-CH_2-Y) = 9.767 - (\sigma_X + \sigma_Y) \qquad 4.8$$

where σ_X and σ_Y are called *effective shielding constants* of groups X and Y, respectively. Values of σ_X and σ_Y for some groups are given in Table 4.3.

Equation 4.8 is often referred to as the Shoolery rule. The mean deviation between measured τ-values and those calculated from Equation 4.8 is ± 0.05 for methylene groups. For methine groups the Shoolery rule does not agree so well with experiments.

The method of successive grouping of chemical shifts discussed previously for methyl groups can in principle be applied to methylene

Table 4.3

Shoolery σ-values

Group	Effective shielding constant
Cl	2.53
Br	2.33
I	1.82
N⟨R_1, R_2	1.57
OR	2.36
SR	1.64
CR=O	1.70
$CR_1 = CR_2R_3$	1.32
C≡CH	1.44
C≡N	1.70
CH_3	0.47

groups, too. However, taking into account N substituents, the number of sub-divisions will be $N(N + 1)/2$, which is a rather high number. The first step in the simple case when two carbon atoms are coupled is the following

$$-\overset{|}{\underset{|}{C}}-\overset{H}{\underset{H}{\overset{|}{\underset{|}{C}}}}-\overset{|}{\underset{|}{C}}- \qquad \tau = 7.97\text{–}9.2$$

$$-\overset{|}{\underset{|}{C}}-\overset{H}{\underset{H}{\overset{|}{\underset{|}{C}}}}-\overset{|}{\underset{|}{C}}= \qquad \tau = 7.58\text{–}8.14$$

$$-\overset{|}{\underset{|}{C}}-\overset{H}{\underset{H}{\overset{|}{\underset{|}{C}}}}-C\equiv \qquad \tau = 7.20\text{–}7.87$$

As shown by the τ-values the local field at the methylene protons depends strongly on the type of bonding of the adjacent carbon atoms. Since the methylene group rotates around the $-C-$ bond, the two protons are equivalent.

Further substitution of carbon atoms results in the following situation

$$-\overset{\overset{\displaystyle H}{|}}{\underset{\underset{\displaystyle H}{|}}{C}}-\overset{|}{\underset{|}{C}}-\overset{|}{\underset{|}{C}}-\overset{|}{\underset{|}{C}}-\quad \tau = 8.46\text{–}9.02$$

$$-\overset{\overset{\displaystyle H}{|}}{\underset{\underset{\displaystyle H}{|}}{C}}-\overset{|}{\underset{|}{C}}-\overset{|}{C}=\overset{|}{C}\quad \tau = 7.88\text{–}8.14$$

$$\overset{|}{C}=\overset{|}{C}-\overset{\overset{\displaystyle H}{|}}{\underset{\underset{\displaystyle H}{|}}{C}}-\overset{|}{C}=\overset{|}{C}\quad \tau = 7.27\text{–}7.30$$

As shown by these simple examples, the chemical shift of the methylene protons is very sensitive to substitutions 2 to 3 bonds away from the group.

A collection of some important methylene groups is shown in Table 4.4 with the corresponding chemical shift ranges.[4.3]

Methine groups. These groups exhibit less intensive proton resonance lines since they only consist of a single proton. The possible coupling of three X-, Y-, Z-groups offers a large amount of combinations for sub-divisions

$$Y-\overset{\overset{\displaystyle Z}{|}}{\underset{\underset{\displaystyle H}{|}}{C}}-X$$

Table 4.4
CHEMICAL SHIFTS OF METHYLENE GROUPS

Group	τ ppm
$>$C—CH$_2$—C$<$	8.0–8.9
$>$C—CH$_2$—C=	7.7–8.2
$>$C—CH$_2$—N$<$	6.8–7.8
=C—CH$_2$—C≡	6.0–7.1
=C—CH$_2$—O—	5.5–6.5
C$_6$H$_5$—CH$_2$—N$<$	5.5–6.5
≡C—CH$_2$—O—	5.0–5.8
C$_6$H$_5$—CH$_2$—O—	4.5–5.3

STRUCTURE DETERMINATION AND ANALYSIS 245

The Shoolery rule is

$$\tau \left(Y - \underset{H}{\overset{Z}{\underset{|}{\overset{|}{C}}}} - X \right) = 9.767 - (\sigma_X + \sigma_Y + \sigma_Z) \qquad 4.9$$

where the shielding constants σ_X, σ_Y, σ_Z are given in Table 4.3.

Aromatic groups. As shown in Section 4.1 the diamagnetic local field of an aromatic ring is of the same direction at the ring protons and opposite to the polarizing field H_0 inside the ring. The resonance line of the ring protons is therefore shifted to *lower fields* (negative shielding), the protons inside the ring current are shifted to *higher fields*. This is illustrated in the proton resonance spectrum of coproporphyrin-I methyl ester. The structure of the compound is the following

The resonance line of the two inner protons H_i is shifted to $\delta = +3.89$ from the benzene line which serves as reference in this case. There is a stronger signal coming from the four outer protons; H_0 is shifted to lower fields by $\delta = -9.96$ ppm from benzene. The intensity ratio of the lines is $1:2$ according to the ratio of protons.[4.4]

Benzene derivatives. As an illustrative example of chemical shifts in aromatic compounds, consider the effect of substitution on the benzene ring proton resonance lines. Upon substituting a ring hydrogen atom by some other group, the diamagnetic ring current will be changed, resulting in a shift of the line of the remaining ring protons. Substitution usually also makes the ring protons non-equivalent, resulting in different splitting of *ortho-*, *meta-* and *para-*protons with respect to the substituted group:

NO_2 substituted benzene: ortho −0.97, meta −0.30, para −0.42

In this case the single line of benzene splits into three lines with an intensity ratio of 2 : 2 : 1, corresponding to the different splitting of *ortho-*, *meta-* and *para-*protons. The numbers given above are the shifts from the pure benzene line in ppm. Since all shifts are negative, the substitution of NO_2 *has increased the ring current* to produce higher local fields at the ring protons.

In some cases positive shifts are observed with respect to benzene

OCH_3 substituted benzene: ortho +0.23, meta +0.23, para +0.23

NH_2 substituted benzene: ortho +0.77, meta +0.40, para +0.13

indicating that substitution has decreased the diamagnetic ring current. There are substituents which do not influence the ring currents so there is no shift with respect to pure benzene:

CH_2Cl: $\delta = 0$ Cl: $\delta = 0$ Br: $\delta = 0$

The effect of substituents on ring current is directly connected with their electronegativity. Substituting groups X which are electron-acceptors results in the formation of the following mesomeric form:

$$\langle\bigcirc^{+}\rangle-X^{-}$$

In these cases the ring current is increased and the shielding of *ortho-*, *meta-* and *para-*protons becomes different. If the substituents are electron-donors the mesomeric form will be the following

$$\langle\bigcirc^{-}\rangle-X^{+}$$

This will result in a decrease of the ring current and a corresponding positive shift of the lines. Groups of neutral electron affinity will cause small changes in the ring current.

Data for benzene derivatives is collected in Table 4.5. The shift of the ring proton resonances is shown in positions *ortho* (*o*), *para* (*p*) and *meta* (*m*) with respect to the substituent group X. The tendency of the shift of the lines towards lower fields with increasing electron affinity of the substituents is clearly seen. Groups NO_2, COCl, $COOCH_3$ are known as strong electron-acceptors; the corresponding negative shift is high. The electron-donor groups NH_2 and $NH(CH_3)$ cause strong positive shifts of the ring proton lines.

Substituting two ring hydrogens results in a characteristic shift of the resonance lines of the remaining ring protons. Substitution of two methyl groups results in the following shifts with respect to that of pure benzene

$$H_3C\diagdown\underset{+0.27}{\underset{|}{\bigcirc}}\diagup\overset{CH_3}{\underset{+0.27}{|}}\quad +0.27$$

o-Xylene

Table 4.5
PROTON RESONANCE LINE SHIFT OF BENZENE DERIVATIVES[4.30]

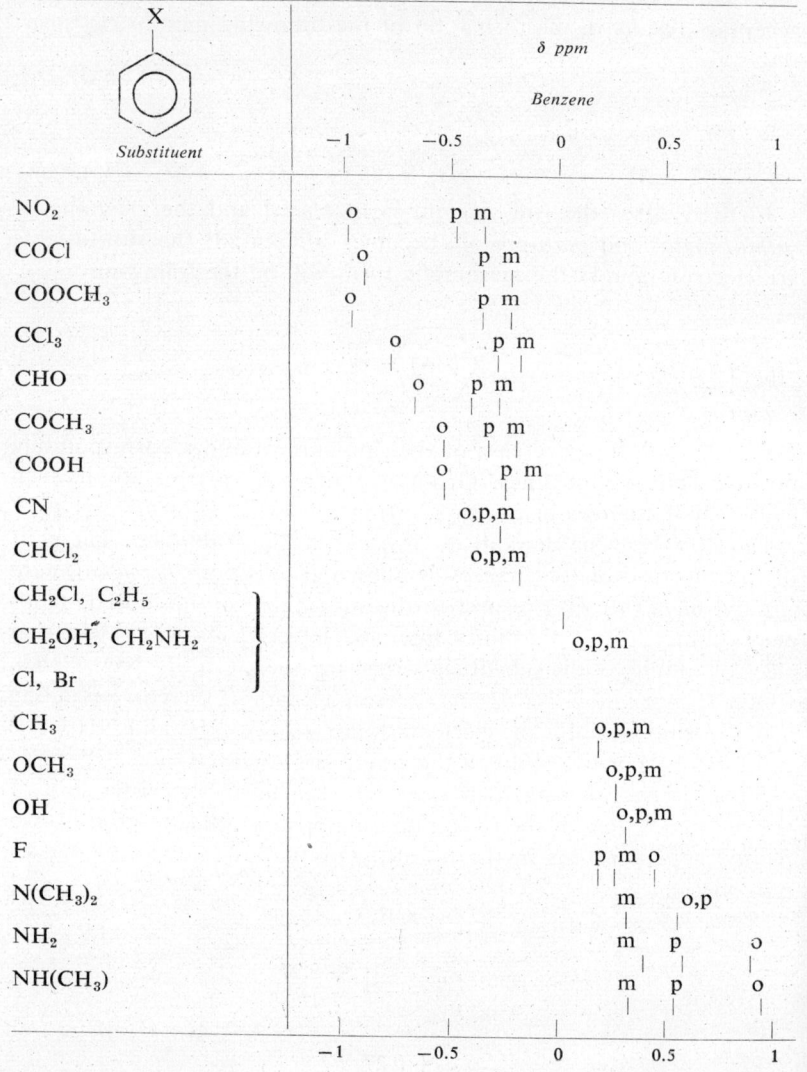

m-Xylene: ring protons shifts +0.40, +0.40, +0.40, +0.40 (with CH₃ groups at two positions)

p-Xylene: ring protons shifts +0.32, +0.32, +0.32, +0.32 (with CH₃ groups para)

the corresponding compound with a single substitution is

Toluene: ring protons shifts +0.17, +0.17, +0.17

These benzene derivatives can be identified by the shift of the ring protons and also by the shift of the methyl groups.

Heterocyclic groups. Ring currents are appreciably changed if a hetero-atom is present. The chemical shifts of the *ortho-*, *para-* and *meta-*protons with respect to the hetero-atom are usually different. The shifts of the ring protons in pyridine e.g. with respect to pure benzene are the following

Pyridine: −0.19, +0.18, +1.33

There is a strong spin–spin coupling between the *ortho-*, *meta-* and *para-*protons.

The position of the hetero-atom can also be identified. Quinoline and isoquinoline for example differ only by the position of the hetero-atom. The shifts with respect to benzene are the following[4,5]

```
          +0.4  +0.4
   + 0.4  ╱═══╲   + 1.02        Quinoline
   + 0.4  ╲═══╱N  − 0.88
          −0.22

          +0.72 +0.72
   + 0.72 ╱═══╲   − 0.58        Isoquinoline
   + 0.72 ╲═══╱N
          +0.72 −1.38
```

The spin–spin coupling constants are also changed by changing the position of the hetero-atom.

4.2.2 SPIN–SPIN COUPLING IN HYDROCARBON GROUPS

Spin–spin couplings between nuclei offer further possibilities of structure determination. As shown above, hydrocarbon groups can be located by measuring the exact position of the proton resonance lines. Spin–spin coupling between nuclei of different groups make proton resonance spectra of molecules far more characteristic. In Fig. 4.8 proton resonance spectra of some simple aliphatic alcohols are shown in order to illustrate how spin–spin couplings make spectra unique. These compounds can be recognized at a glance by their proton resonance spectra.

In Section 4.1 some ideas have been given about the interpretation of spin–spin splitting. In analyzing any spectrum the following main problems are to be solved.

1. *How many spin systems are there in the compound?* The answer to this question is simple in cases when spin–spin coupling is weak,

i.e. $\delta \gg J$ for each system. In such cases the groups are well separated, as in the aliphatic alcohols shown in Fig. 4.8. The $-OH$, CH_2 and CH_3 groups are clearly separated by the relatively large chemical shift.

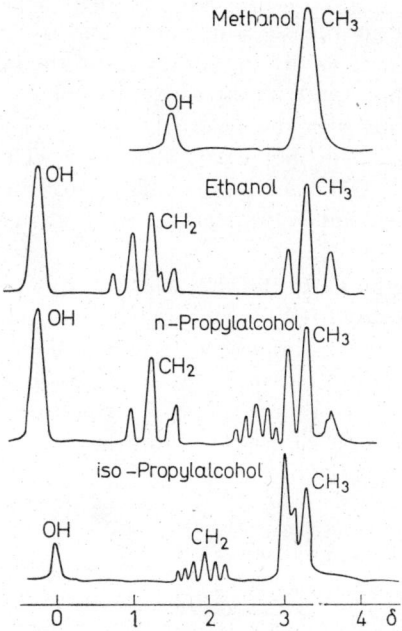

Fig. 4.8 Proton magnetic resonance spectra of some aliphatic alcohols

If the spin–spin coupling is strong with respect to the chemical shifts, the analysis of the spectra is somewhat more difficult. The technique of spin–spin decoupling is very useful in such cases.

2. *How many nuclei are there in each spin system?* There are two ways of solving this problem. One is to measure exactly the intensity ratios of the groups, which should be proportional to the number of protons. The other and more exact solution

is to analyze the spin–spin splittings. As shown in Section 4.1 the neighbouring spin system containing N equivalent spins of $I=1/2$ would cause a splitting to $N+1$ lines with intensity ratios defined by the binomial coefficients. This is very well illustrated in the spectrum of ethanol (Fig. 4.8) where the methyl group is split into three lines with an intensity ratio of $1:2:1$ as a result of the interaction with the 2 methylene protons, and the methylene group is split into four lines with an intensity ratio of $1:3:3:1$ as a result of the interaction with the 3 methyl protons. The intensity ratios are well illustrated in methanol, where the $-CH_3$ line is 3 times as high as the $-OH$ line. In the OH group the spin–spin interaction is averaged out by the rapid proton exchange (see Chapter 5).

It is possible to build up the characteristic spin–spin coupling constants of the main hydrocarbon groups as was done in the case of chemical shifts. Some examples are given as follows.

Coupling between CH groups. Spin–spin coupling between the protons of two $C-H$ groups depend strongly upon the chemical structure. The simplest case is the following

$$\begin{array}{c} 2\text{--}9 \text{ c/s} \\ \overset{\curvearrowright}{H \quad H} \\ | \quad | \\ -C-C- \\ | \quad | \end{array}$$

The actual value of splitting depends on what groups are joined to the carbon atoms. If two adjacent carbon atoms are coupled with *double bonds*, the spin–spin coupling constant between H atoms will be larger

$$\begin{array}{c} 10\text{--}13 \text{ c/s} \\ \overset{\curvearrowright}{H \quad H} \\ | \quad | \quad | \quad | \\ C=C-C=C \\ | \qquad \qquad | \end{array}$$

If the $-CH$ groups are separated by carbon atoms, the coupling constant becomes approximately zero

$$J = 0$$

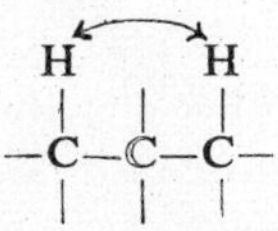

$$\begin{array}{ccc} H & & H \\ | & | & | \\ -C- & C- & C- \\ | & | & | \end{array}$$

Coupling between ethylene protons. Ethylene groups exhibit characteristic spin–spin couplings between the two protons depending on the relative position of the two protons. The three possibilities and the corresponding coupling constants are the following

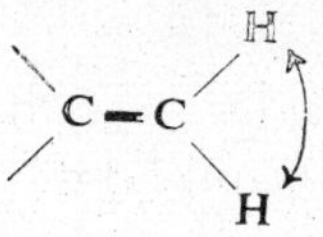

0–3.5 c/s   *geminal*

6–14 c/s   *cis*

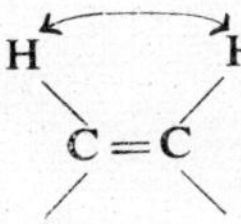

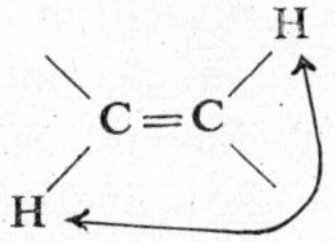

11–18 c/s   *trans*

Coupling between ring protons. The protons of an aromatic molecule are coupled through the π-electrons of the ring. The characteristic coupling constants of different benzene derivatives are the following

1–3 c/s   5–8 c/s 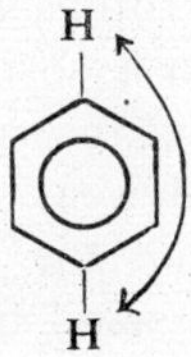 ~ 1 c/s

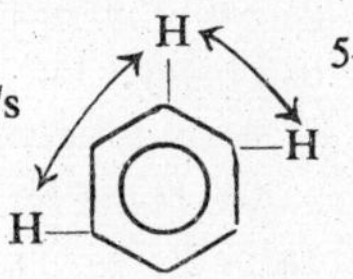

Thus the coupling is very small between the *meta*-protons and large between *ortho*-protons.

4.2.3 ILLUSTRATIVE EXAMPLES

In comparison with such other methods as optical and infrared spectroscopy the interpretation of NMR spectra is relatively easy. The reason is that only nuclei having magnetic moments exhibit spectral lines. The spectrum range of various nuclei are very well separated (see Table 6.1 for the resonance frequencies). It is therefore certain that the lines of a given spectrum have originated from similar nuclei being in different chemical surroundings. The technique of spin–spin decoupling makes it easy to separate chemical shifts and spin–spin couplings.

2,3-Dibromopropene. As an example of an NMR spectrum of a simple organic molecule the spectrum of 2,3-dibromopropene is discussed in some detail:[4,6]

$$H_X-\underset{\underset{Br}{|}}{\overset{\overset{H_X}{|}}{C}}-\underset{\underset{Br}{|}}{C}=\underset{\underset{H_B}{|}}{\overset{\overset{H_A}{|}}{C}}$$

The spin systems in this molecule are H_A; H_B and $2H_X$. The magnetic ^{79}Br, ^{81}Br nuclei in this molecule are present in an abundance of 50.57% and 49.43%, respectively. These nuclei, however, have relatively large quadrupole moments, which makes spin–spin relaxation times shorter (see the discussion in Section 2.3). Such nuclei do not contribute to the spin–spin splitting of the protons; they can be regarded as non-magnetic nuclei.

The proton resonance spectrum of 2,3-dibromopropene is shown in Fig. 4.9. The two groups of equal intensity on the low field side correspond to protons H_A and H_B, the high field group to the two protons H_X with a corresponding double intensity. The spectrum is of ABX type, i.e. the spin–spin coupling is strong between protons H_A and H_B and weak between H_X-H_A and H_X-H_B. The exact chemical shifts of the groups can be obtained by irradiating the system with a second frequency corresponding to the resonance of the H_X group. This would remove couplings H_X-H_A and H_X-H_B and thus the AB structure is obtained. From the decoupled spectrum the chemical shifts and the coupling constants can be determined directly (see the second spectrum of Fig. 4.9).

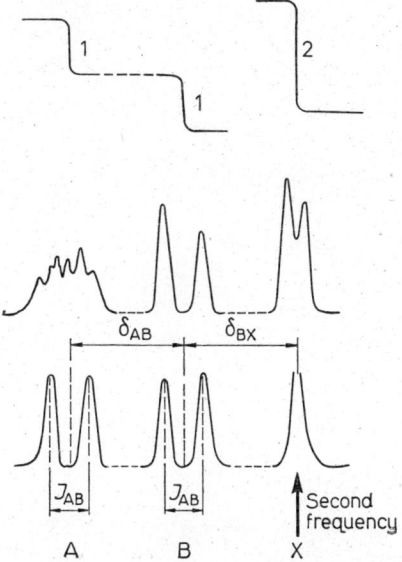

Fig. 4.9 Proton resonance spectrum of 2,3-dibromopropene[4.6]

Styrene. The spectrum of styrene consists of a strong line corresponding to the ring protons at low fields and an ABX-type spectrum from the three ethylene protons. The spin–spin coupling constants are the following[4.7]

The small coupling between the geminal hydrogen atoms is in agreement with the general arguments given above for the spin–spin coupling in ethylene groups. The ring protons do not contribute to the spin–spin splitting, because they are separated from H_X by two

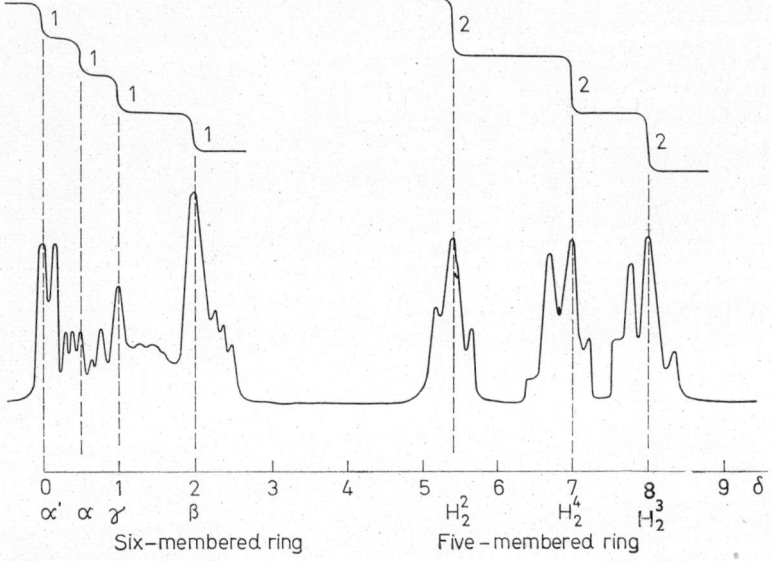

Fig. 4.10 Proton resonance spectrum of the alkaloid myosmine[4.8]

carbon atoms. For determining the exact chemical shifts between groups H_A and H_B, the spin–spin decoupling technique should again be used.

Myosmine. This alkaloid consists of a six-membered and a five-membered heterocyclic ring. The exact structure was first revealed by proton resonance measurements. The structure is the following

The ring proton lines of the five-membered ring and of the six-membered one are well separated, as a result of the different diamagnetic ring currents. The spectrum is shown in Fig. 4.10. The chemical shift of the ring protons are different and also spin–spin coupling exists between them. The four groups with intensity ratio 1 : 1 : 1 : 1 at the low field side of the spectrum correspond to protons $H_{\alpha'}$, H_{α}, H_{γ}, H_{β}, respectively. Each group is split into many lines by the spin–spin coupling between the four protons. Nitrogen nuclei do not contribute to the spin–spin splitting because ^{15}N has a small natural abundance and ^{14}N has a relatively high nuclear quadrupole moment.[4,8]

The groups of five-membered ring protons are shifted by about 6 ppm to higher fields. The intensity ratio of the three groups is 2 : 2 : 2 corresponding to the hydrogens in positions 2, 4 and 3, respectively. In this case the structure can be identified by measuring only the spin–spin splittings and the relative intensities (integrals) of the lines. Analysis of the spin–spin splitting is not necessary.

4.2.4 ^{13}C RESONANCES

Proton resonance provides enough information to determine most hydrocarbon groups. In some cases, however, valuable additional information can be derived from ^{13}C spectra. The natural abundance of the ^{13}C isotope is 1.1%, its spin is $I = 1/2$, magnetic moment 0.70216 nuclear magnetons. The resonance frequency at 15 kG field is 15.35 Mc/s. The relative intensity of the lines in comparison with proton lines is only about 1%. For measuring ^{13}C resonances in natural abundance thus very high sensitivity is needed.

^{13}C resonances are now extensively used to determine carbonyl acetylenic and carboxyl groups

$$\begin{array}{c} O \\ \parallel \\ -C-O- \end{array}$$

$$-C \equiv C-$$

$$\begin{array}{c} -C- \\ \parallel \\ O \end{array}$$

Table 4.6

Some ^{13}C line shifts

Compound	^{13}C line shift ppm
n-Alkanes	107–127
Cycloalkanes	95–105
Cyclohexane	100
Ethylene oxide	92
Dioxan	55
Nitrile groups	−2 to +14
Benzene	0
Aromatic hydrocarbons	−10 to +10
$\left(\underset{\underset{O}{\parallel}}{C}-OR\right)_2$	−31
$R-O-\underset{\underset{O}{\parallel}}{C}-O-R$	−36
HCOOH	−37
$R-\underset{\underset{O}{\parallel}}{C}-OH$	−52
$R-\underset{\underset{O}{\parallel}}{C}-R$	−79
Carbonyl-carbon of α-hydroxy-acid	−90

In such groups ^{13}C is the 'strongest' magnetic nucleus, for the natural abundance of ^{17}O is only 0.04%.

The range of ^{13}C chemical shifts is quite large. Some compounds and groups are collected in Table 4.6 with their ^{13}C shifts with respect to that of pure benzene.[4,5]

As an example of ^{13}C resonances the spectrum of dimethyl acetylene-dicarboxylate is shown in Fig. 4.11. The structure of the compound is the following

$$H_3C-O-\underset{\underset{O}{\parallel}}{C}-C\equiv C-\underset{\underset{O}{\parallel}}{C}-O-CH_3$$

This compound cannot be identified by proton resonance; only the atoms nearest to the methyl groups can be determined. As mentioned above ^{17}O resonances are very difficult to measure, ^{16}O has no magnetic moment. The only possibility is to measure ^{13}C resonances. As shown

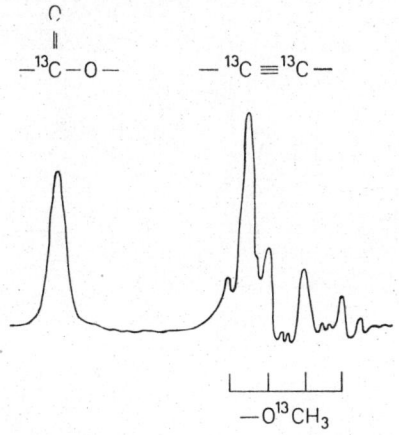

Fig. 4.11 ^{13}C spectrum of dimethyl acetylenedicarboxylate. After Lauterbur[4.9]

in Fig. 4.11 the following ^{13}C groups are present in the system

$$-\overset{\overset{\displaystyle O}{\|}}{^{13}C}-O-$$ A single line at the low field side

$-^{13}C\equiv{}^{13}C-$ A single line at higher fields

$$-\overset{\overset{\displaystyle H}{|}}{\underset{\underset{\displaystyle H}{|}}{^{13}C}}-H$$ Four lines corresponding to the spin–spin coupling between the ^{13}C atom and the three protons of the methyl group

The ^{13}C–H spin–spin coupling can be removed by simultaneous r.f. irradiation at the methyl proton frequencies.

Since ^{13}C nuclei are always present in hydrocarbon groups, proton resonance spectra should also be split by $^{13}C-H$ coupling. Since the abundance of the ^{13}C nuclei is only 1.1%, ^{13}C splittings are generally overwhelmed by the strong proton signals. By using the technique

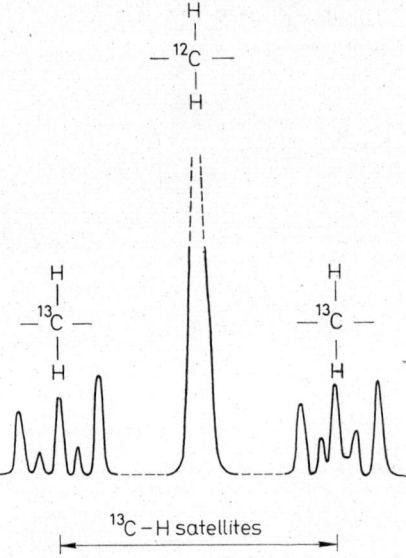

Fig. 4.12 ^{13}C satellites in the proton resonance spectrum of p-dioxan[4.10]

of internuclear double resonance, INDOR, discussed in Section 2.4, it is possible to filter out strong proton signals and detect the side lines caused by $^{13}C-H$ spin–spin coupling.[4.9]

In some special cases $^{13}C-H$ side lines can be observed directly. This is illustrated in Fig. 4.12 where the proton resonance spectrum of p-dioxane is shown. The structure is the following

The 8 protons of this molecule are equivalent, they exhibit a single strong line at the centre of the spectrum in Fig. 4.12. According to the natural abundance of the ^{13}C nuclei about 4% of the molecules contain a $^{13}CH_2$ group. The spin–spin splittings are the following:[4.10]

$$H_2C^{12} - {}^{12}CH_2 - \overset{O}{\underset{}{\diagup}}$$

(structural diagram showing ^{13}C with H couplings: 6.1 c/s and 2.7 c/s)

These coupling constants are average values of the carbon configurations of the molecule. The averaging is good at room temperatures where the conformational configurations are changing at a high rate.

4.3 DETERMINATION OF OTHER GROUPS

There are many possibilities for analyzing chemical compounds other than via hydrocarbon groups by high resolution NMR. The only requirement is that the group to be identified should contain magnetic isotopes having sufficiently large resonance signals. As seen in Table 6.1 in the Appendix there are many possibilities to choose from. A few of them will be discussed below in some detail.

4.3.1 DETERMINATION OF GROUPS X–H

When an arbitrary atom or group X other than C is connected with hydrogen the proton resonance line will be shifted in the range of about $\delta = 20$ ppm according to the nature of the group. The chemical shifts are usually referred to methane.

Groups X–H can be determined by the characteristic chemical shifts of the proton resonance, the spin–spin splitting arising from magnetic nuclei in group X, and/or by measuring the spectra of the nuclei in group X.

Shielding of protons in X−H groups. The local fields induced by the polarizing field H_0 in group X result in positive or negative shielding

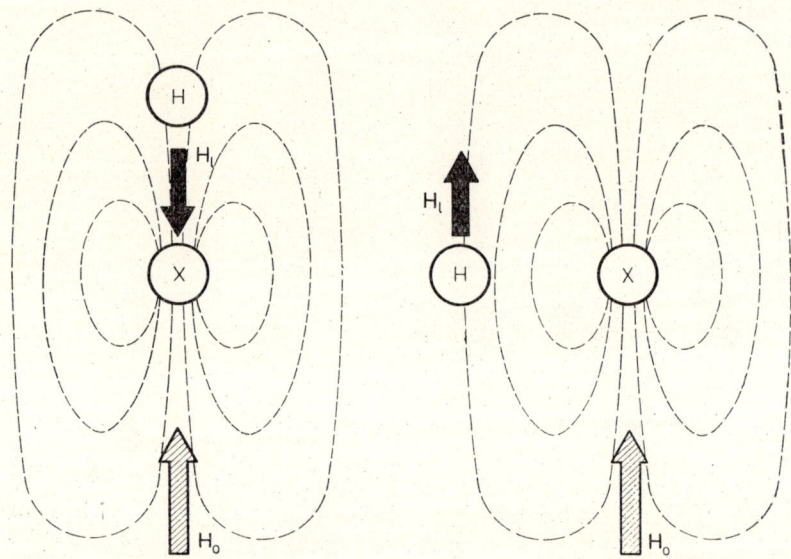

Fig. 4.13 Magnetic shielding in X–H groups

at the proton, as shown in Fig. 4.13. If the main field H_0 is oriented parallel to the symmetry axis of group X−H the local field H_1 will be opposite to H_0 at the proton. The shielding in this orientation is positive. In orientations perpendicular to the main field the local field has the same direction as H_0, and thus the shielding is negative. As shown in Section 4.1 the average shielding, i.e. chemical shift in gaseous or liquid phase, depends on the diamagnetic anisotropy of the group and on the steric position of the proton. Generally, when group X has no definite symmetry axis, the shift due to the diamagnetic anisotropy of group X is given by

$$\delta = \frac{1}{3} r^{-3} \sum_{i=1}^{3} \chi_i (1 - 3 \cos^2 \alpha_i) \qquad 4.10$$

where δ is the shift in ppm, r is the distance of the proton from group X, χ_i the diamagnetic susceptibility of group X in direction i, α_i is the angle between direction i and the bond X−H.

As well as diamagnetic fields of group X and those of atoms H there is a third important factor which contributes to the total shielding of the group. This is the paramagnetic susceptibility caused by intermixing of the electronic states (hybridization). Although most groups X are not paramagnetic, i.e. do not have unpaired electrons to result in net electronic magnetic moments in the ground state, the hybridization of electronic states often results in small residual paramagnetic susceptibilities which contribute to the total shielding of the H atom. The total susceptibility of the system is thus the following

$$\chi \text{ (total)} = \chi \text{ (paramagnetic)} - \chi \text{ (H atom)} - \chi \text{ (X group)}$$

The chemical shifts observed are thus determined by the magnitude and anisotropy of the total susceptibility χ (total) and by the distance and angle between group X and H. Some data are collected in Table 4.7. The reference is methane, CH_4. Large negative shifts are observed in OH groups, especially in inorganic acids. The highest positive shifts are observed in halogen groups HI, HBr.

The large shift ranging to about 20 ppm offers good possibilities for identification of atom or group X by locating the chemical shift of the proton. Additional information can be derived from spin–spin splittings.

Illustrative example: pyrrole. The proton resonance spectrum of X−H groups can generally be observed without any difficulties. In some cases, however, when X is a magnetic nucleus having large quadrupole moments, the spin–spin coupling between X and H is averaged out by the quadrupole effects. This is well illustrated in the proton resonance of pyrrole, where proton resonance of the N−H

Table 4.7
PROTON RESONANCE SHIFTS IN X–H GROUPS*

Compound	X atom	δ-values ppm
Organic acids	O	−8 to −7
Inorganic acids	O	−7.5 to −7
Phenols	O	−8 to −2
Alcohols	O	−2 to −0.5
H$_2$	H	≈ −4.5
SiH$_4$	Si	≈ −3
PH$_3$	P	−2.5 to −2
H$_2$O	O	−6 to −1 (hydrogen bonded)
CH$_4$	C	0
NH$_3$	N	−0.5 to 0
H$_2$S	S	−0.5 to 0
HCl	Cl	−0.5 to 0
Aryl amides	N	0.5 to 1.5
Thiophenols	S	1 to 1.5
Alkyl amides	N	3 to 4
HBr	Br	2 to 3
HI	I	10 to 12

*Hatched columns indicate shifts due to hydrogen bonding.

group cannot be observed at all at ordinary conditions. The structure is the following

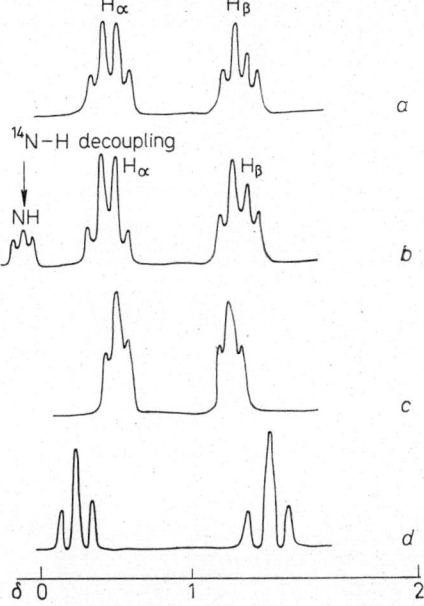

The ordinary proton resonance spectrum consists of two groups of lines corresponding to the ring protons H_α and H_β. The spectrum is shown in Fig. 4.14a. The splitting of the lines suggests the existence of the proton in the $N-H$ group, for from the ring protons 3 + 3

Fig. 4.14 Proton resonance spectrum of a—Pyrrole; b—Pyrrole with ^{14}N decoupling; c—Deuterated pyrrole[4.11]; d—Furan[4.12]

lines an A_2X_2-type spectrum should be observed. The proton resonance line of the N—H group appears immediately if the ^{14}N—H coupling is decoupled by double resonance. This is shown in Fig. 4.14b. The pure A_2X_2-type spectrum of the ring protons can be observed by changing the N—H hydrogen into deuterium (Fig. 4.14c)

$$\underset{H_\alpha}{}\underset{\underset{D}{N}}{\overset{H_\beta \quad H_\beta}{}}\underset{H_\alpha}{}$$

The resolution is not very good, because there is a small spin–spin coupling between the deuteron nucleus and the ring protons.[4.11]

By substituting the hetero-atom with oxygen (Fig. 4.14d)

$$\underset{H_\alpha}{}\underset{O}{\overset{H_\beta \quad H_\beta}{}}\underset{H_\alpha}{}$$

a well-resolved A_2X_2-type spectrum is observed from the ring protons H_α and H_β.[4.12]

Hydrogen bonding. Intra- and intermolecular hydrogen bonding of groups X—H has a very large effect on proton shielding. In such cases the X—H group cannot be considered separately, since it is only part of the whole system

$$X-H \cdots H-Y$$

$$X-H \cdots H-X$$

$$Y-H \cdots H-Y$$

The chemical shifts are determined by the following main factors.

1. *Direct charge transfer.* The magnetic field generated by the currents due to transfer of charges makes the local field at the protons change. This would result in a positive shift.

2. *Magnetic interaction.* If the total susceptibility of group X–Y is anisotropic there will be a mutual contribution to the shielding of the protons from groups X and Y. The sign of the shift depends on the steric orientation, as shown in Section 4.1.

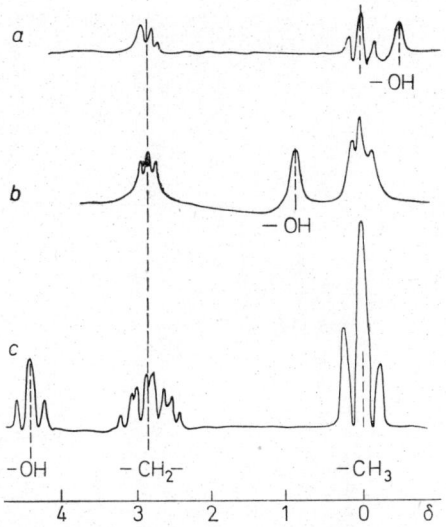

Fig. 4.15 Effect of hydrogen bonding to the OH line in ethanol

3. *Electric interaction.* If group Y is highly polar or ionic, the electron configuration of the X–H group can be distorted by the electrostatic forces. This would result in a change of the total susceptibility and thus in the shift of the group.

The effect of hydrogen bonding on the chemical shift of OH groups is illustrated in Fig. 4.15. In the gaseous phase practically no hydrogen bonding is present. The spectrum is shown in Fig. 4.15a. The OH group is shifted far away to high fields. In dilute solution in CCl_4 the number of hydrogen bonds is increased, resulting in a shift of the OH line to lower fields as shown in Fig. 4.15b. In water-free pure ethanol the number of hydrogen bonds is very high; the OH line is shifted further to lower fields. The total shift from gaseous phase to the pure liquid is −4.58. This is a general observation; *hydrogen bonding causes proton resonance lines to be shifted to lower fields* (negative

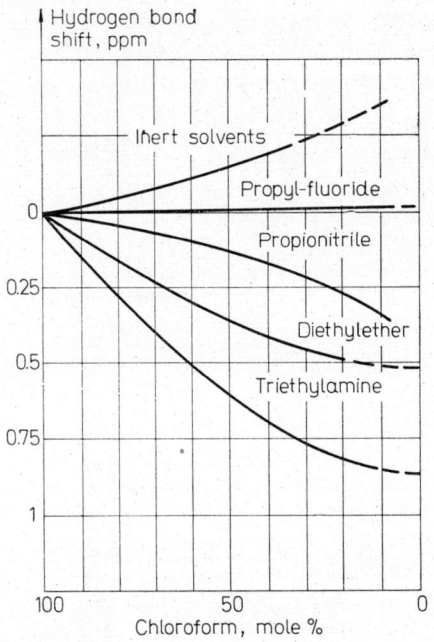

Fig. 4.16 Hydrogen bonding line shifts of chloroform in different solvents[4.31]

shielding). The total shift due to hydrogen bonding can be measured by measuring the spectrum at high temperatures in the gaseous phase, where hydrogen bonding is practically absent and at low temperatures in the liquid state near the melting point, where hydrogen bonding is considered to be complete. Some hydrogen bonding shifts are indicated in Table 4.7 by shading. In HCl, for example, the range of shifts is −1.5 to +0.5 ppm, depending on the extent of hydrogen bonding; similar ranges are indicated for HBr, HI, H_2S, NH_3 etc.

It is not absolutely necessary to measure in the gaseous phase to study hydrogen bonding. One can take proton resonance spectra in neutral solutions, varying the concentration. Extrapolation of the observed shifts to zero concentration results in a fair approximate value for δ in the absence of hydrogen bonding.

In this way it is relatively easy to decide if hydrogen bonding is present or not. The groups not taking part in the association do not exhibit any shifts as concentration is varied. This is also illustrated in Fig. 4.15 where the position of groups $-CH_3$ and $-CH_2-$ remain unchanged in the gaseous phase, in dilute solution, and in the pure material.

For investigating hydrogen bonds the sample must be dissolved in an inert solvent. The chemical shift in these solutions is increased as the concentration of the sample is decreased. This is illustrated in Fig. 4.16 where the chemical shift of chloroform protons is plotted against chloroform concentrations. In such inert solvents as n-hexane, cyclohexane and cyclopentane the extrapolated shift is $+0.3$ ppm with respect to the line of pure chloroform. Thus the chemical shift due to hydrogen bonding between chloroform molecules and the solvent molecules is approximately -0.3 ppm.

In electron-donor solvents the shift is mainly determined by the donor–acceptor interaction between chloroform groups X and solvent groups Y

$$X^+H\cdots Y^-$$

In this case decreasing the chloroform concentration results in a decrease of the chemical shift, as in triethylamine or diethyl ether shown in Fig. 4.16.

The hydrogen bonds between molecules are not stable at ordinary temperatures. Thus during the course of the measurement a large number of hydrogen bridges are formed and broken. If the rate of formation and disintegration of the hydrogen bonds is large in comparison with the chemical shift, the spectrum line will represent the time average of situations

$$XH\cdots Y$$

and

$$X \cdots HY$$

Near the freezing point, however, the exchange of protons between groups X and Y is slow in comparison with the chemical shift and the lines X−H and Y−H are separated. This phenomenon known as chemical exchange can be studied very effectively by NMR. Some examples are given in Chapter 5. From the viewpoint of structure determination, hydrogen bonding and chemical exchange phenomena provide additional information to identify certain groups by high resolution NMR.

4.3.2 GROUPS CONTAINING FLUORINE

The natural isotope of fluorine ^{19}F has a nuclear spin of 1/2 and a magnetic moment of 2.7927 nuclear magnetons, slightly different from that of protons. The resonance frequency is therefore about 5% lower than proton frequencies, the sensitivity is almost the same as that of proton resonances. These parameters make fluorine resonances quite easy to observe.

A very attractive feature in fluorine resonance spectra is that the chemical shift of F−X groups is spread over a range of about 300 ppm which is considerably larger than the 20 ppm band observed for H−X groups. Groups containing fluorine can therefore be determined even more accurately by measuring the chemical shifts than those containing hydrogen. There is a definite connection between the electronegativity of group X and the fluorine chemical shift. The magnetic shielding of fluorine nuclei in F−X groups has been found to be strictly determined by the ionic character of the bond. The more the ionic-to-covalent percentage is, the larger the shielding of the fluorine nucleus will be. This is well illustrated by substituted fluorobenzenes, where a direct correlation between the Hammett σ-values and the fluorine chemical shifts is observed; Hammett σ-values are defined as follows

$$\sigma = \ln k_i - \ln k_i^0 \qquad 4.11$$

where k_i is the ionization constant of the substituted fluorobenzene i, and k_i^0 is that of the corresponding non-substituted fluorobenzene. The dependence of the chemical shifts from unsubstituted fluorobenzene on the corresponding Hammett σ-value is described by the following equations

$$\delta_m = -5.92\,\sigma_m$$

$$\delta_p = -17.9\,\sigma_p + 4.84$$

4.12

Subscripts m and p refer to the position of the substituting group with respect to the fluorine atom

```
        F
        |
   H——⬡——H
   H——⬡——R  meta
        |
        R  para
```

The absolute values of the shifts are much higher than those of the corresponding normal benzenes. Substituting with an OH group the corresponding monofluoro-benzenes would exhibit the following fluorine chemical shifts, depending on the relative position of the OH group with respect to the fluorine

```
        OH
        |
       ⬡  + 25.0
          − 0.9
       +10.6
```

The numbers denote the chemical shifts of the compound in ppm, which has its fluorine atom at the position of the number. Thus the largest shift from unsubstituted fluorobenzene is found when the

OH group is in *ortho*-position with respect to the fluorine atom:

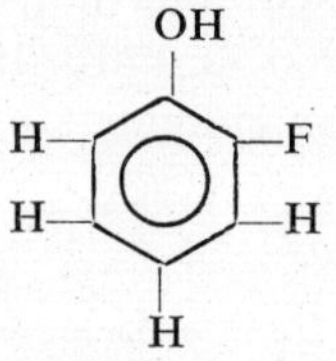

$$\text{2-fluorophenol}\qquad \delta = +25.0 \text{ ppm}$$

The shifts are much lower if chlorine is substituted:

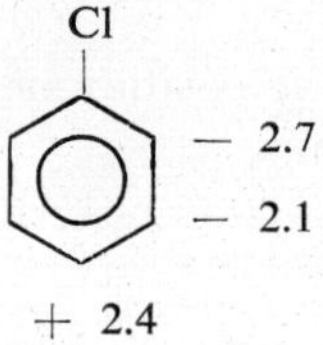

Cl-substituted benzene:
- ortho: -2.7
- meta: -2.1
- para: $+2.4$

Here again the largest positive shift is observed when fluorine is in *ortho*-position with respect to chlorine. Further halogen substitution results in negative shifts. In the case of Br

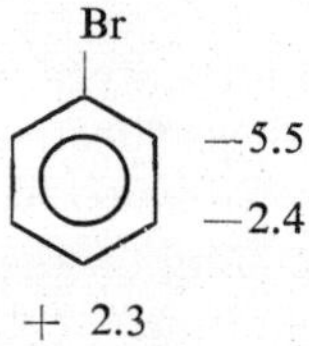

Br-substituted benzene:
- ortho: -5.5
- meta: -2.4
- para: $+2.3$

By substituting iodine a fairly large negative shift is observed in *ortho*-position

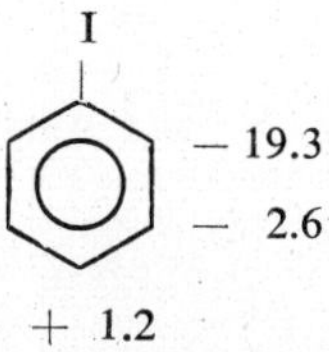

I-substituted benzene:
- ortho: -19.3
- meta: -2.6
- para: $+1.2$

In accordance with Equation 4.16 the compounds with substituted groups in *para*-position always give positive shifts, although the absolute values decrease upon increasing the σ-values. In *meta*-substituted compounds negative shifts are always observed.

Substitution acts in two ways on the shielding of the fluorine nucleus. One is the simple magnetic induction conducted by the σ-type bonds of the ring. This is described by the 'induction' Hammett factors $\sigma(i)$. The other way is the so-called 'resonance' effect conducted by the π-electrons of the ring. This is described by the resonance Hammett factor $\sigma(r)$. The total chemical shift caused by multiple substitution can generally be described as follows

$$\delta(meta) = -5.83 \sum_{k=1}^{N} \sigma_k(i) + 0.20$$

$$\delta(para) = -5.83 \sum_{k=1}^{N} \sigma_k(i) + 18.80\,\sigma(r) + 0.80$$

4.13

Here N is the number of substitutions. Values of $\sigma(i)$ and $\sigma(r)$ are tabulated.[4.13] Thus the substituted fluorobenzene shifts can be predicted theoretically, which makes their identification quite easy. The mean error between the values calculated from Equation 4.13 and of those measured experimentally is ± 0.7 ppm.

Spin–spin couplings. Fluorine resonance spectra are split mainly by F–F and F–H spin–spin interactions. Fluorine–fluorine splitting constants are usually higher than those of H–H. In difluorobenzenes for example the following F–F and F–H coupling constants are observed

6–10   20 c/s   *ortho*-difluorobenzene

6–8 c/s   2–4 c/s   *meta*-difluorobenzene

2 c/s — [para-F,H benzene]   [para-F,F benzene] — 12–15 c/s

The spin–spin splittings give valuable additional information for locating groups containing fluorine nuclei.

The examples above are given only to illustrate the possibilities. An enormous quantity of fluorine containing compounds have been investigated and their NMR spectra interpreted.

4.3.3 GROUPS CONTAINING PHOSPHORUS

The natural phosphorus isotope is ^{31}P with a nuclear spin of 1/2 and magnetic moment of 1.1305 nuclear magnetons. ^{31}P resonance frequencies are therefore much lower at the same magnetic field than proton frequencies. The sensitivity is about 6% of that of proton resonances. This is, however, quite enough for structure determinations. The chemical shifts and spin–spin splitting constants are favourably high. Thus groups containing phosphorus atoms are relatively easy to analyse.

As an example of phosphorus resonances the ^{31}P spectrum of polyphosphates is discussed in some detail. These compounds contain groups of oxygen and phosphorus only. The general structure is the following

$$\sim O-\underset{\underset{O}{\|}}{\overset{\overset{O}{\|}}{P}}-O-\underset{\underset{O}{\|}}{\overset{\overset{O}{\|}}{P}}-O \sim$$

Two main groups can be considered in this polymer: the end groups

$$^-O-\underset{\underset{O}{\|}}{\overset{\overset{O}{\|}}{P}}-O \sim$$

and the intermediate group

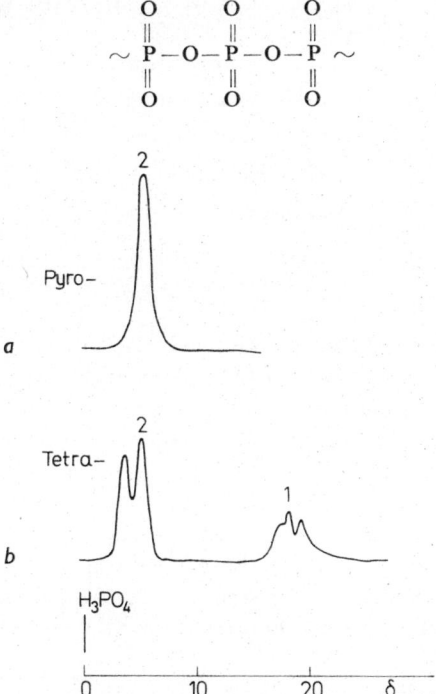

Fig. 4.17 ^{31}P resonances in pyro- and tetra-polyphosphate. After Cullis, van Wazer and Shoolery[4.14]

The local field is different at the ^{31}P nuclei in the end groups from that in the intermediate groups, because the chemical surroundings are different. In Fig. 4.17a the ^{31}P resonance of pyrophosphate is shown[4.14]

This compound consists of two end groups and exhibits a single ^{31}P line, as shown in the figure. In triphosphate there are two end groups and a single intermediate group. This is why two groups of lines appear (Fig. 4.17) with an intensity ratio of 2 : 1 corresponding to the ratio of the ^{31}P nuclei. The spin–spin splitting is due to the ^{31}P–^{31}P coupling:

$$\begin{array}{c} \text{O} \text{O} \text{O} \\ \| \| \| \\ ^-\text{O}-\text{P}-\text{O}-\text{P}-\text{O}-\text{P}-\text{O}^- \\ \| \| \| \\ \text{O} \text{O} \text{O} \end{array}$$

end intermediate end

The line corresponding to the intermediate group splits into three with an intensity ratio of 1 : 2 : 1 as a result of the interaction with the two equivalent end group nuclei. The end group signal at the low field side is, on the other hand, split into a doublet by the interaction with the single intermediate group ^{31}P atom. The spectrum is thus of AX_2 type.[4.14]

This example has shown that groups containing ^{31}P atoms can generally be identified even in those unfavourable cases, when the only magnetic nucleus in the system is ^{31}P. If other magnetic nuclei are present as well, the structure determination is more accurate.

4.3.4 GROUPS CONTAINING BORON

Natural boron has two magnetic isotopes. One is ^{10}B with an abundance of 18.8% having a nuclear spin of 3, and magnetic moment of 1.8006 nuclear magnetons. The other natural magnetic isotope is ^{11}B with an abundance of 81.2% having a nuclear spin of 3/2 and magnetic moment of 2.6880 nuclear magnetons. The nuclear quadrupole moment of ^{10}B is about 10 times as much as that of ^{11}B. According to the higher abundance, higher nuclear magnetic moment and smaller nuclear quadrupole moment ^{11}B is much more favourable for NMR measurements. Its sensitivity relative to protons is 16%.

^{11}B spectra are rather complex as a result of the high spin $I = 3/2$. Spin–spin coupling between ^{11}B and H nuclei would result in $2I + 1 =$

4 lines, that between $^{11}B-^{11}B$ in case of weak coupling in 8 lines. Here the spin–spin decoupling technique is essential.

For illustrating ^{11}B resonance spectra the following compound is discussed

$$H-\underset{\underset{CH_3}{|}}{\overset{\overset{CH_3}{|}}{P}}-BH_3$$

The spin systems of this molecule are as follows:

2 methyl groups	6 equivalent protons
P	^{31}P nucleus
P–H group	1 proton
B	^{11}B nucleus
BH_3 group	3 equivalent protons

First it is advisable to record the spectrum at proton resonance to locate the methyl groups, BH_3 groups and PH. From the observed spin–spin splitting of the proton resonance line the following couplings can be identified[4,15]

$$\begin{array}{c} CH_3 \quad H_b \\ 6\ c/s \nearrow \quad \uparrow \\ \quad \quad 12\ c/s \\ \quad \quad \downarrow \\ H_a \leftrightarrows P \quad - \quad B \quad H_b \\ 350\ c/s \quad \quad \quad 90\ c/s \\ \quad \quad \downarrow \\ CH_3 \quad H_b \end{array}$$

The spectra are shown in Fig. 4.18.

The different couplings can be separated by ^{11}B and ^{31}P double resonances. The ^{11}B resonance spectrum exhibits four doublets with an intensity ratio of 1:3:3:1. This corresponds to the spin–spin coupling of the ^{11}B nucleus with three equivalent protons, evidently with protons H_b. The doublet splitting of the lines is due to the interaction with the 1/2 spin phosphorus nucleus. Thus the ^{11}B

resonance spectrum reveals the following spin–spin couplings

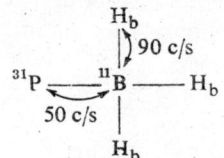

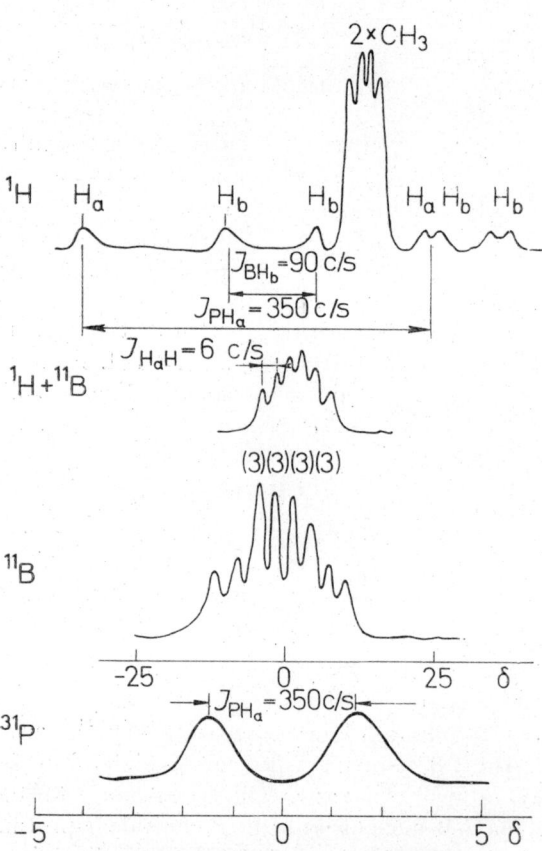

Fig. 4.18 Proton, boron and phosphorus resonance in $(CH_3)_2PHBH_3$. After Shoolery[4.15]

This confirms the $^{11}B-H_b$ constant of 90 c/s derived from proton resonance and also gives the new coupling $^{31}P-^{11}B$. Finally it is possible to take the ^{31}P resonance spectrum, although it is not necessary since the whole structure is accurately determined by proton and ^{11}B spectrum. The ^{31}P spectrum consists of two lines separated by 350 c/s according to the coupling constant between $P-H_a$ already derived from the proton resonance spectrum.

As shown by this example molecules containing many magnetic nuclei can be analysed very effectively by measuring each resonance spectrum, observing the chemical shifts and spin–spin couplings and using the method of spin–spin decoupling.

The amount of information gained from these spectra is in most cases more than enough for identifying even much more complicated structures.

4.3.5 GROUPS CONTAINING OTHER NUCLEI

It is not possible to mention all the nuclei which can be measured by high resolution NMR. A glance at Table 6.1 in the Appendix will show how many possibilities there are for structure determination and analysis. In planning NMR experiments with new nuclei it is advisable to consider the following problems:

1. The NMR set-up must be equipped with a new head for the new resonance frequencies. It is always advisable to work at the highest magnetic field level of the given spectrometer, so changing from one nucleus to the other involves changing heads.

2. The signal intensity relative to that of protons should be higher than the sensitivity limit of the spectrometer. In quantitative analysis a good signal-to-noise ratio is required in order to assure intensity (integral) measurements with the required accuracy.

3. Nuclei having high quadrupole moments are very difficult to measure because of the broadening of the lines. If nuclei of high quadrupole moments are coupled to the spin system investigated, the decoupling technique should be used.

4. If the system to be analyzed contains paramagnetic atoms or ions, the lines may be broadened seriously by the electron-nuclear spin interaction. In these cases the high paramagnetic local fields are usually higher than the diamagnetic fields and thus chemical shifts and spin–spin coupling cannot be observed at all. Some examples of unusual nuclei are given as follows.

Nitrogen. The natural nitrogen isotope of high abundance (99.6%) has a nuclear spin 1 and an electric quadrupole moment of 2×10^{-2}. The magnetic moment is 0.40357 nuclear magnetons. The measurement is difficult because of the quadrupole effects; ^{14}N resonances have been measured in such compounds and ions as NH_3; NH_4OH; N_2H_4; $(NH_2)_2CO$; $(SCN)^-$; $(CN)^-$; HNO_3; pyridine; N_2; NO_2. The resolution is not good: the lines are broadened by the quadrupole interaction. The other stable isotope ^{15}N has a nuclear spin of 1/2, a magnetic moment of -0.28304 nuclear magnetons. The abundance is only 0.37%, and thus the signal intensities are too low to be detected.

Silicon. The magnetic Si isotope, ^{29}Si, has a nuclear spin $I = 1/2$. a magnetic moment of -0.55477 nuclear magnetons. The abundance is 4.7%. The chemical shifts are high enough for structure determinations and analysis. Data on quartz, various kinds of glasses, semiconductors and on a number of organic silicon compounds are available in the literature.

Tin. The magnetic isotope of highest abundance is ^{119}Sn (8.68%). Its nuclear spin is 1/2, magnetic moment -1.0409 nuclear magnetons. The signals are relatively strong; ^{119}Sn spectra can be taken without any difficulties. The chemical shifts are extremely large in comparison with those of proton and fluorine. Between SnI_4 and $(CH_3)_4Sn$ the ^{119}Sn lines are shifted by 1698 ppm. The spin–spin coupling with hydrogen is readily observed. The only difficulty in observing ^{119}Sn spectra is that Sn compounds generally cannot be dissolved in suitable inert solvents. Thus in most cases additional shifts due to association appear.

Lead. The magnetic isotope of lead, ^{207}Pb, has a nuclear spin of $I = 1/2$, a magnetic moment of 0.5837 nuclear magnetons. Spectra are easy to measure. The chemical shifts are very large. A shift of 3,000 ppm is observed between ionic and covalent Pb compounds.

The resonance line of metallic lead and those of semiconducting compounds are shifted strongly to low fields as a result of the negative shielding of the conduction electrons. The line of solid $PbSO_4$ is shifted by 14,700 ppm to higher fields from that of pure metallic lead.

4.4 STEREOCHEMICAL ANALYSIS

As shown in Section 4.1 the magnetic shielding depends not only on the chemical surroundings of the spins but on the steric configuration as well. The reason for this is that most chemical groups are diamagnetically anisotropic, i.e. the local fields are not averaged out by the random motion of the molecules. The steric distribution of the diamagnetic shielding of a group $X-Y$ is given by Equation 4.7 for axially symmetric groups. The chemical shift regions are shown in Fig. 4.7. According to the arguments given in Section 4.1 for absolute determination of steric structures the following main factors should be known:

1. *The diamagnetic symmetry of the groups.* This can be derived from the electronic structure or experimentally determined by using model compounds.
2. *The absolute value of the average diamagnetic moment $\langle \mu \rangle$.* This factor can in principle be derived theoretically from the electronic structure of the group. However, it is much easier to use model compounds of known steric structure and measure values of $\langle \mu \rangle$ experimentally.

In most cases the possible steric conformations are already known from chemical arguments. NMR measurements are used in these cases to identify them. This is much easier than absolute structure determination, only a few general rules derived from simple arguments such as are given in Section 4.1 are needed.

Besides chemical shifts, spin–spin couplings can also be used to identify stereoisomers. Spin–spin couplings between groups of nuclei are conducted by the molecular electrons and correspondingly are rather sensitive to steric configurations. On the basis of the measured chemical shifts and spin–spin coupling constants steric structures can easily be identified.

Steric isomers are usually not stable at ordinary temperatures. Different possible steric configurations are usually present in different amounts. A very attractive feature of high resolution NMR is that it is possible not only to determine the relative amount of isomers in the system, but the rates of interchange between them as well. In the case of very rapid interchange between steric forms the local fields are averaged and the isomers can no longer be distinguished. By lowering the rate of interchange (lowering the temperature) the lines will be broadened as a result of incomplete averaging of the local fields. Further lowering of the interchange rates results in separation of the isomer spectra. Thus by measuring NMR spectra at different temperatures the process of interchange between certain steric forms can easily be followed.

The separation between stereoisomer NMR lines are in the order of 0.1–1 ppm. At operating frequencies of 100 Mc/s, this corresponds to absolute shifts of 10–100 Mc/s. Thus for separating the isomer lines, the rate of interchange must be lower than about 10 c/s.

In NMR analysis of stereoisomers the following main groups are considered.

Group I. The activation energy of the interchange is so great that only a single isomer is present at the temperature range of the measurement. In these cases a single NMR spectrum appears which practically does not change with temperature. The lifetime of these isomers (τ_i) must be much higher than the inverse chemical shift $\Delta\delta_i$

$$\tau_i \gg \frac{1}{\Delta\delta_i}$$

$\Delta\delta_i$ in these cases cannot be measured directly.

Group II. The activation energy of the interchange is low enough to allow different isomeric forms to be present at the same time. The NMR spectrum in these cases consists of superimposed spectra of the stereoisomers. The shifts $\Delta\delta_i$ can be directly measured and the relative abundance of the isomers can be determined readily from the relative intensities of the lines. Lines widths and relative intensities are changed upon changing the temperature. For the lifetime of the isomers in the temperature range of the measurement:

$$\tau_i \approx \frac{1}{\Delta\delta_i}$$

Group III. The activation energy of the interchange is so low that even at the lowest possible temperature in the liquid state

$$\tau_i < \frac{1}{\Delta\delta_i}$$

In such cases the isomers cannot be distinguished by NMR. Interchange in such systems can only be lowered in the solid state, where the resolution of NMR is too low to resolve $\Delta\delta_i$.

Rotational isomers having low activation energies can be studied by microwave molecular spectroscopy in the gaseous phase.

4.4.1 ROTATIONAL ISOMERS

A good example of isomers of Group III is the *cis–trans-* isomerization about $-C=C-$ double bonds. The chemical shift of ethylene derivatives for example is strongly dependent on the steric position of the protons. The interchange between *cis-* and *trans*-configurations is very slow even at higher temperatures.

Substitution of one ethylene proton results in a shift of the remaining three with respect to pure ethylene. In propylene, for example, the following shifts are observed

$$\underset{H_3C}{\overset{-0.16}{\diagdown}} C = C \underset{-0.9}{\overset{-0.10}{\diagup}}$$

The numbers represent the shifts of the corresponding protons in ppm from the pure ethylene line.

In the case of double substitution as

$$\begin{array}{c} Br \\ \diagdown \\ CH_3 \end{array} C=C \begin{array}{c} H \\ \diagup \\ H \end{array}$$

the shift of the protons in *cis*- and *trans*-positions with respect to Br is

$$\delta_{trans} - \delta_{cis} = 0.20 \text{ ppm}$$

Generally it is possible to identify the following isomer pairs

$$\begin{array}{c} X \\ \diagdown \\ R \end{array} C=C \begin{array}{c} H \\ \diagup \\ R \end{array} \quad cis$$

$$\begin{array}{c} X \\ \diagdown \\ R \end{array} C=C \begin{array}{c} R \\ \diagup \\ H \end{array} \quad trans$$

by measuring the shift $\delta_{trans} - \delta_{cis}$ of the proton with respect to group X. It is also possible to identify *cis–trans* isomers by measuring resonances of group X. There is a shift of the methyl proton resonances in the following isomer pair

$$\begin{array}{c} H_3C \\ \diagdown \\ R \end{array} C=C \begin{array}{c} H \\ \diagup \\ COCH_3 \end{array} \quad trans$$

$$\begin{array}{c} R \\ \diagdown \\ HC_3 \end{array} C=C \begin{array}{c} H \\ \diagup \\ COCH_3 \end{array} \quad cis$$

In these pairs the methyl resonances are separated by about $\delta_{trans} - \delta_{cis} \approx 0.2$ depending on the nature of group R.

Spin–spin splitting constants between ethylene protons have already been discussed in Section 4.2. As shown there, coupling constants between geminal hydrogens are small (0–3 c/s), between vicinal hydrogens are large (6–14 c/s and 11–18 c/s for the *cis*- and *trans*-configurations, respectively). The steric position of substituting groups can thus be confirmed by the spin–spin splitting constants. In *trans*- and *cis*-propenylbenzene, e.g., the following spin–spin coupling constants are observed[4.16]

11.4 c/s

Ph–C(H)=C(H)–CH₃, 1.4 c/s, 6.7 c/s   cis

Ph–C(H)=C(CH₃)–H, J=0, J=0, 6 c/s   trans

The chemical shift between the ethylene protons is $\delta_{trans} - \delta_{cis} = 0.1$ ppm. As shown, the coupling constants are strongly dependent on the steric form.

Cis–trans isomerization around partial double bonds. The next group of stereoisomers, rather stable at higher temperatures, are those having partial double bonds. Groups containing OCN in many respects behave as though they had C=N double bonds. This is interpreted by the partial transfer of the C=O bond to C–N, resulting in the following structure

$$\diagdown_{/}C=N^+\diagup^{O^-}_{\diagdown}$$

The second bonding electron in these systems is shared between O–C and C–N forming *partial double bonds*

$$\begin{array}{c} O \\ \diagdown \\ C \cdots N \\ \diagup \quad \diagdown \end{array}$$

The rotation about C–N is much more hindered than that about single bonds. This is why *cis-* and *trans-*isomers in amides are rather stable even at room temperatures. Even in the simplest compound of such type, formamide, the differences in the spin–spin splitting constants indicate hindered rotation around the C–N partial double bond:[4,17]

$$\begin{array}{c} O \quad \xrightarrow{13 \text{ c/s}} \quad H_3 \\ C \cdots N \\ H_1 \xrightarrow{2.1 \text{ c/s}} H_2 \end{array} \quad J_{23}=0$$

The exact proton–proton coupling in this molecule can only be determined by ^{14}N–H decoupling. The difference in coupling between *trans-* and *cis-*hydrogens indicates that rotation is greatly hindered. At higher temperatures the difference in splitting constants J_{12} and J_{13} vanish.

By substituting two hydrogens in formamide by methyl groups

$$\begin{array}{c} O \quad \quad CH_3 \\ \diagdown \quad \diagup \\ C \cdots N \\ \diagup \quad \diagdown \\ H \quad \quad CH_3 \end{array}$$

a shift can be observed between the methyl resonance lines in *cis-* and *trans-*position with respect to the oxygen. The shift is

$$\delta_{trans}(CH_3) - \delta_{cis}(CH_3) = 0.25 \text{ ppm}$$

the lines are well separated in the temperature range of 0–30 °C and are broadened between 30 and 100 °C. At 150 °C a single sharp line is observed from the two methyl groups indicating that rotation is fast enough to average the local fields.

Rotational isomerism about partial double bonds can also be investigated in compounds of the following type

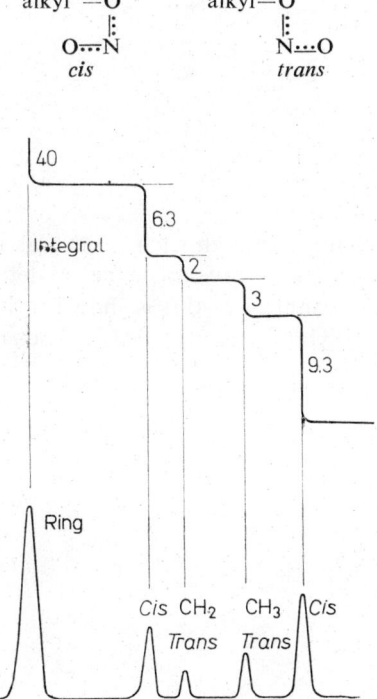

Fig. 4.19 *Cis — trans* isomer shift in benzylmethylnitrosamine[4,32]

Most of these isomers are separated even at room temperature. The proton resonance lines of the alkyl group are shifted depending on the *cis*- or *trans*-position of the N—O group. The relative abundance of these isomer-pairs can be measured directly by comparing the intensities of the alkyl proton resonance signals. Partial double bonds can also be formed in N_2O groups of the following type

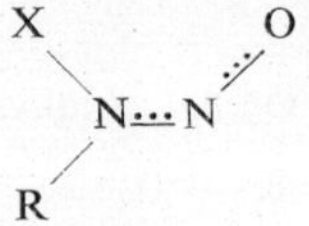

$$\begin{array}{c} X \\ \diagdown \\ N\cdots N \\ \diagup \diagdown\!\!\!\cdot \\ R O \end{array} \quad cis$$

$$\begin{array}{c} X \\ \diagdown \\ N\cdots N \\ \diagup \diagdown\!\!\!\cdot \\ R O \end{array} \quad trans$$

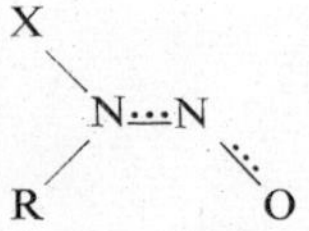

The rate of interchange between such isomers is usually intermediate at room temperature. It is thus possible to detect both isomers at the same time. An example of this is shown in Fig. 4.19. The compound is benzylmethylnitrosamine; the two isomers are the following

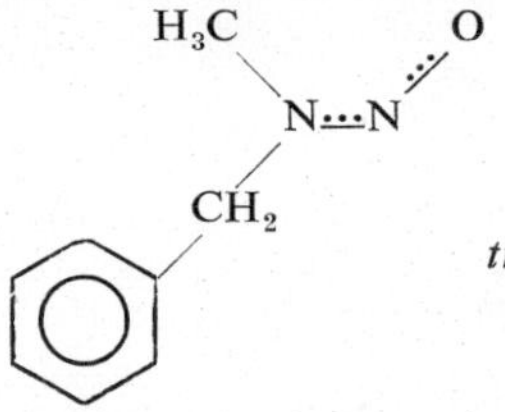

trans

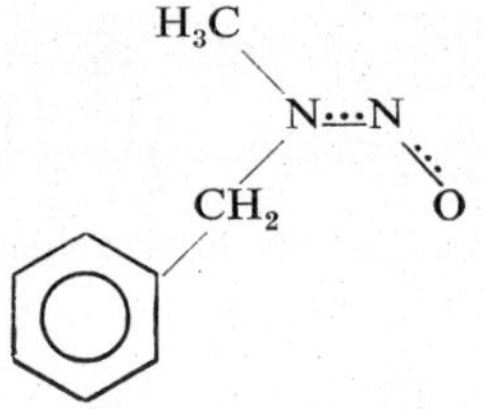

cis

There are three proton spin systems in this molecule, the 5 ring protons, the 2 methylene protons and the 3 methyl protons. As shown in Fig. 4.18 ring protons are not sensitive to steric configuration, but methyl and methylene protons are. Upon integrating the spectrum the relative concentration of the isomers can readily be determined. The intensity ratio of the *cis–trans* methyl and methylene protons is 3.1 : 1; this is the ratio of *cis–trans* isomers. The shifts are rather large:

$$\delta_{trans}(CH_2) - \delta_{cis}(CH_2) = 0.47 \text{ ppm}$$
$$\delta_{trans}(CH_3) - \delta_{cis}(CH_3) = -0.75 \text{ ppm}$$

Rotation isomerisation around single bonds. Rotation around single bonds is generally less hindered than that around double or partial double bonds. The rotational isomers usually cannot be distinguished at room temperature by the NMR technique. In these cases the compounds must be dissolved in solvents having very low freezing temperature in order to take spectra at the lowest possible temperature. As an example of this the rotational isomers of ethane derivatives are discussed.

According to the conventional notation of conformational analysis the steric structures of ethane derivatives are illustrated as follows

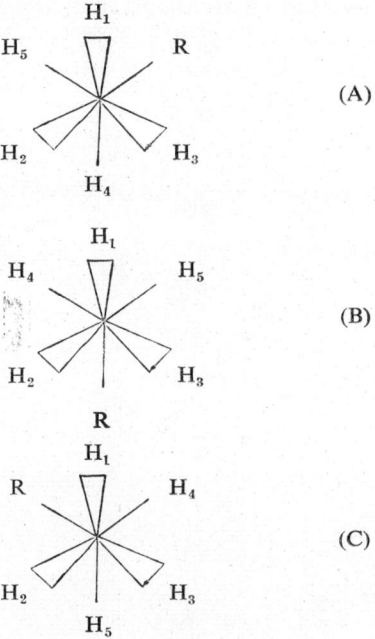

The molecule is viewed from the $H_1H_2H_3$ methyl group, the C–C bond being perpendicular to the surface of the paper. In the case of a single substitution by group R the conformations (A) (B) and (C) are possible according to the relative positions of the $H_1H_2H_3$ methyl and H_4H_5 methylene groups.

If the methyl group did not rotate, the shielding of protons H_1 and H_3 would be different from that of H_2 in configuration (A). In this case the proton resonance spectrum would be of A_2B_2 or A_2X_2 type (provided R contained no magnetic nuclei), depending on the strength of the spin–spin coupling. The spin systems are the following

A_2 protons H_1H_3

B_2 protons H_4H_5

The observed spectra are of A_2B_3 type even at low temperatures, indicating that methyl groups are rotating rapidly.

Rotational isomers of ethane derivatives can only be distinguished at low temperatures provided that the substitution is asymmetric:

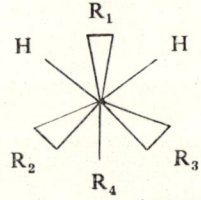

Every 120° rotation results in different shielding of the methyl protons in this molecule. As an example let us consider the molecule CF_2BrCBr_2CN. There are two steric forms of this molecule

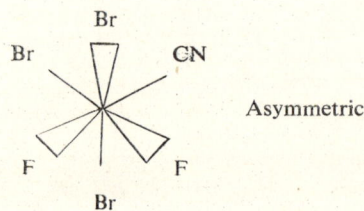

Asymmetric

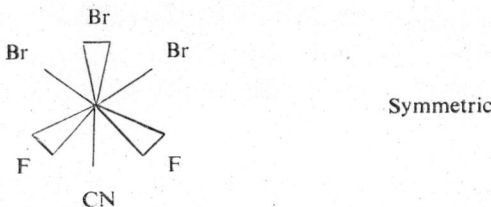

Symmetric

In the symmetric case only a single spin system is present containing two fluorine atoms. Br and CN are considered to be non-magnetic

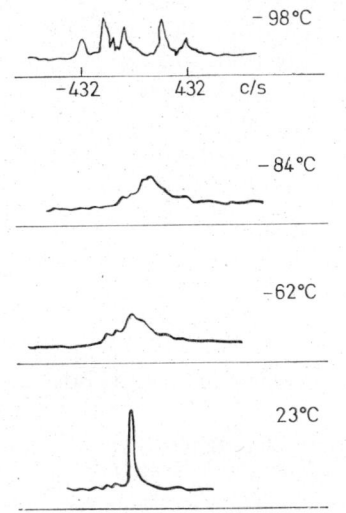

Fig. 4.20 Rotational isomers in CF_2BrCBr_2CN. ^{19}F resonance spectra[4.32]

because of the quadrupole interaction. The asymmetric isomer contains two spin systems, the two non-equivalent fluorine nuclei. This should result in AX or AB type spectra depending on the relative strength of the F—F coupling.

The observed ^{19}F resonance spectrum is shown in Fig. 4.20. At very low temperatures the two isomers exhibit separate ^{19}F resonance spectra. The asymmetric isomer exhibits an AB-type quartet, the symmetric one a single line. The intensity ratio is asymmetric : symmetric = 4 : 1. Upon increasing the temperature the lines become broadened and a single broad line is observed at -62 °C. At this temperature the rate of interchange between symmetric and asymmetric form is of the order of the chemical shift. Further increasing of the temperature results in a decrease of the line width and a corresponding increase of the amplitude. In this case the interchange is fast enough to average the local field at the fluorine nuclei.

Aromatic isomers. As shown in Section 4.1 the local field generated by the diamagnetic ring current has a peculiar steric distribution. It is parallel to the polarizing field outside the ring and opposite inside. In a way similar to the arguments on the steric effects of the shielding of group $X-Y$, discussed in Section 4.1, it can be shown that the shielding of an aromatic ring should depend strongly on its steric position. This is illustrated by the example of diphenyl derivatives. Diphenyl consists of two rings coupled together

The shielding of the ring protons is evidently influenced by the angle between the planes of the rings. In diphenyl in the gaseous or liquid phase the angle between the planes of the two rings is 45°. By substituting some of the ring protons this angle can be changed and the corresponding change in the shielding can be observed by NMR. Substituting chlorine and bromine in *ortho*-position results in the following shift with respect to that of pure diphenyl

Iodine or fluorine substitution results in a slightly smaller shift of the ring proton resonance

```
        0.18
    I,F
   ⬡ — ⬡
    I,F
        0.18
```

A theoretical plot of the chemical shift of the ring protons against the angle between the ring planes is given in Fig. 4.21.

Ring currents affect the chemical shifts of aliphatic groups connected to the ring. The shielding depends on the steric position of the ring. This is illustrated by the following example

```
   Ph      H
     \   /
      C=C
     /   \
    H     Ph
```

```
    H     H
     \   /
      C=C
     /   \
    Ph    Ph
```

the chemical shift of the ethylene protons is different in *cis*- and *trans*-configuration

$$\delta_{cis} - \delta_{trans} = 0.5 \text{ ppm}$$

In the *trans*-isomer the planes of the rings are nearly parallel, the local field generated by the ring protons is thus added. This results in a high negative shift of the ethylene proton line. In the *cis*-isomer the rings cannot be coplanar and thus the ethylene protons are less shielded. The shielding of aliphatic protons in the presence of aromatic rings is thus determined not only by the chemical groups

$$\begin{array}{c}\overset{\|}{-\mathrm{C}}\qquad\mathrm{H}\\ \diagdown\;\;\diagup\\ \mathrm{C}=\mathrm{C}\\ \diagup\;\;\diagdown\\ \mathrm{H}\qquad\mathrm{C}=\\ \qquad\;\;|\end{array}$$

but also by the relative steric position of the rings.

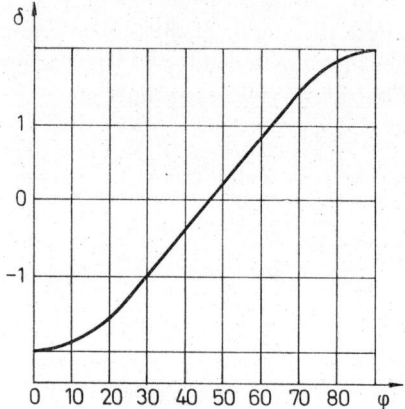

Fig. 4.21 Angular dependence of the chemical shift in two aromatic rings

As shown by the examples above high resolution NMR offers a relatively simple straightforward method for studying any kinds of rotational isomers. The only requirement is that the compound to be investigated can be dissolved in a solvent having a freezing temperature low enough to make rotational frequencies much lower than chemical shifts between the isomers. This can easily be done in isomerization around double bonds or partial double bonds. Rotational isomers around single bonds are somewhat more difficult to separate because of the small activation energies involved. Besides qualitative and quantitative identification of rotational isomers, high resolution NMR provides information about the rate of interchange and about activation energies of the rotation.

4.4.2 CYCLOHEXANE-TYPE STEREOISOMERS

The dependence of the diamagnetic shielding on steric configuration can be applied very effectively to identify conformational isomers of cyclohexane derivatives. The structure of a stereoisomeric form of cyclohexane, the chair form, is the following

Six protons are joined by C−H bonds nearly parallel to the surface of the ring; these are the equatorial protons H_e. The other six are connected by nearly perpendicular bonds; these are the axial protons H_a. In cyclohexane there is a rapid conformational inversion and therefore interchange between axial and equatorial protons, so a single proton resonance line is obtained. Upon substituting some hydrogen atoms by non-magnetic groups the shielding of the remaining protons will become strongly dependent on the steric configuration.[4.18]

Let us consider the hexachlorocyclohexane stereoisomers. In this case six protons are substituted by chlorine, which is taken as non-magnetic because of its large nuclear quadrupole moment.

Substituting chlorine for five equatorial and one axial hydrogens the following isomer is formed

δ-isomer

The corresponding proton resonance spectrum consists of two lines with the intensity ratio of 5 : 1 separated by $\delta = 0.51$ ppm. In this case the rotation of CH_2 groups is hindered by asymmetric substitution and thus the shielding of the axial protons becomes different from that of equatorial ones. As a result of the diamagnetic anisotropy of the cyclohexane ring the shielding of the axial protons is always larger than that of equatorial protons.

By substituting chlorine for two equatorial and four axial hydrogens the ε-isomer is formed

ε-isomer

The proton resonance spectrum consists of two lines again, but with an intensity ratio of 2 : 1 corresponding to the ratio of the equatorial and axial protons left in the molecule.

Symmetric substitution of 3 axial and 3 equatorial hydrogen atoms results in a single line with a chemical shift different from that of pure cyclohexane (β-isomer). Following this procedure all isomers can be interpreted. The chemical shifts of hexachlorocyclohexane stereoisomers are illustrated in Fig. 4.22. The shifts are referred to that of pure cyclohexane. As shown, axial proton lines are always shifted to higher fields with respect to equatorial lines. The magnitude of the shift is in the order of

$$\delta(\text{axial}) - \delta(\text{equatorial}) \approx 0.4 \text{ ppm}$$

This figure is in agreement with the calculations based on the simple model of Fig. 4.23. The diamagnetic local field of each C–C bond of the ring is substituted by that of the corresponding dipoles. The sum of the local fields of these 6 dipoles is then calculated at the position of the axial and equatorial proton.

Dependence of spin–spin coupling constants on bond angles. Analysis of the spin–spin coupling constants in cyclohexane-type isomers has shown that there is a direct correlation between bond angles and spin–spin coupling constants. The main ranges of proton–proton

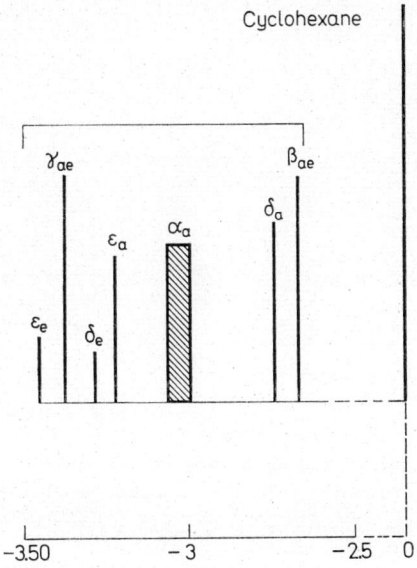

Fig. 4.22 Chemical shifts in hexachloro-cyclohexane stereoisomers[4.18]

spin–spin coupling constants in cyclohexane derivatives are the following

axial–axial	6–14 c/s
equatorial–equatorial	0– 6 c/s
equatorial–axial	~ 0 c/s

The general dependence of the coupling constant on bond angles is illustrated in Fig. 4.24. In simple CH_2 groups with 120° between the two bonds the coupling constant is rather small, 2–3 c/s. Upon decreasing the bond angle the coupling constant is strongly increased. In Fig. 4.24a the calculated values of coupling constants J_{HH} are plotted against bond angles ϕ in groups

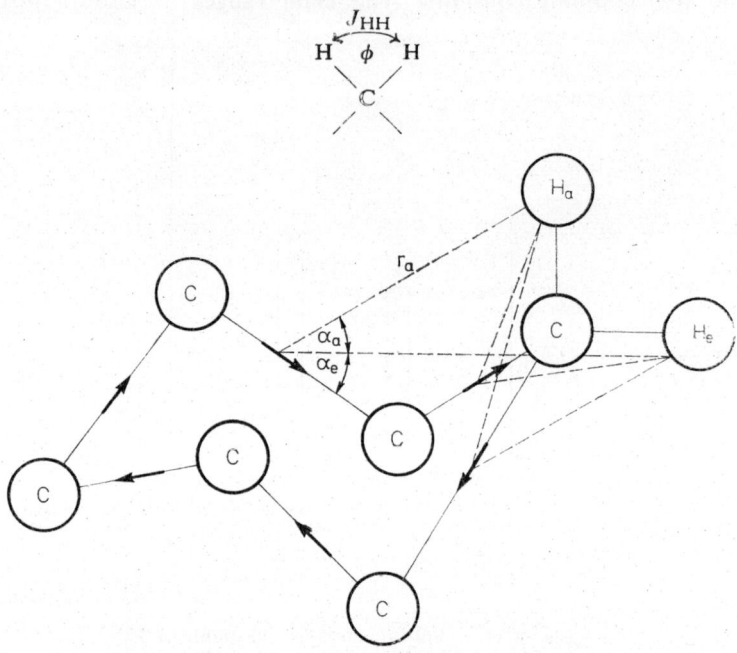

Fig. 4.23 Local magnetic fields in cyclohexane

The measured values are generally in good agreement with those plotted in Fig. 4.24a except in cases when groups of high electronegativity are coupled to the group CH.

Couplings between geminal H atoms in C_2H_2 groups are plotted against bond angles in Fig. 4.24b. Here bond angles are defined as follows

$\varphi = 0$ corresponds to the *cis*-formation

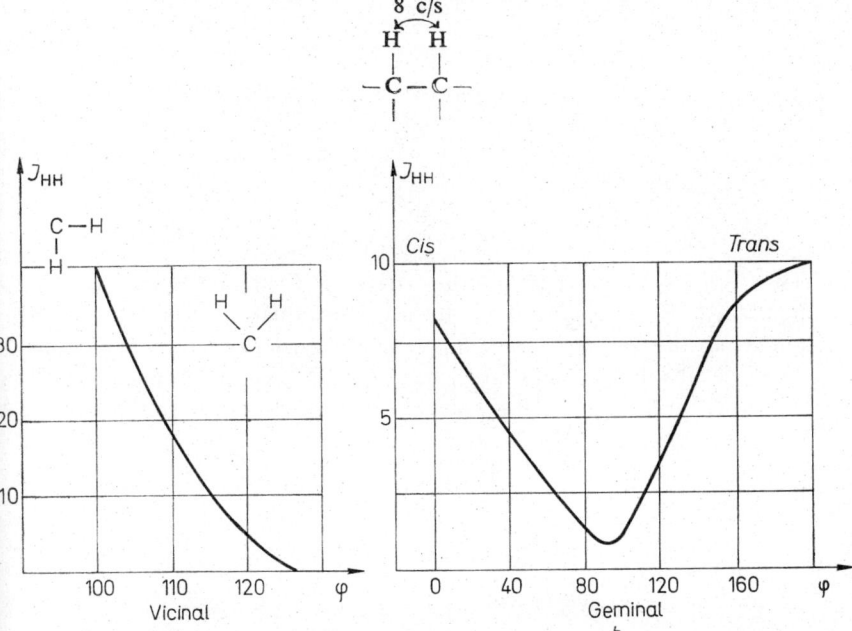

Fig. 4.24 Dependence of the spin–spin coupling on bond angles

while $\varphi = 180°$ corresponds to the *trans*-position

As shown in Fig. 4.24b in the intermediate cases when the two bonds are perpendicular to each other the J_{HH} coupling constant is very small. This is why axial–axial ($\phi = 0°$, $180°$) coupling constants in cyclohexane-type compounds are large and axial–equatorial ($\phi \sim 90°$)

coupling constants are small. The measured values are the following[4,18]

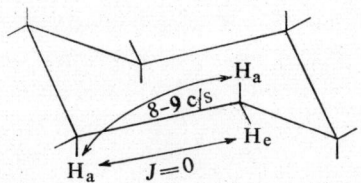

Thus spin–spin can help in the analysis of stereoisomers. The dependence of the spin–spin splitting constants on the relative bond angles is of the same type in cases of nuclei other than protons, only the absolute values are changed. In cyclohexyl fluoride, for example, at $-90\,°C$ the following $^{19}F-H$ spin–spin coupling constants are observed[4,19]

Axial isomer

Equatorial isomer

The chemical shift between these two isomers is fairly large, about 90 ppm.

4.4.3 STEROID STRUCTURES

In structure determination of such large molecules as steroids by high resolution NMR new problems arise. The large molecular framework, being less mobile in solution, exhibits wide resonance bands instead of sharp resonances. Sharp lines originate from smaller groups coupled to the skeleton of the molecule, which can rotate freely. The situation is illustrated by the well-known steroid, cholesterol

The spin systems in this molecule are the following

1. The ring protons.
2. The proton in the OH group.
3 – 4. The two methyl groups connected to carbons 10 and 13 of the rings.
5. The proton connected at carbon (6).
7. Methyl, methylene and methine protons in the side chain.

It is evident that the protons coupled rigidly to the skeleton would not exhibit sharp lines, because their local fields are not averaged sufficiently. Indeed, ring protons exhibit an absorption band shown in Fig. 4.25. Fortunately there are some sharp lines in the spectrum corresponding to the 18 and 19 methyl groups, the 3 OH group, the 6 = CH group and some to the groups in the side chain.

The chemical shift of the 18 and 19 methyl groups is sensitive to the steric structure of the steroid, i.e. the relative orientation of rings A, B, C, and D. It is also sensitive to substitution. From spectra of several hundred steroids it was concluded that the structure can be identified by considering the shifts of the 18 and 19 methyl groups only. It was shown that the shift due to different substitutions is additive, i.e. the *Shoolery rule* discussed in Section 4.2 for methylene and methine groups holds.[4.20]

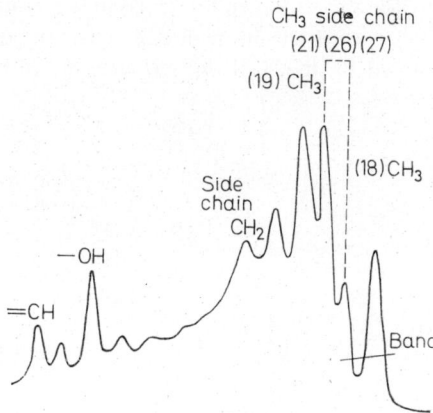

Fig. 4.25 Proton resonance spectrum of cholesterol. After Zürcher[4.20]

Additional information about the structure can be derived from the lines of other groups connected to the ring system.

In analyzing steroid structures much care should be taken with the solvents; usually highly purified deutero-chloroform $CDCl_3$ is used. The only magnetic nuclei in this solvent are ^{13}C with an abundance of 1.2%. To avoid hydrogen bonding and allow maximum motional freedom for the molecules preferably low concentrations are used with a correspondingly high sensitivity level. The usual concentrations are of the order of 0.1 mole/litre. The resolution for the methyl groups at this concentration is about 5×10^{-9}. The range of shifts is within 0.1 ppm, which means that methyl proton shifts can be measured with an accuracy of about 10%.

Despite these difficulties NMR technique has become the most widely used tool for identifying steroid structures. It is now being used for routine analysis in pharmacology.

4.4.4 METHYL-GROUP SHIFTS IN STEROIDS

5α-Steroids. This type of steroid contains a hydrogen connected to carbon 5 in α-position. Rings A and B are in *trans*-position. Two sub-divisions can be made depending on the position of the hydrogen coupled to carbon 14. The structures are the following

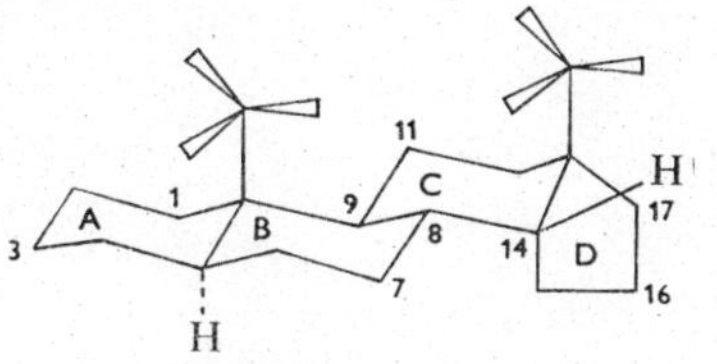

5α, 14β

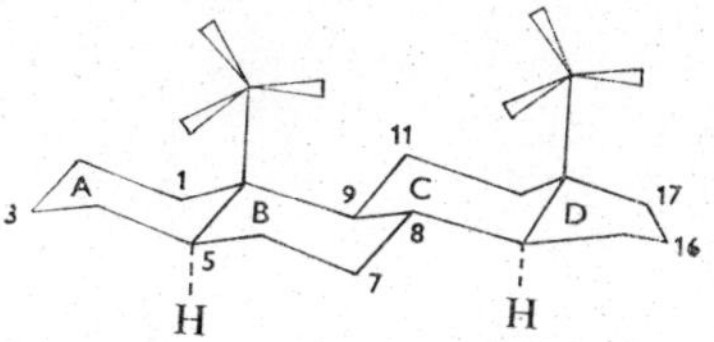

5α, 14β

In the 5α, 14α-isomer rings C and D are in *trans*-linkage, in the 5α, 14β one, in *cis*-linkage. The problem is how the shifts of the 18 and 19 methyl groups are changed upon substituting groups in the rings of 5α, 14α and 5α, 14β-steroids. As an example of this the shifts of 17-oxo-5α,14α-androstane are given in comparison with 5α,14α-androstane:

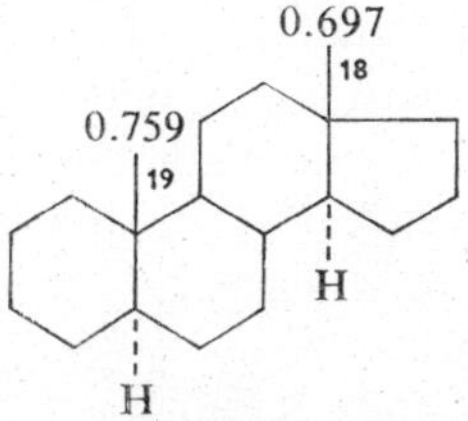

5α, 14α-androstane

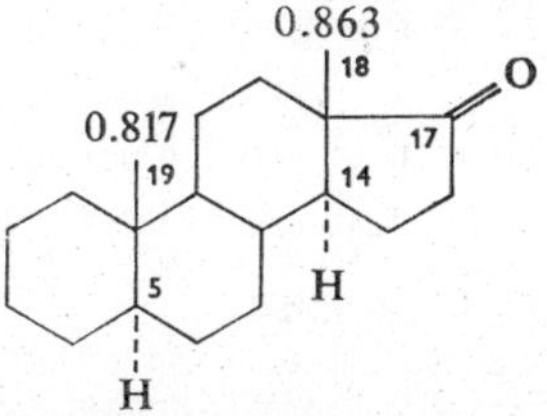

17-oxo-5α, 14α-androstane

The numbers given beside the 18 and 19 methyl groups represent their proton chemical shifts in ppm with respect to tetramethylsilane.

The effect of a single substitution has thus changed the diamagnetic shielding of the 18 and 19 methyl groups appreciably. The stereoisomers 5α,14α and 5α,14β can also be distinguished by the shift of the methyl resonances.

5β-Steroids. In these groups hydrogen is in the β-position at carbon 5. The two basic forms are as follows

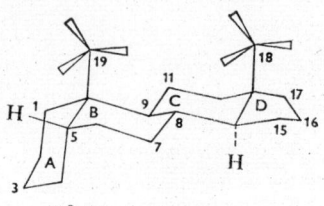

5β, 14α-androstane

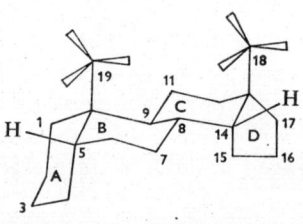

5β, 14β-androstane

Rings A and B are in *cis*-linkage; rings C and D are linked *trans* in the 5β,14α-isomer and *cis* in the 5β,14β-isomer. The change in the position of the hydrogen at carbon 5 causes the 18 and 19 methyl resonances to be shifted. The difference in shifts of the 19 methyl group is

$$\delta_{(19)}(5\beta,14\alpha) - \delta_{(19)}(5\alpha,14\alpha) = 0.1333 \text{ ppm}$$

The effect of substitution is illustrated by the methyl group shifts of 3,17-dioxo-5β,14β-androstane

5β, 14β-androstane (0.90 at 19, 0.990 at 18, H positions shown)

3,17-dioxo-5β, 14β-androstane (1.027, 1.110 at 18, with O at 3 and 17)

The shifts caused by substitution are

$$\Delta\delta_{(19)} \text{ (dioxo)} = 0.127 \text{ ppm}$$

$$\Delta\delta_{(18)} \text{ (dioxo)} = 0.083 \text{ ppm}$$

Effect of substitution. According to the Shoolery rule the shifts caused by multiple substitution are additive. This is illustrated by the example of 3β,17β-dihydroxy-Δ5-androstene. The measured shifts of the 18 and 19 methyl protons are 0.770 and 1.035 ppm, respectively. The partial shifts for single substitutions are the following:

	18-methyl δ ppm	19-methyl δ ppm
5α,14α-androstane	+0.692	+0.792
3β-OH substitution	+0.008	+0.033
17β-OH substitution	+0.033	0
Δ5 double bond	+0.042	+0.233
total shift	0.775	1.058
measured shifts	0.770	1.035

The mean deviation of the 18 and 19 methyl resonance shifts from that calculated by the Shoolery rule in multiple substituted steroids is ±0.015 ppm. Thus the Shoolery rule can be used for predicting the shifts in new types of steroids.

The examples given above have indicated the possibilities of identifying steroids by measuring their 18 and 19 methyl group shifts. The other well-resolved lines of the side groups are also suitable for NMR analysis. Also some information can be derived from the shape of the band of the ring protons.

4.5 POLYMERS

The methods of radio frequency and microwave spectroscopy have proved to be useful in the study of physical and chemical structure of polymers. Determination of the chemical structure of polymers usually involves stereochemical analysis. It can be done efficiently by high resolution NMR, as shown in the previous section, provided the samples can be dissolved in suitable inert solvents. In polymers determination of the steric structure of the chain-segments (stereo-regularity or tacticity) is usually done by comparing high resolution NMR spectra with those obtained for suitable model compounds.

The physical structure of polymers can be studied by ESR, NMR and dielectric spectroscopy. Although there are many other ways of studying crystallinity, orientation, molecular and segment motion in polymers, the information gained from radio-spectroscopic methods is of considerable interest. The following main problems can be studied.

1. *Microcrystalline structure of polymers.* Radio frequency spectroscopic methods, especially dielectric spectroscopy and NMR, are sensitive to the crystalline electric fields no matter how small the microcrystalline zones are. Small crystallites which cannot be detected by the X-ray diffraction method can be studied by measuring the electronic and nuclear relaxation times.

2. *Oriented polymeric structures.* As shown in Chapter 3, the ESR spectra of irradiated oriented polymers are anisotropic. Consequently, it is possible to study orientation effects by irradiating the polymers with a small radiation dose, in order to avoid physical damage, and to study the oriented radicals by ESR. As an example of this the oriented structure formed by stretching polypropylene and poly(tetrafluoro-ethylene) has been discussed in some detail in the previous chapter.

3. *Phase transitions in polymers.* The symmetry of the molecular order in polymers is often changed many times if the temperature is raised to the softening point. Such second-order transitions can also occur during the course of reactions in the solid state. Dielectric spectroscopy offers an excellent, yet rather undeveloped, way of investigating such transitions. It is also possible to use wide line NMR technique to follow the changes in the rotational degree of freedom (spin–spin relaxation time T_2) of the polymer molecules, or segments.

4. *Molecular motion in polymers.* The study of the motion of polymer molecules or segments is a very important problem in polymer physics. In principle, wide line NMR and dielectric spectroscopy can be used to investigate such problems. Nuclear relaxation times and dielectric dipole relaxation times, being sensitive to motional degrees of freedom, can be useful in doing dynamic studies, for example, or the molecular motion itself can be followed.

Some illustrative examples on these possibilities will be given below.

NMR study of phenol-formaldehyde resins. As an example of the direct way of deriving analytical information from polymeric structures the high resolution proton resonance of phenol-formaldehyde resins is discussed by Woodbrey, Higginbottom and Culbertson.[4.21]

There are many reaction schemes describing the synthesis of phenol-formaldehyde resins. A typical one is the following:

$$\text{PhOH} + CH_2O \longrightarrow \text{HO-C}_6H_4\text{-}CH_2OH$$

$$\text{HO-C}_6H_4\text{-}CH_2OH + CH_2O \rightleftharpoons \text{HO-C}_6H_4\text{-}CH_2OCH_2OH$$

By subsequent reaction with CH_2O the following structure is formed:

$$\text{HO-C}_6\text{H}_4\text{-(CH}_2\text{O)}_n\text{CH}_2\text{OH} \quad n = 2, 3 \ldots$$

There is a possibility of methylene bridge formation

$$\text{C}_6\text{H}_5\text{OH} + \text{HO-C}_6\text{H}_4\text{-CH}_2\text{OH} = \text{HO-C}_6\text{H}_4\text{-CH}_2\text{-C}_6\text{H}_4\text{-OH} + H_2O$$

or

$$\text{HO-C}_6\text{H}_4\text{-CH}_2\text{-OH} + \text{HO-C}_6\text{H}_4\text{-CH}_2\text{-OH} = \text{HO-C}_6\text{H}_4\text{-CH}_2\text{OCH}_2\text{-C}_6\text{H}_4\text{-OH}$$

By the use of high resolution NMR it is possible to identify characteristic groups in the pre-polymers. From the relative intensities of the groups quantitative analysis of different resins can be performed.

In order to illustrate the position of the functional groups, in Table 4.8 the spectral bands of acylated phenol-formaldehyde resins are shown in comparison with tetramethylsilane ($\tau = 10$). As seen there, most of the bands are well resolved except those of $AcOCH_2(OCH_2)_n$-OCH_2OAc and $ArCH_2OCH_2OAc$ and bridged $ArOAc$ and $Ar(CH_2O)_{1,2}Ac$ groups. An actual spectrum is shown in Fig. 4.26 for acylated resol–N measured by Woodbrey et al.[4.21] From the integral curve the relative concentration of the groups can readily be determined.

As shown in Fig. 4.26 the ArH and ArOAc groups exhibit rather well-resolved spin–spin splittings; this also helps to identify the groups.

Different types of commercial phenol-formaldehyde resins exhibit

Table 4.8

FUNCTIONAL GROUPS OF ACETYLATED PHENOL-FORMALDEHYDE RESINS IN CCl_4 SOLUTION[4.21]

Group	τ-value
ArH	2.4–3.3
$(AcO)_2CH_2$	4.5
$AcOCH_2(OCH_2)_nOCH_2OAc$	4.7
$AcOCH_2(OCH_2)_nOCH_2OAc$	5.2
$ArCH_2OCH_2OAc$	4.5–4.7
$ArCH_2OAc$	4.9–5.1
$ArCH_2OCH_2OAc$	5.4
$ArCH_2OCH_2Ar$	5.6–5.9
$ArCH_2Ar$	6.0–6.2
ArOAc	7.8
Bridged ArOAc	7.9
$Ar(CH_2O)_{1,2}Ac$	8.0
$AcO—(CH_2O)_n—Ac$	8.0

different proton resonance spectra. Detailed analysis of the resins can be made by comparing the measured chemical shifts and splittings with those measured in simple model compounds. In the case of acylated resins such compounds as $(AcO)_2CH_2$, $C_6H_5CH_2OAc$, etc., are used.

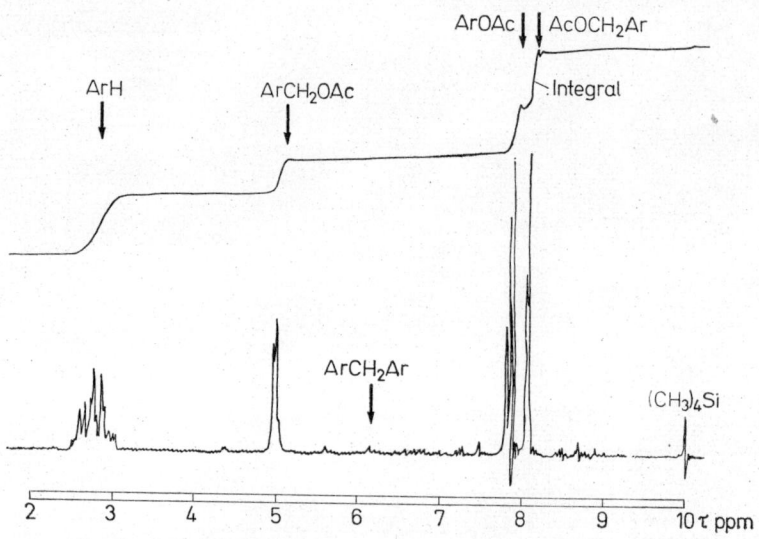

Fig. 4.26 Proton resonance spectrum of an acylated phenol formaldehyde resin. After Woodbrey[4.21]

4.5.1 STEREOCHEMICAL STUDY OF POLYMERS

As shown in Section 4.5 the chemical shifts and spin–spin couplings of functional groups depend on their steric structures as a result of the diamagnetic anisotropy. This principle can also be applied to study the tacticity of polymers, although in the large polymer molecules the resolution is usually not good. A typical spectrum of polyvinyl chloride at 60 Mc/s is shown in Fig. 4.27a measured by U. Johnsen[4.22] in 20% solution of chlorobenzene. As shown, the CH_2 and $CHCl$ groups are well separated but the spin–spin splitting is only partially resolved. The structures can be simplified by the spin–spin decoupling technique. The resulting groups are shown in Fig. 4.27b. The three lines appearing between $\tau = 5$ and 6 are interpreted by Bovey et al.[4.23] as coming from syndiotactic, atactic and isotactic triads, respectively. Further analysis of the spectrum can be made by choosing suitable low molecular weight model compounds

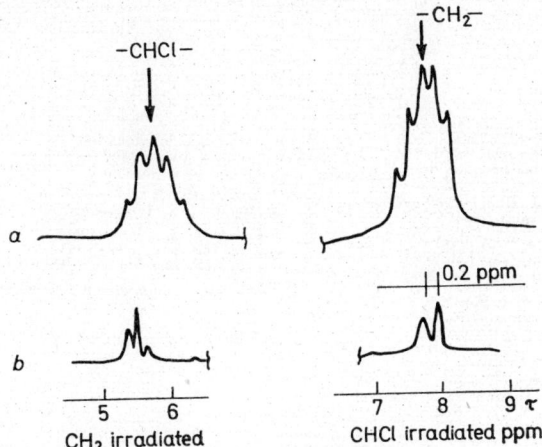

Fig. 4.27 The proton resonance spectrum of polyvinyl chloride.
a — Undecoupled; b — Decoupled. After Johnsen[4.22] and Bovey, Anderson and Douglas[4.23]

and comparing their spectra with those of the corresponding polymers. For polyvinyl chloride and other polymers, for example, various conformations of 2,4-dichloropentanes have been used as model compounds.[4.24] These models can be used for studying the steric surroundings of the methylene group in the following compounds:

$$\sim \underset{R}{\overset{H}{\underset{|}{C}}} - \underset{H}{\overset{H}{\underset{|}{C}}} - \underset{R}{\overset{H}{\underset{|}{C}}} \sim$$

The model compound 2,4-dichloropentane has two stable d,l-rotational isomers at room temperature

These steric forms are analogous to those in syndiotactic diads in polyvinyl chloride. The *meso*-isomers of 2,4-dichloropentane are

These forms are analogous to those found in isotactic diads in polyvinyl chloride. Upon mixing the two *meso*-isomers in a ratio of 1 : 1, at lower resolutions spectra similar to Fig. 4.27a are obtained.

4.5.2 DIELECTRIC SPECTROSCOPY OF POLYMERS

Dielectric spectroscopy has recently gained interest again in connection with the study of physical structure determination of polymers.

The dielectric absorption peaks observed in polymers correspond either to the rotation of the whole molecule or to that of a segment or bead. The absorption maximum corresponding to the rotation of the whole molecule lies at low frequencies and is dependent on

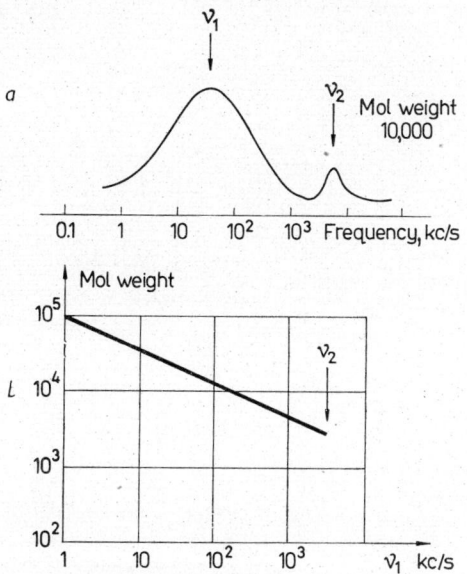

Fig. 4.28 The dependence of the dielectric absorption on the molecular weight of polymers. Cellulose acetate

the average molecular weight of the polymer. The absorption due to segment rotation lies at higher frequencies and is independent of the molecular weight. A typical dielectric absorption spectrum of cellulose acetate in dioxan solution is shown in Fig. 4.28a. The molecular weight of the polymer is 10,000. Upon decreasing the molecular weight the low frequency peak v_1 is shifted to higher frequencies, the high frequency peak v_2 is unchanged. The frequency of the low frequency peak as a function of the molecular weight is plotted in Fig.

4.28b. It is determined by the following equation

$$v_1 \propto \frac{1}{\langle M \rangle^{2.5}}$$

where $\langle M \rangle$ is the average molecular weight of the polymer.

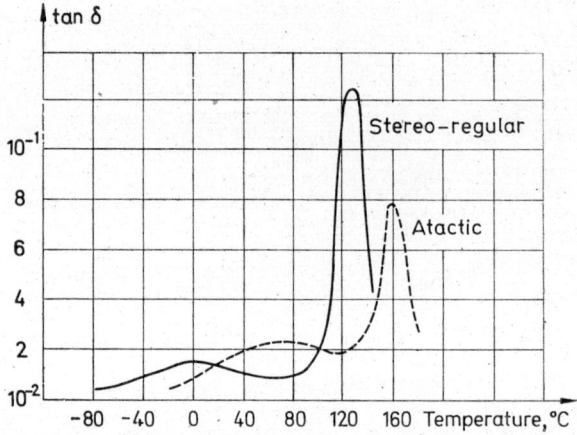

Fig. 4.29 Dependence of the dielectric absorption on the tacticity of polymers. Poly(butyl methacrylate). After Michailov, Burshtein and Krasnev[4.26]

Dielectric spectra may depend on the tacticity of the polymers. In poly(butyl methacrylate) the effective dipole moments of the atactic and stereo-regular forms are different. According to the measurements of Michailov, Burshtein and Krasnev[4.26] the dielectric spectral bands of the different stereoisomeric forms are separated. In Fig. 4.29 the dielectric loss of atactic and a stereo-regular form are plotted against temperature at a fixed frequency of 200 c/s. The peaks corresponding to stereo-regular forms are shifted to lower temperatures with respect to that of the atactic polymer. On the basis of the additivity of the dielectric polarization, it is possible to calculate the fraction of the stereo-regular forms from the observed effective dipole moment μ_{eff}, derived from the dielectric spectrum, provided the dipole moments corresponding to the stereoisomers are known. If isotactic and syndiotactic forms are present, for example, with dipole moments of μ_i and μ_s, respectively, the total effective dipole

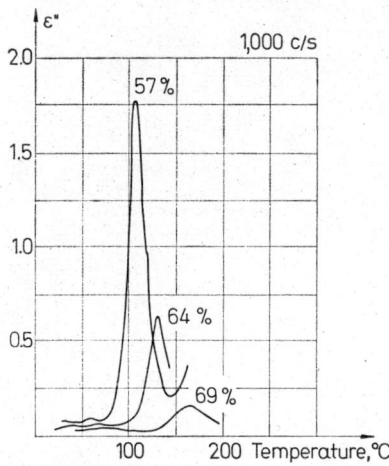

Fig. 4.30 Dependence of the dielectric loss on chlorine content of chlorinated polyvinyl chloride. After Reddish[4.27]

moment can be expressed as follows

$$\mu_{\text{eff}}^2 = \alpha_i \mu_i^2 + (1 + \alpha_i)\mu_s^2$$

where α_i is the mole concentration of the isotactic form.

According to Michailov et al.[4.26] the dipole moment of isotactic poly(butyl methacrylate) is $\mu_i = 1.73$ Debye, that of atactic and syndiotactic forms is $\mu_s = 1.54$ Debye.

Chlorinated PVC. The dielectric spectroscopy of PVC was extensively studied by Reddish[4.27] using the low frequency polarization method described in Section 2.5 and the usual high frequency technique as well. As an example, the dielectric loss peaks of chlorinated PVC are shown in Fig. 4.30 with different chlorine content. The spectra were taken at a fixed frequency of 1 kc/s as a function of the temperature. As shown, the increase of the chlorine content results in a drastic change in the dielectric loss and in a shift of the band to higher temperatures. The reason is a physical one, the introduction of a second

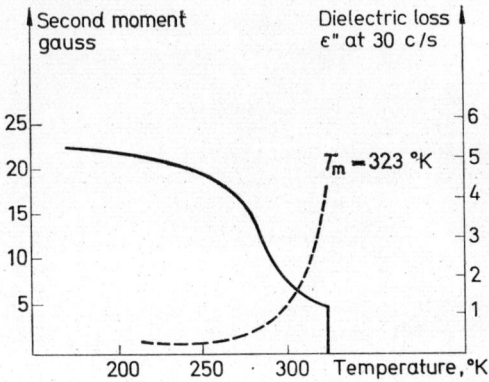

Fig. 4.31 Molecular motion in polyethylene glycol studied by wide line NMR and dielectric spectroscopy. After Hikichi and Furuichi[4.28]

chlorine atom makes the corresponding segment stiff and therefore the absorption band is shifted to higher temperatures.

Molecular motions in polymers. The physical properties of polymers are known to depend on the location of the operating temperature with respect to the melting point and to the glass transition temperature. It is recognized that besides glass transition and melting there are other second-order transition regions in the polymer corresponding to certain changes in the motional degrees of freedom of the molecules or segments. These multiple transition regions can be studied by wide line nuclear magnetic resonance and by dielectric spectroscopy.

Wide line NMR spectra of polymers usually exhibit a broad and a superimposed narrow line, corresponding to the crystalline and amorphous parts, respectively. In the crystalline region, the orientation-dependent spin–spin coupling is not averaged out and therefore the spin–spin relaxation time T_2 is short and the corresponding line is broad. In the amorphous part the spin–spin coupling is partially averaged out by the rotation of groups and the corresponding line is narrow. As the temperature is increased the rotation of the segments gets less and less hindered, which results in a narrowing of the NMR line. An illustrative example of this is shown in Fig. 4.31 (after K. Hikichi and I. Furuichi[4.28]) for polyethylene glycol (molecular weight 4,000). The second moment (see Section 1.4) of the NMR line is plotted against temperature. As shown, there is a significant decrease in the second moment well before the melting point is reached. At higher temperatures a narrow NMR line appears superimposed on the broad one. The decrease in the second moment of the broad NMR line (increase of the spin–spin relaxation time T_2) is accompanied by a sharp increase of the low frequency dielectric loss as indicated in Fig. 4.31. It is evident that between 250 and 300 °K the rotational freedom of groups

$$\sim \underset{\underset{H}{|}}{\overset{\overset{H}{|}}{C}} - O - \underset{\underset{H}{|}}{\overset{\overset{H}{|}}{C}} \sim$$

is continuously released. At higher frequencies (≥ 100 kc/s) the dielectric absorption does not change between 250 °K and 300 °K, indicating that quite large parts or segments take part in the rotation.

Together with the methods of mechanical relaxation and nuclear magnetic resonance, dielectric spectroscopy has been applied to study transition phenomena in many polymers. A recent review of this field has been given by W. P. Slichter.[4.29]

REFERENCES

4.1 Nukada, K., Yamamoto, O. and Suzuki, T., *Anal. Chem.* **35**, 1892 (1963).
4.2 Dailey, B. P. and Shoolery, J., *J. Amer. Chem. Soc.* **77**, 3977 (1955).
4.3 Tiers, G. V. D., *Characteristic NMR shielding values*. Minnesota Mining Co. (1958).
4.4 Becker, E. D. and Bradley, R. B., *J. Chem. Phys.* **31**, 1413 (1959).

4.5 Pople, J. A., Schneider, W. G. and Bernstein, H. J., *High Resolution NMR*, McGraw Hill, New York, 268 (1959).
4.6 Anderson, W. A., *Phys. Rev.* **102**, 151 (1956).
4.7 Fessenden, R. W. and Waugh, J. S., *Amer. Chem. Soc. Conference*, New York, (1957).
See also ref. 4.5, p. 238.
4.8 Pople, J. A., Schneider, W. G. and Bernstein, H. J., *ref.* 4.5, p. 281.
4.9 Lauterbur, P. C., *Ann. N. Y. Acad. Sci.* **70**, 841 (1958).
4.10 Cohen, A. D., Sheppard, N. and Turner, J. J., *Proc. Chem. Soc.* 118 (1958).
4.11 Abraham, R. J. and Bernstein, H. J., *Canad. J. Chem.* **37**, 6, (1959).
4.12 Corey, E. J. G., Tobinga, S. D. and Glazier, E. R., *J. Amer. Chem. Soc.* **80**, 1204 (1958).
4.13 Gutowsky, H. S., McCall, D. W., McGarvey, B. R. and Meyer, L. H., *J. Amer. Chem. Soc.* **74**, 4809 (1952);
Taft, R. W., *J. Amer. Chem. Soc.* **79**, 1045 (1957).
4.14 Callis, C. F., Van Wazer, J. R. and Shoolery, J. N., *Analyt. Chem.* **28**, 269 (1956).
4.15 Shoolery, J. N., *Discuss. Faraday Soc.* **19**, 215 (1955).
4.16 Pople, J. A., Schneider, W. G. and Bernstein, H. J., *ref.* 4.5, p. 238.
4.17 Gutowsky, H. S., McCall, D. W. and Slichter, C. P., *J. Chem. Phys.* **21**, 279 (1953).
4.18 Lemieux, R. U., Kullnig, R. K., Bernstein, H. J. and Schneider, W. G., *J. Amer. Chem. Soc.* **80**, 6098 (1958).
4.19 Bovey, E. A., Anderson, E. W., Hood, F. P. and Kornegay, R. L., *J. Chem. Phys.* **40**, 3099 (1964).
4.20 Zürcher, R. F., *Helv. Chim. Acta* **44**, 1380 (1961); **46**, 2055 (1963).
4.21 Woodbrey, J., Higginbottom, H. P. and Culbertson, H. M., *J. Polymer Sci.* **3**, 1079 (1965).
4.22 Johnsen, U., *J. Polymer Sci.* **54**, S6 (1961).
4.23 Bovey, F. A., Anderson, E. W., Douglas, D. C. and Manson, J. A., *J. Chem. Phys.* **39**, 1199 (1963).
4.24 Doskocilova, D. and Schneider, B., *Polymer Letters* **3**, 213 (1965).
4.25 Doskocilova, D. and Schneider, B., *Collect. Czech. Chem. Commun.* **29**, 2290 (1964).
4.26 Michailov, G. P., Burshtein, L. L. and Krasnev, L. V., *Vysokomol. Soed.* **7**, 870 (1965).
4.27 Reddish, W., *Amer. Chem. Soc., Polymer Preprints* **6**, 571 (1965).
4.28 Hikichi, K. and Furuichi, I., *J. Polymer Sci.* A **3**, 3003 (1965).
4.29 Slichter, W. P., *Symposium on Multiple Transition in Polymers*, Chicago, (1964). *Amer. Chem. Soc., Polymer Preprints* **6**, 632 (1965).
4.30 Corio, P. L. and Dailey, B. P., *J. Amer. Chem. Soc.* **78**, 3043 (1956).
4.31 Korinek, G. and Schneider, W. G., *Can. J. Chem.* **35**, 1157 (1957).
4.32 Phillips, W. D., *Ann. N. Y. Acad Sci.* **70**, 817 (1958).

5
Reactions

Investigation of reactions chiefly involves the quantitative analysis of the products formed. Conventional chemical analysis, however, in most cases cannot be carried out in such a short time and with such accuracy, as required for following reaction kinetics. Radical intermediates, for example, are very difficult to analyze by conventional chemical methods. These methods do not usually provide any information about radical structures. As shown in Chapter 3, the method of electron spin resonance provides a very sophisticated way of determining radical concentrations and structures. The application of ESR to the study of radical reactions has indeed become one of the most interesting fields of chemistry today.

In studying radical reaction kinetics one generally meets the following basic problems:

1. What are the initiating centres of the reaction?
2. What intermediates are formed, in what concentrations?
3. What are the products?
4. What are the rates of formation of the intermediates and that of the products?
5. Are there any changes in the physical structure of the system?

These problems generally cannot be solved without introducing a radio frequency method, in particular ESR, but other radio frequency spectroscopic methods can also be applied. Chemical exchange, for example, can be conveniently studied by nuclear magnetic resonance, especially by the 'spin–echo' method of measuring nuclear spin relaxation times (Chapter 2). Nuclear magnetic resonance can also be applied to analyze non-radical intermediates and products.

The following sections will provide some examples of the capabilities of studying chemical reactions by using radio frequency spectroscopic methods.

5.1 SIMPLE RADICAL REACTIONS. OXIDATION

A simple radical oxidation can be described schematically as follows

$$\dot{R} + O_2 = RO\dot{O}$$
$$RO\dot{O} + \dot{R} = ROOR \qquad 5.1$$

Radical $\dot{R}$ reacts with oxygen to produce a new radical $RO\dot{O}$, known as the peroxy radical. It is the intermediate radical product of the oxidation. In the second step radical $RO\dot{O}$ reacts with $\dot{R}$ to produce the final product ROOR.

This general although simplified scheme of oxidation can in many cases be verified by ESR, because the spectrum of $\dot{R}$ is usually quite different from that of $RO\dot{O}$. Thus the reaction can be followed either by measuring the decay of the R radicals or by following the kinetics of formation of $RO\dot{O}$ radicals. Schimmel and Heineken[5.1] for example found in 1957 that by oxidation of tri-*p*-nitrophenylmethyl a new ESR line appears during the course of the reaction indicating the appearance of $RO\dot{O}$ type intermediates. The initial radical is the following

5.2

At the time of this early measurement, the resolution was not high enough to produce the hyperfine splitting caused by the ring protons. Only a single ESR line could be observed with a g-factor of 2.0036. By admission of oxygen, however, a second line appeared with a g-factor of 2.006, corresponding to the radical

$$O_2N-C_6H_4-\overset{\overset{\displaystyle C_6H_4-NO_2}{|}}{\underset{\underset{\displaystyle C_6H_4-NO_2}{|}}{C}}-OO\cdot \qquad 5.3$$

The difference in the g-factors observed is due to the different extent of delocalization of the unpaired electron. Radical 5.2 has an unpaired electron delocalized over the three benzene rings, and the corresponding g-factor is closer to the free electron value. In the case of the peroxy radical 5.3, the extent of delocalization is smaller, and thus the deviation of the g-factor from the free electron value is greater. In this case the radical could be identified by measuring the g-value of the ESR-lines. The concentration of the intermediates was high enough to be detected by simple technique.

5.1.1 OXIDATION OF HYDROQUINONES

During the course of slow reduction of quinones or oxidation of hydroquinones certain types of intermediates are formed, known as *semiquinones*. The structure of these intermediate radical products can be very thoroughly analyzed by ESR.

The scheme of the reduction of quinones is the following

$$\text{quinone} \xrightleftharpoons{e^-} \dot{R}_S^- \xrightleftharpoons{H^+} \dot{R}_{HS} \xrightleftharpoons{H^{\cdot}} \text{hydroquinone} \qquad 5.4$$

According to this scheme two intermediate semiquinone radicals are formed, one is the benzo-semiquinone radical anion $\dot{R}_S^-$, the other is the undissociated semiquinone radical $\dot{R}_{HS}$. The equilibrium between these intermediates depends on the pH of the medium.

The scheme for the oxidation of hydroquinones is the following

$$\text{hydroquinone} \xrightarrow{\text{laccase, peroxidase or } Ce^{4+}} \dot{R}_S^- \xrightleftharpoons{H^+} \dot{R}_{HS} \longrightarrow \text{quinone} \qquad 5.5$$

Again the dissociated $\dot{R}_S^-$ anion and the undissociated $\dot{R}_{HS}$ semiquinone radicals are formed.

Using the method of ESR, the structure of the semiquinone intermediate radicals can be readily determined, and the equilibrium between the dissociated and undissociated semiquinones can also be studied. The simplest case is reaction 5.4 in basic solutions. In the case of slow reduction a considerable amount of $\dot{R}_S^-$ radical ions are formed. These intermediates can be detected simply by performing the experiment right in the sample holder of the spectrometer. The ESR spectrum obtained consists of 5 lines corresponding to the hyperfine interaction of the unpaired electron with the four ring protons

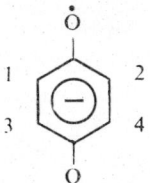

Owing to the symmetry of this structure, the spin density is the same at each ring proton and thus one gets the binomial 1 : 4 : 6 : 4 : 1

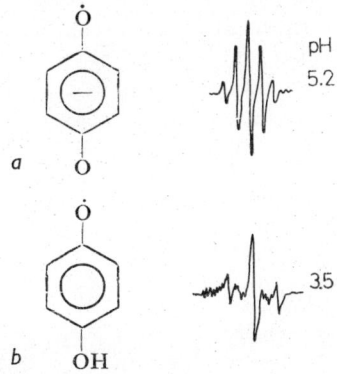

Fig. 5.1 ESR spectra of intermediate radicals formed by oxidation of hydroquinone. After Yamazaki and Piette[5.2]

intensity distribution of the lines, corresponding to the equal HFS interaction with four spins (see the discussion in Chapter 4).

It is also possible to measure the decay of these intermediate radical ions simply by repeatedly sweeping through the spectrum. The decay curves are simply given by the intensities of the $\dot{R}_S^-$ lines as a function of reaction time.

The undissociated $\dot{R}_{HS}$ radicals are more difficult to analyze. They are formed mostly at low pH values, i.e. in acidic media. Since in this case the intermediate radical concentration is not high enough to be measured directly, the *flow method* described in Chapter 2 should be used. Yamazaki and Piette[5.2] have recently studied reaction 5.5 very thoroughly using the flow technique. They could follow the change in the equilibrium

$$\dot{R}_S^- \rightleftarrows \dot{R}_{HS} \qquad 5.6$$

by measuring the ESR spectra as a function of the pH. Some of their results are shown in Fig. 5.1. At high pH values the usual 5-line spectrum can be observed indicating that the radical ions $\dot{R}_S^-$ are stabilized (Fig. 5.1a). By lowering the pH value of the medium the hyperfine structure is changed rather drastically to give a 3-line spectrum in acidic media. This spectrum is interpreted as corresponding to the undissociated radicals $\dot{R}_{HS}$

and the HFS components are interpreted to be the results of interaction with protons 1 and 2 (Fig. 5.1b). The protonation of $\dot{R}_S^-$ makes the structure unsymmetrical, and as a result of this the unpaired electron is more localized. So the spin density at the remaining three protons is very small.

The experiments have been repeated with such unsymmetrical hydroquinones as toluhydroquinone or t-butylhydroquinone. The intermediates in the case of t-butylhydroquinone are

$$\qquad 5.7$$

Here again protonation results in a change in the spin-density distribution of the radicals. At low pH values the unpaired electron is practically localized to position 1, giving the doublet of Fig. 5.2b. Since a doublet is observed at low pH values, the position of protona-

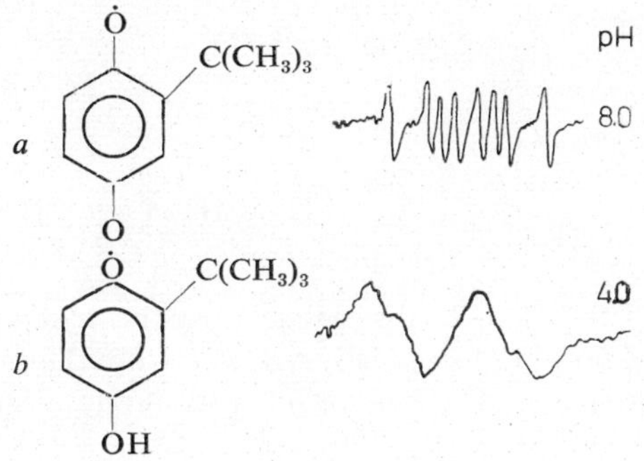

Fig. 5.2 The ESR spectrum of t-butyl-benzosemiquinone. After Yamazaki and Piette[5.2]

tion can be identified as being at position 4. If the other oxygen were protonated, the spectrum should consist of a triplet as a result of interaction with the ring protons at C_3 and C_5.

The spectral lines of the semiquinone intermediates show a significant broadening at intermediate and low pH values. This is a result of the exchange reaction

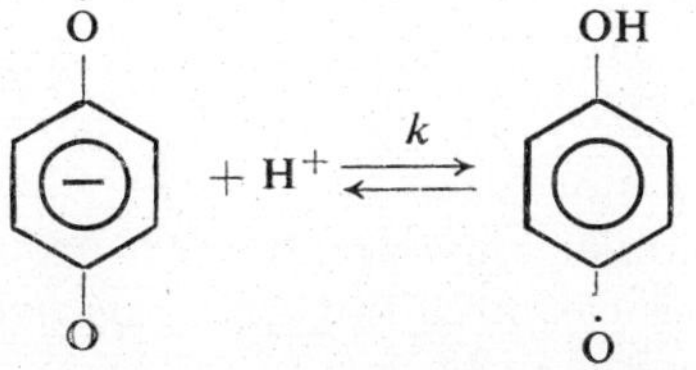

$$\text{semiquinone}^- + H^+ \underset{}{\overset{k}{\rightleftarrows}} \text{hydroquinone radical} \qquad 5.8$$

As will be discussed later in Section 5.2, the rate of exchange is directly linked with the line widths. At very slow exchange the average lifetime of the species is high, and the corresponding line width is small. This is the case of high pH values where the $\dot{R}_S^-$ radical ions are stabilized. At intermediate pH values

$$\frac{d[\dot{R}_S^-]}{dt} = k[\dot{R}_S^-][H^+] \qquad 5.9$$

The mean lifetime is (see Section 5.2)

$$\frac{1}{\tau} = \frac{\dfrac{d[\dot{R}_S^-]}{dt}}{[\dot{R}_S^-]} = \frac{\sqrt{3}}{2}[\Delta H - (\Delta H)_0] = k[H^+] \qquad 5.10$$

Thus by plotting the line widths ΔH against the hydrogen ion concentration $[H^+]$ a straight line is observed and from the slope the value of k can be calculated. In the case of p-benzosemiquinone the k-value is found to be 1.5×10^{10} M^{-1} sec^{-1}.

This illustrative example has shown how much information can be obtained by measuring ESR spectra during the course of a radical reaction.

5.1.2 OXIDATION OF RADICALS TRAPPED IN THE SOLID PHASE

It has been shown in Chapter 3 that trapped radicals can be formed in solids by different ways: by illumination of the solids with ultraviolet light, by irradiation with X-, γ- or β-rays or heavier particles, by mechanical treatment or by gaseous discharges.

The study of such reactions by ESR is relatively easy, since in the solid phase the intermediate radical products are also trapped and thus high radical concentrations are formed. The trapped radicals $\dot{R}$ exhibit characteristic ESR spectra, and the reaction can be followed by observing the transformations of $\dot{R}$ into other types of radicals. In the case of oxidation, for example, intermediate peroxy radicals are formed. These are oxidized further to give peroxides, according to the general scheme given by Equation 5.1.

The different extent of delocalization of $\dot{R}$ and $R O\dot{O}$ radicals always makes it possible to measure their concentrations separately. The reaction kinetics can thus be followed by simultaneous measurement of [Ṙ] and [ROȮ] concentrations. Bamford and Ward,[5.3] e.g., have investigated the oxidation of methacrylate salts. They generated radicals by gaseous discharge on the surface of the solid salts (powders) and followed the decay of these radicals and the formation

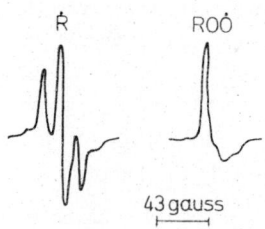

Fig. 5.3 Oxidation of radicals formed in sodium methacrylate by gaseous discharge. After Bamford and Ward[5.3]

of ROȮ-type radicals by ESR. The ESR spectra of the initial radicals and the final ROȮ radicals are shown in Fig. 5.3.

A lot of similar investigations have been made with irradiated materials, especially with polymers. Tsvetkov, Lebedev and Voevodsky[5.4] have studied in detail the oxidation of irradiated poly(tetrafluoroethylene). Upon irradiation with X-rays or with γ-rays a fairly large concentration of radicals is formed in poly(tetrafluoroethylene). In samples irradiated *in vacuo* the following radicals are formed

$$\sim \underset{F}{\overset{F}{C}} - \underset{F}{\overset{\cdot}{C}} - \underset{F}{\overset{F}{C}} \sim \qquad 5.11$$

These radicals are very stable even at temperatures up to 200 °C. The ESR spectrum of this radical consists of 10 lines. The four equivalent fluorine atoms in β-position produce a quintet with a splitting of 33 gauss. The fluorine atom in α-position to the radical doubles this 5-line spectrum by a splitting of 90 gauss (Rexroad and Gordy[5.5]).

Upon admission of oxygen the spectrum is gradually transformed into an asymmetric single line corresponding to the following peroxy radical

$$\sim \underset{F}{\overset{F}{C}} - \underset{F}{\overset{\overset{\displaystyle O}{|}}{\underset{|}{C}}} - \underset{F}{\overset{F}{C}} \sim \qquad 5.12$$

An intermediate case is shown in Fig. 5.4. The sharp line in the middle of the spectrum corresponds to the peroxy radicals. The kinetics of

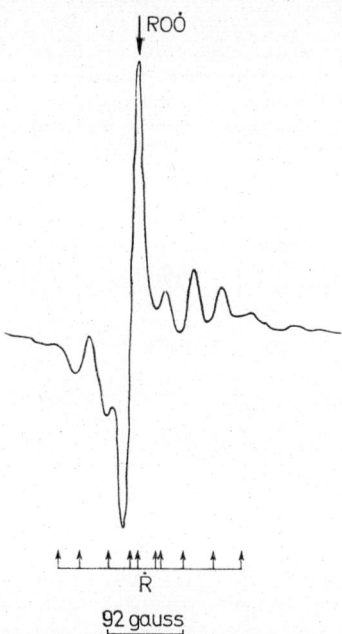

Fig. 5.4 Oxidation of the fluorocarbon radicals formed in irradiated poly(tetrafluoroethylene). ESR spectrum measured during the course of oxidation

the reaction can be followed very easily by measuring the line intensities as a function of time. At the end of the reaction only the peroxy line remains. Repeating the measurements at different temperatures, the activation energies can also be calculated. It has been found that the transformation of the fluorocarbon radicals 5.11 into the peroxy radicals 5.12 is a reversible process

$$\dot{R} + O_2 \underset{2}{\overset{1}{\rightleftarrows}} R\dot{O}\dot{O} \qquad 5.13$$

If an oxidized sample is re-evacuated and kept at elevated temperatures, the sharp peroxy line is gradually transformed back into the 10-line spectrum indicating the existence of process 2. This transformation can only be observed at temperatures as high as 160–250 °C.

Fig. 5.5 represents the observed variation of [ROȮ] radical concentrations as a function of time. It can be shown by kinetic analysis that at the beginning of the reaction

$$\frac{d[ROO]}{dt} = -k_1 [RO\dot{O}] \qquad 5.14$$

In the first stage of the reaction in thin sheets of PTFE the effect of diffusion can be neglected and thus from the slope of the initial part of the decay curve an approximate value of the constant k_1 can be obtained. The second part of the decay curve can be approximated by the following kinetic equation

$$\frac{d[RO\dot{O}]}{dt} = -k_1 [RO\dot{O}] \frac{k_d}{k_2 [\dot{R}] + k_d} \qquad 5.15$$

where k_d is the rate constant of the diffusion and k_2 is the rate constant of reaction 2. Since the radical concentrations [ROȮ] and [Ṙ] are measured by ESR, the only term to be given is the diffusion rate constant k_d. This can be evaluated by repeating the experiments with samples of different sizes. The decay of ROȮ radicals is shown in Fig. 5.5b in a cylindrical sample with a diameter of 3.5 mm. In the decay curves, log ([ROȮ]/[Ṙ]) is plotted against time at different temperatures. From the temperature dependence of rate constants k_1, k_2 and k_d the activation energies of process 1, process 2 and that of the diffusion can be calculated.

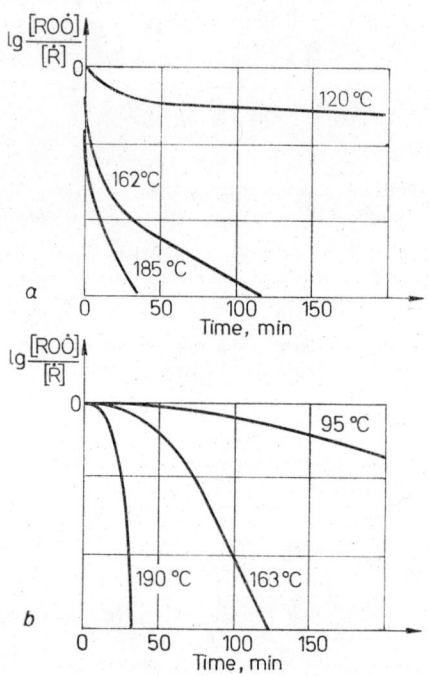

Fig. 5.5 Kinetic curves of the radical oxidation in irradiated polytetrafluoroethylene. a — In thin sheets; b — In cylinders of 3.5 mm diam. After Tsvetkov, Lebedev and Voevodsky[5.4]

This method can be applied in every case when relatively stable radicals react with gases. Similar techniques can be used when the initiating radicals are in the liquid state or even in the case when the reactant is liquid.

5.2 CHEMICAL EXCHANGE

The methods of radio frequency spectroscopy can be applied not only to analyze the reaction products; some of the methods are suitable to measure such dynamic properties as electron and proton exchange and electron and proton transfer. The principle of these methods is that the widths of the ESR and NMR spectrum lines depend on the lifetime of the configuration, in which the electron or nuclear spin is. In the case of inter- or intramolecular exchange of protons or electrons the corresponding ESR or NMR lines are broadened. The higher the rate of exchange, the broader the corresponding resonance lines become. In principle this phenomenon allows the possibility of telling which proton or electron is taking part in the exchange, and in many cases the rate constants of the exchange can also be calculated.

5.2.1 ELECTRON EXCHANGE REACTIONS

It has been shown in Chapter 3 that in a free radical the unpaired electron is usually smeared out to a considerable area of the molecule and the distribution of the electron density can be determined by measuring the hyperfine interaction with the magnetic nuclei of the system. In certain systems, however, radical electrons can be transferred to, or exchanged between, separate molecules. A classical example of this is the naphthalene radical ion–naphthalene molecule system. If naphthalene radical ions are present together with undissociated naphthalene molecules in solutions of e.g. tetrahydrofuran, the unpaired electron of the radical ion can be exchanged with an electron of the naphthalene molecule. The scheme of this reaction is the following

$$(\dot{C}_{10}H_8)^- + C_{10}H_8 \rightarrow C_{10}H_8 + (\dot{C}_{10}H_8)^- \qquad 5.16$$

The electron exchange reaction can be followed by measuring the line widths of the $(\dot{C}_{10}H_8)^-$ spectrum as a function of the $[C_{10}H_8]$ concentration. If no naphthalene molecules are present, a very well resolved ESR spectrum can be observed (25 lines), as a result of the hyperfine interaction with the protons. The radical ions in this solution are stable. By adding naphthalene molecules to the solution a broadening of the lines is observed. This is a result of shortening of the average lifetime of the radicals. The lifetime is expressed as follows

$$\tau \propto \frac{1}{\Delta H - (\Delta H)_0}$$

where $(\Delta H)_0$ and ΔH are the original and the actual line widths, respectively.

On the other hand, the rate constant of the exchange reaction is

$$k = \frac{1}{\tau_0 [C_{10}H_8]} \qquad 5.17$$

Thus by measuring ΔH as a function of $[C_{10}H_8]$ the rate constant k can be calculated.

In most cases it is more practical to measure the decrease in the *amplitude of the lines* instead of line widths. Since the total concentration of the paramagnetic centres is unchanged, there is a definite correlation between the line widths and amplitudes. For Lorentzian lines approximately

$$\sqrt{\frac{A_0}{A}} = 1 + 6.5 \frac{[C]}{\Delta H_0} k \, 10^{-8} \qquad 5.18$$

where A_0, ΔH_0 are the amplitude and width of the stable radical ion lines, A is the measured amplitude, and $[C]$ is the concentration of the molecules which take part in the exchange, i.e. in this case $[C] = [C_{10}H_8]$, k is the rate constant of the exchange reaction.

Using Equation 5.18 the rate constant k can be calculated simply by plotting the measured amplitudes against $[C_{10}H_8]$.

5.2.2 ELECTRON TRANSFER

In certain systems known as charge transfer complexes, there is a possibility for electrons to be transferred from one system of molecules to another. If certain molecules having low ionization energies (electron-donor molecules) are brought together with high electron affinity (electron-acceptor) molecules, a special type of interaction occurs, known as donor–acceptor interaction. This intermolecular interaction makes the electrons move from the donor molecules to the acceptors. This transfer of charge can be observed by optical spectroscopy, by measurement of electrical conductivities and also by ESR.

Example. The electron transfer reaction of chloranil and N,N,N′,N′-tetramethyl-p-phenylenediamine (TMPD).[5,6]

Upon mixing these two materials in solution in acetonitrile the following observations can be made:

1. In the absence of oxygen two kinds of radicals can be observed by ESR. One is the chloranil–semiquinone radical ion (one sharp line), the other is the TMPD positive radical ion (39 lines). At the same time at 9240 Å an optical absorption band appears, which is characteristic of the donor–acceptor complex formed. The existence of the chloranil–semiquinone and the TMPD ion is also proved by optical absorption spectroscopy.

2. A few minutes after mixing the reactants the amplitude of the single ESR lines decreases with the simultaneous increase of the 39-line spectrum. This indicates that during the course of the electron transfer reaction the concentration of the chloranil–semiquinone radicals is decreasing and that of the TMPD positive radical ions is increasing.

3. After a few hours the chloranil line vanishes, and the other lines reach a steady-state amplitude. At this time the optical absorption band at 9240 Å also vanishes, indicating that the electron transfer reaction has been finished.

4. After several weeks a slow decrease in the TMPD ESR line amplitudes can be observed. This radical decay, however, is not connected with the charge transfer reaction.

ESR is thus a powerful method for analyzing radical intermediates formed during the course of charge transfer reaction. It can be used simultaneously with such other methods as the measurement of electrical conductivities and optical spectroscopy.

5.2.3 PROTON EXCHANGE

In certain systems there is a possibility for protons to be exchanged between molecules. The general form of these reactions is the following

$$HA + \overline{H}B \underset{k}{\overset{k_1}{\rightleftharpoons}} \overline{H}A + HB \qquad 5.19$$

where A and B are the molecules containing the protons H and $\overline{H}$.

The chemical shift and spin–spin coupling of proton magnetic resonance (see Chapter 2) makes it possible to distinguish between configurations AH and BH. If there is no proton exchange, these two protons exhibit different NMR spectrum lines at frequencies of v_A and v_B. The shift $|v_A - v_B|$ between the two proton lines depends on the local magnetic field (shielding), i.e. on the structure of molecules A and B. In the case of proton exchange, however, the residence time of the protons is shared between molecules A and B, and thus in the case of fast exchange the local fields are averaged out to give a single sharp line from both protons H and $\overline{H}$. The condition for this averaging is

$$|v_A - v_B| \ll \frac{1}{\tau_A}, \; \frac{1}{\tau_B} \qquad 5.20$$

where τ_A, τ_B are the residence times of protons in molecules A and B, respectively. So if the rate of exchange is high in comparison with the chemical shift, NMR cannot distinguish between protons H and $\overline{H}$ any more. At intermediate cases when

$$|v_A - v_B| \approx \frac{1}{\tau_A}, \; \frac{1}{\tau_B} \qquad 5.21$$

the averaging of the local fields is incomplete, which results in a broadening of the spectrum lines v_A and v_B.

This effect of averaging of the local magnetic fields by chemical exchange provides a means of investigating these reactions by nuclear magnetic resonance. By altering the rate of exchange in some way and recording continuously the NMR spectrum of the system the reaction can be followed very easily and the rates and activation energies of the exchange can be calculated.

Slow proton exchange. In this case the effective magnetic fields are not averaged out, just a slight increase in the line widths can be observed as a result of the limited lifetimes of the protons. The exchange rate is

$$\frac{1}{\tau_A} = \frac{1}{[AH]} \frac{d[AH]}{dt} \qquad 5.22$$

where τ_A is the residence time of the proton at molecule A, [AH] is the concentration of molecules AH in the system. The width of the ν_A line is the following

$$(\Delta \nu)_A = \frac{1}{\pi} \left(\frac{1}{T_2^A} + \frac{1}{\tau_A} \right) \qquad 5.23$$

where T_2^A is the spin–spin relaxation time of the protons in molecule A. This is either known or can be measured in a separate system, where no exchange is possible. The spin–spin relaxation time can also be measured independently by the 'spin–echo' method described in Chapter 2.

Thus by measuring the individual line widths $(\Delta \nu)_A$ and $(\Delta \nu)_B$ the residence times τ_A and τ_B can easily be calculated in the case of slow exchange.

Intermediate exchange. In this case the residence times τ_A and τ_B cannot be determined individually, because the spectrum lines almost coincide. If the original line widths (without exchange) are small compared with the chemical shift $|\nu_A - \nu_B|_0$, the measured chemical shift $|\nu_A - \nu_B|_m$ can be expressed as follows

$$\frac{|\nu_A - \nu_B|_m}{|\nu_A - \nu_B|_0} = \left(1 - \frac{2}{\tau_{AB}^2 |\nu_A - \nu_B|_0^2} \right)^{1/2} \qquad 5.24$$

From this the average rate of exchange $1/\tau_{AB}$ can be calculated. If the two lines coincide, i.e. $\nu_A \approx \nu_B$, the residence time is simply given by Equation 5.23.

Fast proton exchange. In this case only a single resonance line can be observed ($\nu_A = \nu_B$) and the line gets narrower as the rate of exchange increases, because the averaging is better the higher the exchange rate. The decrease in the line width, is then, a measure of the increase in the exchange rate. If the average residence time $\tau_{AB} \ll T_2$, the spin–spin relaxation time, the line width can be expressed as follows

$$(\Delta v)_{AB} = \frac{1}{\pi}\left[\frac{1}{T_2} + (M_2^A + M_2^B)\,\tau_{AB}\right] \qquad 5.25$$

where M_2^A, M_2^B are the second moments of the lines v_A and v_B, measured in the case where there is no exchange (see Chapter 2 for the definition of the second moments).

Illustrative example: Proton exchange in the ethanol–water system. In the ethanol–water system, the following proton exchange reactions can occur, depending on the pH of the medium:

$$\text{ROH} + \text{OH}^- \underset{}{\overset{k_1}{\rightleftarrows}} \text{RO}^- + \text{H}_2\text{O}$$

$$\text{ROH} + \text{RO}^- \underset{}{\overset{k_2}{\rightleftarrows}} \text{RO}^- + \text{ROH}$$

$$\text{ROH} + \text{H}_3\text{O}^+ \underset{}{\overset{k_3}{\rightleftarrows}} \text{ROH}_2^+ + \text{H}_2\text{O} \qquad 5.26$$

$$\text{ROH} + \text{ROH}_2^+ \underset{}{\overset{k_4}{\rightleftarrows}} \text{ROH}_2^+ + \text{ROH}$$

$$\text{ROH} + \text{H}_2\text{O} \underset{}{\overset{k_5}{\rightleftarrows}} \text{ROH} + \text{HOH}$$

These reactions can be studied by measuring the proton magnetic resonance spectra of the system at different pH values. In the neutral ethanol–water system the exchange reaction is very slow; thus the line for H_2O can be observed separately.

Fig. 5.6 shows the NMR spectrum of pure, water-free ethanol (*a*), the neutral ethanol–water system (*b*) and that for acidic ethanol–water system (*c*). In pure ethanol the methyl group is a triplet, the methylene group is a quartet with some indications of further splitting, the OH group is a triplet, according to the spin–spin coupling among these groups. In the presence of water the lines of the OH group and those of the $-CH_2$ group are somewhat broadened, and the water line can

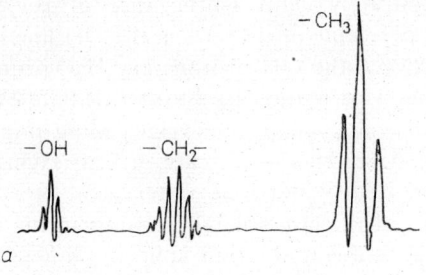
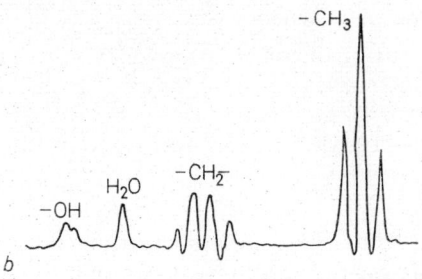
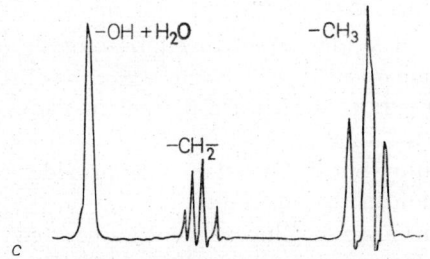

Fig. 5.6 Proton exchange in the ethanol–water system as investigated by NMR.
a — Water-free ethanol; *b* — Neutral mixture of ethanol and water; *c* — Acidic mixture of ethanol and water

be observed separately. This is the case of very slow exchange. By lowering the pH value of the system the OH line is first broadened and the shift between the OH protons and H_2O protons decreases till the lines coincide. This is the case of intermediate exchange. Further lowering of the pH value results in the sharpening of the $OH - H_2O$ line, as shown in Fig. 5.6c. This is the case of fast exchange. The experiment shows very clearly that the water protons exchange with the hydroxyl protons of the ethanol. The change in the line width of the methylene protons is due to the spin–spin coupling with the OH group. The methyl group remains unaffected during the course of the exchange.

Using the method described above it is possible to determine the rate constants k_1 to k_5 of the reactions 5.26.

5.2.4 INTERMOLECULAR CATION EXCHANGE INVESTIGATED BY ESR

In the case of electron exchange and electron transfer ESR has been applied to identify the radicals formed during the course of the charge transfer reaction and also to measure directly the electron exchange by observing the variation in the ESR line widths.

Recently ESR has been applied to study the formation of ion pairs in cases where radical ions are involved. In solutions of aromatic radical anions in some cases hyperfine interaction has been observed with the alkali metal cations which are present in the system. The interaction between the unpaired electron of the radical anions and the magnetic moment of the alkali metal nuclei provides direct evidence of the ion association in these systems.

Atherton and Goggins[5.7] have recently observed that the widths of the hyperfine lines corresponding to ion association show a peculiar variation as a function of temperature. This variation is interpreted as being due to the intermolecular cation exchange.

The high resolution ESR spectrum of pyrazine radical ion-Na consists of several lines corresponding to the following hyperfine interactions:

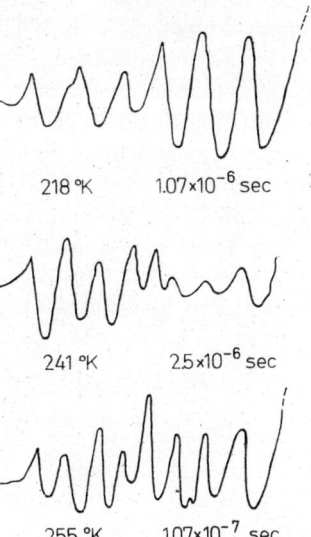

Fig. 5.7 Intramolecular cation exchange in pyrazine containing sodium. A part of the ESR spectrum representing the coupling between sodium and pyrazine radicals. After Atherton and Goggins[5.7]

1. The interaction with the nitrogen nucleus of pyrazine

$$A_N = 7.14 \pm 0.03 \text{ gauss}$$

2. The interaction with the protons

$$A_H = 2.70 \pm 0.03 \text{ gauss}$$

3. The interaction with metallic sodium

$$A_{Na} = 0.55 \pm 0.03 \text{ gauss.}$$

The splitting constants are given at room temperature. The variation of the spectrum is illustrated in Fig. 5.7. As shown, the lines responsible for the coupling with Na completely vanish at 241 °K, and reappear at higher temperatures (255 °K). The exchange rates $1/\tau$ calculated from the observed line broadening are also given in the figure. For the calculation it is necessary to know the line widths and the coupling constants in the absence of exchange. The coupling constants are, however, difficult to measure. To avoid the difficulties Atherton and Goggins used mixed solvents to lower the temperature of collapse of the lines. The results are represented by the following equation

$$\frac{1}{\tau} = 10^{13.3} \exp(-7{,}400/RT) \text{ sec}^{-1}.$$

The activation energy of the exchange is thus 7.4 kcal/mole. The mean deviation is ± 1 kcal/mole.

The calculation of the exchange lifetimes τ is generally not very easy. It involves a very thorough mathematical analysis of the line shapes and sometimes electronic computers must be used to obtain the numerical values. Equations 5.23, 5.24 and 5.25 given above are oversimplified, and serve only to illustrate the possibilities.

Recently the spin–echo technique described in Chapter 2 has also been applied to study chemical exchange. It has been shown[5.8] that the multiple pulse technique for measuring the spin–spin relaxation time T_2 developed by Carr and Purcell[5.9] can be applied to the case of chemical exchange too. By measuring the envelope of the echo pulses as a function of some parameters of the exchange, the exchange rates can be calculated.

5.3 PHOTOLYSIS AND RADIOLYSIS

Irradiation of organic or inorganic compounds usually results in the formation of paramagnetic centres of varying stability.

These centres may be formed immediately under the action of the radiation or may arise during the course of the radiochemical or photochemical reaction. Investigation of the concentration and structure of these products is one of the most important problems in the study of radiolysis and photolysis.

If the radiation energy is high enough to break chemical bonds, free radicals are formed. Owing to their high reactivity these radicals may initiate reactions which through several steps lead to the transformation of the compound. Visible light in most cases is not energetic enough to produce radicals. If, for example, benzyl chloride is illuminated by visible light, no radicals can be observed. If, however, it is illuminated by 2540 Å ultraviolet light, very high concentrations of radicals are formed, which can easily be detected by ESR. This indicates that in this system the radiation energy must be higher than 4 eV (92 kcal/mole) to produce radicals, i.e. to break chemical bonds.

In the case of high-energy irradiation (X-rays, γ-, β-rays, neutrons or irradiation with heavier particles) the radiation energy is always high enough to break chemical bonds. In this case the primary action of the radiation is always connected with a formation of paramagnetic centres, in most cases free radicals.

In some cases, the paramagnetic centres formed by radiolysis are not radicals. In alkali halide crystals, for example, irradiation results in a characteristic coloration, which is caused by electrons, holes or ions trapped in defects of the crystal lattice. If, for example, an NaCl crystal is irradiated by X-rays it becomes yellow and an ESR spectrum line appears with a splitting factor very close to the free electron value. The paramagnetism of this crystal is explained by the unpaired electrons trapped in crystal defects. These centres are called *F-centres*. There are many other kinds of colour centres formed in ionic crystals by irradiation. They can be investigated by the methods of ESR and ENDOR (see Chapter 2).

For chemists working in the field of photolysis and radiolysis the following general problems are of basic interest.

1. *Primary radical products of radiolysis and photolysis.* These products can be analyzed by the method of ESR at very low temperatures. The measurements are usually performed in an inert gas matrix, e.g. in solid argon. At liquid helium temperatures the reactivity of the primary radicals is negligible. These radicals are thus trapped in the rigid network during irradiation and can be measured quite easily.

2. *Transformation of primary radicals.* This problem can also be investigated by ESR. Irradiation is made at very low temperatures to get primary radicals in a frozen matrix. The ESR spectrum is taken and recorded continuously while increasing the temperature of the sample. The reactions of the primary radicals can thus be followed by observing the transformation of the ESR spectra.

3. *Measurement of reaction kinetics.* During the course of radiolysis and photolysis some of the primary radicals react immediately to form intermediate products. At steady state conditions the concentration of certain intermediate radicals may be high enough to be detected by ESR. In these cases, using the appropriate kinetic equations, the rate constants and activation energies can be calculated. In some cases the kinetic process leading to steady state can also be followed.

4. *Changes in the physical structure of the system.* A lot of experimental evidence has recently been published about the effect of physical structure on chemical reactions. The study of reactions in the solid phase always involves the study of the effects of physical structure. During the course of radiolysis and photolysis of crystalline solids, for example, the crystalline structure is gradually changing. This change has a remarkable effect on reaction kinetics. The changes in physical structures can also be followed by such radio spectroscopic methods as dielectric spectroscopy and nuclear magnetic resonance. An example of this will be given in Section 5.4.

5. *Energy transfer and charge transfer processes.* It is known that the molecules of a solid or liquid system cannot be treated individually, because of the strong interactions existing between them. These interactions cause the energy of excitation of one molecule to be transferred to others. In solids and even in liquids this transfer of energy may form *exciton waves*, which can pass through macroscopic distances in the system. The study of such energy transfer processes is very important in radiolysis and photolysis. In some cases paramagnetic excited states (triplet states) can be investigated by ESR and the process of energy transfer can be studied.

Charge transfer may also affect the kinetics of photochemical and radiochemical reactions. It can be investigated by electrical conductivity measurements, and in some cases by ESR and NMR (see the examples in Section 5.2).

5.3.1 DETECTION OF PRIMARY RADICALS

As mentioned above, the primary radicals formed by photolysis or radiolysis are usually too reactive to be studied at ordinary temperatures. A possible way of avoiding the difficulties is to perform irradiations at very low temperatures, where even the atomic products are frozen in. This condition, of course, is different from that where the actual reaction takes place. The ESR signals become broadened because

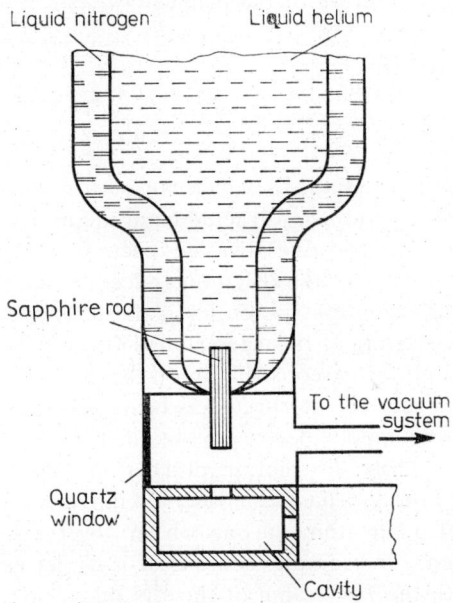

Fig. 5.8 Experimental set-up for detecting primary radicals in photolysis at 4.2 °K

of the magnetic anisotropy of the system. The inert matrix (e.g. solid argon) also has a slight effect on the spectra. The experimental technique is rather difficult because of the very low temperatures needed. Despite these difficulties much fruitful information has been obtained by this technique.

Fig. 5.8 shows an example of how photochemically generated radicals can be measured by ESR at liquid helium temperatures (4.2 °K). The material to be investigated is mixed with argon gas and this mixture is deposited on a sapphire rod cooled to the temperature of liquid helium. The solid layer obtained in this way is illuminated by ultraviolet light and inserted in the cavity of the ESR spectrometer. In this way primary radicals formed by irradiation are dissolved in the solid argon matrix and thus the dipole–dipole interaction is decreased. This results in a relatively high resolution which makes it possible to determine the structure of the primary radicals.

5.3.2 ESR SPECTRA OF ATOMS ABSTRACTED BY IRRADIATION

The simplest primary radicals which can be formed by radiolysis or photolysis are single atoms abstracted from the molecules. In radiolysis of water, for example, one always expects formation of hydrogen atoms in the first stage of the reaction. Indeed, irradiation of water at 4.2 °K in a solid argon matrix results in the ESR spectrum shown in Fig. 5.9a. The splitting of this spectrum is 507.9 gauss, which is very close to that found for gaseous H atoms (see Chapter 4). The additional splitting of 5 gauss is interpreted as being a matrix effect. In other noble gas matrices similar spectra are obtained, only the splittings and g-values differ slightly. Formation of H atoms in the first stage of radiolysis and photolysis has been observed in many other compounds.

Formation of other atoms has also been observed. The ESR spectrum of nitrogen, for example, consists of a triplet as a result of the interaction with the $I = 1$ spin of the ^{14}N nucleus (Fig. 5.9b). Phosphorus atoms have also been detected by photolysis of PH_3 and P_2O_3. The spectrum is shown in Fig. 5.9c.

According to V. V. Voevodsky et al.,[5,10] in the first stage of radiolysis of hydrocarbons generally atomic hydrogen is formed. This means that in most cases C—H bonds are broken rather than C—C bonds. In the case of larger molecules the remaining part of the molecule,

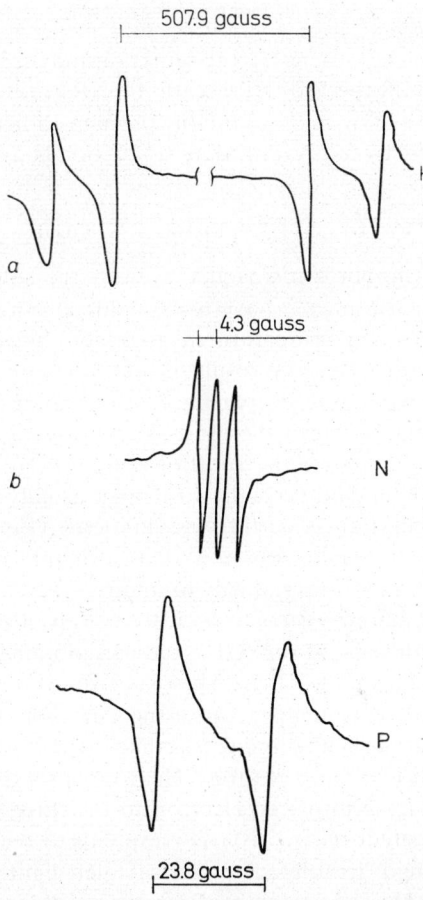

Fig. 5.9 ESR spectra of atoms formed in the first stage of photolysis, trapped in a solid argon matrix at 4.2 °K

from which the hydrogen atom has been abstracted, is trapped even at much higher temperatures.

According to this scheme the primary action of radiolysis of hydrocarbons is the following

$$RH \xrightarrow{} \dot{R} + H \qquad 5.27$$

where the symbol $\xrightarrow{}$ denotes the radiation-induced transformation. Although most of the low temperature ESR results support this scheme, there are a few cases where the primary radicals observed can only be explained by supposing that C−C bonds are broken.

5.3.3 ESR SPECTRA OF OTHER PRIMARY RADICALS

Atoms are very mobile and reactive even in the solid phase. This is why the ESR measurements have to be made at temperatures as low as 4.2 °K. At such low temperatures, however, bigger radical molecules cannot even rotate. This results in a broadening of their spectral lines and a corresponding decreasing of the resolution. More complex radical compounds can therefore be investigated at slightly elevated temperatures, where the rotation is not frozen, but diffusion is. By increasing the temperature, the ESR lines usually become sharper, indicating that rotation is getting less hindered. Further increase in the temperature usually results in a change of the spectra, indicating that transformation of the radicals has started.

Examples of different primary radicals are given in Fig. 5.10. The first spectrum corresponds to $\dot{C}H_3$ radicals obtained by photolysis of CH_3I in solid argon at 4.2 °K. The reactivity of the $\dot{C}H_3$ radicals is so great that this low temperature is needed to fix them. The resolution is rather good, which indicates that the methyl radical can rotate even at such low temperatures. The four lines correspond to the equal coupling of the unpaired electron to the three protons.

The second spectrum (Fig. 5.10b) corresponds to the radicals formed in methacrylic acid irradiated by ultraviolet light at 77 °K; measured at 243 °K. This spectrum is also rather well resolved, indicating that the radicals can almost freely rotate, and thus the spin–spin interaction is averaged out. The radical is stable at this temperature. The spectrum corresponds to the radical

$$CH_3-CH_2-\underset{\underset{COOH}{|}}{\overset{\overset{CH_3}{|}}{C}}\cdot$$

Large, long chain radicals, like those formed in irradiated polymers, may be stabilized even at room temperature. Irradiation of polyethylene in vacuum, e.g., results in the formation of relatively stable radicals

$$\sim\underset{\underset{H}{|}}{\overset{\overset{H}{|}}{C}}-\underset{\underset{H}{|}}{\overset{\overset{}{|}}{\dot{C}}}-\underset{\underset{H}{|}}{\overset{\overset{H}{|}}{C}}\sim \qquad 5.28$$

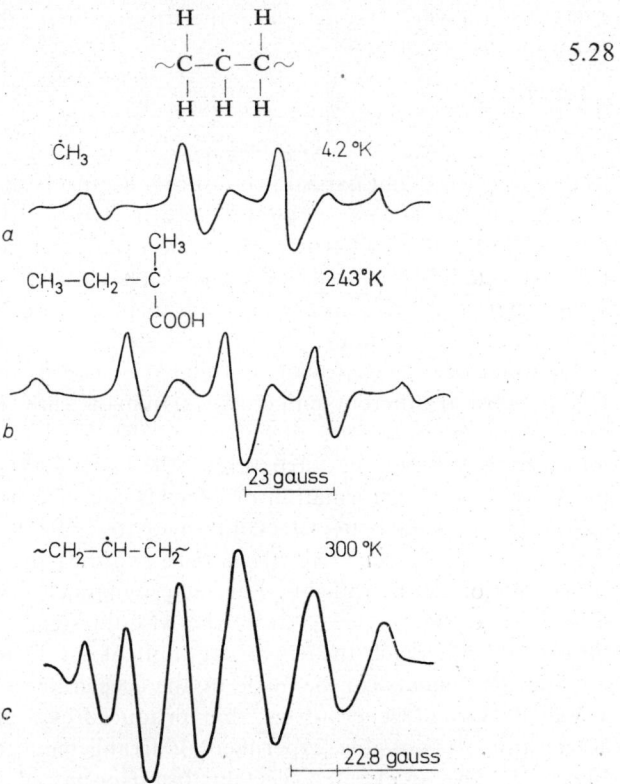

Fig. 5.10 ESR spectra of trapped primary radicals of radiolysis

The corresponding spectrum is shown in Fig. 5.10c. These radicals are formed by abstraction of an H atom from the hydrocarbon chain.

5.3.4. DETECTION OF RADICAL INTERMEDIATES

According to Equation 5.27, the first step in radiolysis of hydrocarbons is the formation of $\dot{R}$ radicals and atomic hydrogen. More generally, the very first action of the irradiation is the formation of two radicals

$$R \leadsto \dot{R}_1 + \dot{R}_2 \qquad 5.29$$

The problem is, what happens to radicals $\dot{R}_1$ and $\dot{R}_2$ during the course of the reaction? The more mobile radical is expected to react immediately to form radical intermediates or non-radical products. The other part is usually less mobile and thus reacts slowly. At the end, after several steps, non-radical products are formed, which can be analyzed by conventional methods.

The intermediate steps can in principle be investigated by measuring ESR spectra at different temperatures. In many cases the intermediate radicals can be trapped in this way, and their structures can be determined from the hyperfine structure of the ESR spectra. In many cases the kinetics of transformation of radicals can also be studied.

Another approach to detect intermediate radicals is to take ESR spectra *during the course of irradiation*, and measure the steady state concentration of the radicals. For such 'dynamic' measurements high irradiation doses are needed, for otherwise the steady state concentration would not reach the sensitivity limit of the ESR spectrometers. Such measurements can be made by the accelerated electron beam of a Van de Graaff generator or that produced by a microwave linear accelerator. A possible experimental arrangement is illustrated in Fig. 5.11. The accelerated electron beam is passed through a hole bored in the axis of the magnet pole-pieces. The cavity must have a thin-walled window in order to minimize the loss in the electron dose rate. The cavity is thermostated in the usual way (see Chapter 2). Since the heating effect of the electron beam is quite appreciable at dose rates of about 500–1000 Mrad/hr, a second, low frequency modulation and phase-sensitive detection is used in order to avoid serious

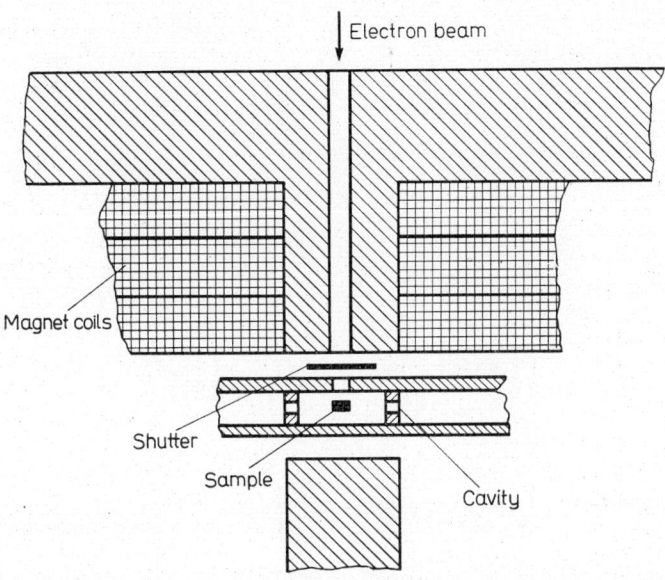

Fig. 5.11 Experimental arrangement for measuring ESR spectra during the course of high energy electron irradiation. After the scheme used in the Institute of Physical Chemistry, Moscow

drifts in the base line of the spectra. The spectra taken by this technique represent the second derivatives of the absorption curves.[5.11]

This 'dynamic' method for measuring intermediate radicals shows certain advantages over the 'freezing in' technique. The radicals are in the liquid state, and thus the resolution is much better than in the frozen-in systems. This is illustrated in Fig. 5.12, where the ESR spectra of $\dot{C}_2H_5$ radicals are shown measured at 4.2 °K in a solid argon matrix (Fig. 5.12a) and that measured at 97 °K during irradiation. As can be seen, the resolution is tremendously improved at the higher temperature. The variation of the splittings is due to the effect of the argon matrix. The spectrum of Fig. 5.12b is the second derivative of the absorption line.

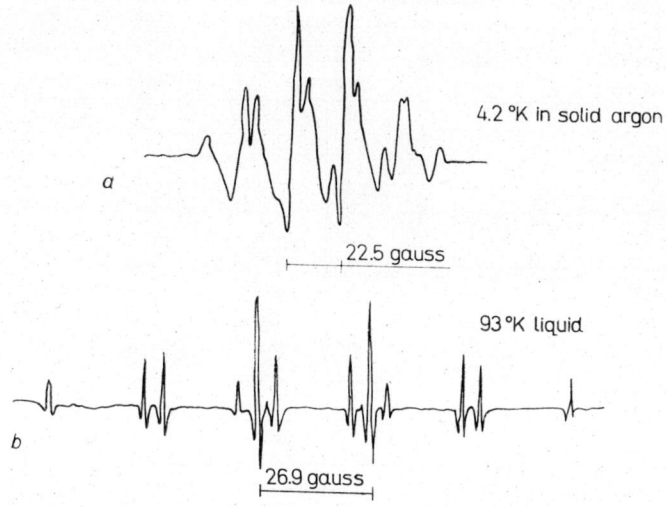

Fig. 5.12 ESR spectra of radicals formed by radiolysis of ethane. a — Radicals trapped at 4.2 °K; b — Measured in the liquid state during the course of irradiation. After Fessenden and Schuler[5.11]

5.3.5 RADICALS FORMED BY ADDITION OF HYDROGEN

According to Equation 5.27 radiolysis of hydrocarbons in most cases gives atomic hydrogen in the first stage of the reaction. In some cases the reaction of atomic hydrogen may result in radical intermediates, which can have an appreciable effect on the reaction kinetics. Using ESR the possibilities of formation of radicals by *addition of hydrogen* atoms can be investigated and the structure of the radicals can be analyzed.

Chachaty and Schmidt[5.12] have investigated a series of organic compounds having $C=C$ and $C=O$ double bonds. They found that hydrogen atoms could be added to such double bonds to form radicals.

The experiments were made using gaseous discharges to produce H atoms. The ESR spectra obtained in this way were compared with those appearing in the first stage of radiolysis and photolysis of the same compounds.

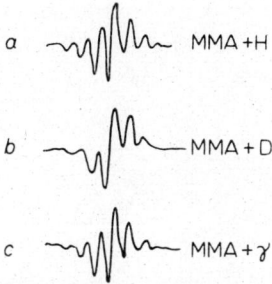

Fig. 5.13 ESR spectra of radicals formed by addition of hydrogen and deuterium to methyl methacrylate. After Chachaty and Schmidt[5.12]

An example is shown in Fig. 5.13. Spectrum (a) has been obtained by addition of hydrogen to methyl methacrylate (MMA). The reaction scheme is the following

$$CH_2=C\begin{smallmatrix}CH_3\\COOCH_3\end{smallmatrix} + \dot{H} \rightarrow CH_3-\dot{C}\begin{smallmatrix}CH_3\\COOCH_3\end{smallmatrix} \qquad 5.30$$

The corresponding ESR spectrum consists of seven lines with intensity ratios of 1 : 6 : 15 : 20 : 6 : 15 : 1. Reaction 5.30 has also been verified, by letting MMA react with deuterium atoms

$$CH_2=C\begin{smallmatrix}CH_3\\COOCH_3\end{smallmatrix} + \dot{D} \rightarrow CH_2D-\dot{C}\begin{smallmatrix}CH_3\\COOCH_3\end{smallmatrix} \qquad 5.31$$

The corresponding ESR spectrum consists of six lines with intensity ratios of 1 : 5 : 10 : 10 : 5 : 1. This indicates that the unpaired electron interacts with the five protons of the CH_2 and CH_3 group. The hyperfine interaction with the deuterium atom is not resolved.

By irradiation of pure MMA at 77 °K a spectrum similar to that of 5.13a can be observed. This indicates that by radiolysis of MMA the

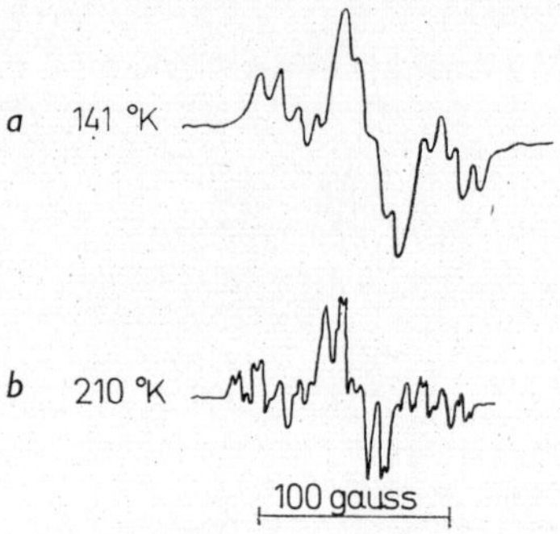

Fig. 5.14 ESR spectra of radicals formed by addition of hydrogen to benzene. After Voevodsky[5.10]

hydrogen atoms formed in the first stage of radiolysis react further with the MMA molecules to form radicals $(CH_3)_2\dot{C}-COOCH_3$.

Similar results were obtained by irradiation of benzene.[5.13] The ESR spectrum at 141 °K consists of 3×4 lines as shown in Fig. 5.14a. At elevated temperatures (210 °K) the spectrum is changed appreciably to result in 36 lines as indicated in Fig. 5.14b. Each line of the spectrum of Fig. 5.14a is further resolved to 3 lines. The ESR spectra can be interpreted by supposing that the H atoms formed in the first stage of radiolysis are added to the benzene ring

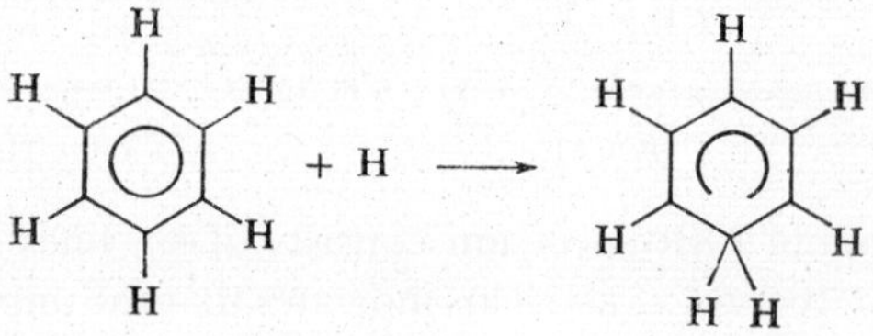

5.32

It seems quite general that H atoms can be added to multiple bonds to form new types of radicals, which have a considerable influence on the reaction. The ESR technique has proved to be a basic method for investigating such problems.'

5.3.6 KINETIC MEASUREMENTS

ESR measurements at steady state. Besides the identification of primary and intermediate radicals ESR can also be used to study radical formation and decay kinetics. One possible way of doing this is to measure the steady state concentration of radicals during the course of irradiation. In the simplest case, when only a single radical species is present, the kinetic equation is the following

$$\frac{d[\dot{R}]}{dt} = P_R - 2k\,[\dot{R}]^2 \qquad 5.33$$

where P_R is the production rate of the radicals and k is the second-order rate constant of the reaction. The radical concentration at steady state where $d[\dot{R}]/dt = 0$ is

$$[\dot{R}]_S = \left(\frac{P_R}{2k}\right)^{1/2} \qquad 5.34$$

At sufficiently high irradiation dose rates $[\dot{R}]_S$ can usually be detected by ESR. The rate of formation of the radicals is expressed as follows

$$P_R = \frac{10 I G_R}{N} \qquad 5.35$$

where I is the dose rate in eV/ml sec, N is the total concentration and G_R is the radical yield in molecules per 100 eV. G_R is either known or can be calculated from the absolute concentration of the radicals measured at low temperatures.

According to Equations 5.34 and 5.35 the steady state concentration $[\dot{R}]_S$ should depend on the square root of the dose rate I. This is illustrated in Fig. 5.15, where the steady state concentration of $\dot{C}_2H_5$ radicals formed by radiolysis of ethane is plotted against dose rate in double logarithmic scale. The radicals are formed during the course of irradiation with an accelerated electron beam. Dose rates are expressed in terms of electron beam current.

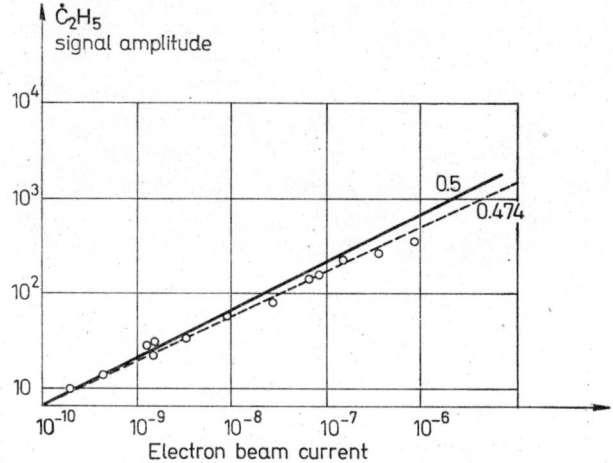

Fig. 5.15 The dose rate dependence of the steady state radical concentration [C_2H_5] formed during the course of radiolysis of ethane. After Fessenden and Schuler[5.11]

The experimental points in Fig. 5.15 determine a straight line with a slope of 0.474 ± 0.014. The slight deviation from the expected value of 0.5 is considered to be insignificant. The deviations occurring at very high dose rates are caused by the increased ionization within the cavity resonator of the spectrometer. This ionization results in an increase in dielectric losses within the cavity and thus the sensitivity is reduced.

For calculating the absolute value of the second-order rate constant k the absolute concentration of radicals has to be determined. This can be done by using one of the methods described in Chapter 2. As shown there, one must always compare the ESR lines with those of some standard material, whose spin concentration is known. In this process there are many sources of error involved, and thus absolute concentrations can only be determined to an accuracy of about ±20%. The second-order rate constant obtained from Equations 5.35 and 5.34 is the following

$$k = \frac{10 G_R I}{2N [\dot{R}]_S^2} \qquad 5.36$$

Unfortunately the least accurate value $[\dot{R}]_S$ is squared in this expression and thus the probable error in k is rather large. In fact, only the order of magnitude of k can be given with certainty. The value obtained by Fessenden and Schuler[5.11] for radiolysis of liquid ethane is $k(-175\ °C) = 3.0 \pm 3 \times 10^8$ litre/mole sec.

The *mean lifetime* of the radicals is defined by the following equation

$$\tau = \frac{[\dot{R}]_S}{P_R} = \frac{1}{2k[\dot{R}]_S} \qquad 5.37$$

For $\dot{C}_2H_5$ radicals in radiolysis of ethane $\tau = 6.8 \times 10^{-3}$ sec was found at $-175\ °C$ and a dose rate of 5.0×10^{17} eV/ml sec. This is a better defined value to find the second-order rate constant k, for it depends on the first power of $[\dot{R}]_S$.

The activation energy of the reaction can also be determined by taking the ESR spectra during irradiation as a function of temperature. If the rate of formation of the radicals P_R is independent of temperature, the steady state radical concentration is

$$[\dot{R}]_S \sim T^{-1/4} \exp(E_a/2RT) \qquad 5.38$$

where T is the absolute temperature in °K, E_a is the activation energy in cal/mole, R is the gas constant.

By plotting ln $[\dot{C}_2H_5]$ against $1/T$ a straight line is indeed obtained over a relatively large temperature interval, 94–154 °K. The activation energy is $E_a = 780 \pm 50$ cal/mole.

The scheme described above for measuring the kinetics of radical disappearance during the course of radiolysis deals with the simplest case of second-order reaction between like radicals. Generally several radical species are present in the system with many possibilities of reactions among them. Since 'dynamic' ESR measurements can be made in the liquid state the resolution is high enough to identify several radicals and for measuring their concentrations separately. Using the appropriate kinetic equations the rate constants, which characterize the different reactions, can be calculated.

5.3.7 NON-STEADY STATE MEASUREMENTS

ESR can, in principle, be applied to the study of non-steady state processes too. This can only be done at high dose rates in such systems where the radical yield is high enough. The usual technique of displaying an ESR spectrum, however, takes about 5–10 minutes. This technique can only be used if the steady state is reached very slowly. There are, however, two possible ways to take ESR spectra fast enough to follow rapid changes in radical concentrations. One is to set the magnetic field on a peak of a spectrum line and let the spectrometer record directly the change of this amplitude as a function of reaction time. In this case the rate of the process is limited only by the time constant of the spectrometer. Kinetic curves taken by this technique are shown in Fig. 5.16. Curve a represents the radical response in the photolysis of t-butylhydroperoxide.[5.14] The sample placed in the cavity resonator is illuminated by ultraviolet light. After switching on the illumination the rise in the radical concentration is recorded until a steady state is reached. The whole spectrum is taken at the steady state by the usual technique (curve b). After switching the light off the radical concentration decreases, as indicated in curve a. By performing the experiment at low temperatures, where the reaction is slow, the kinetics of formation of the radicals can be studied. In this case the radical concentration does not decrease appreciably after the light has been switched off, because the radicals are frozen in.

A more sophisticated way of following non-steady state kinetics is the use of the spectrum accumulating technique described in Chapter 2. Instead of using a large time constant this technique involves repeated rapid scanning through the spectrum. The information is stored in the memory unit of an electronic computer and the sum of these individual spectra is recorded. The first advantage of this method is that the signal–noise ratio is increased. The other is that very fast scanning rates can be used. It is possible by this technique to take radical response curves of 0.1–10 sec without serious reduction of the nomi-

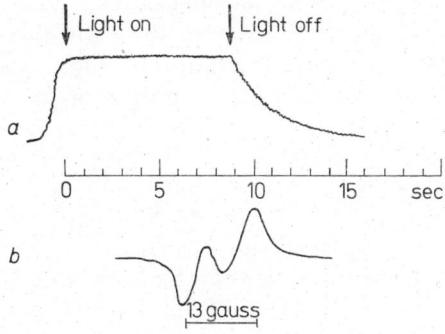

Fig. 5.16 The formation and disappearance of radicals formed by photolysis of t-butyl-hydroperoxide. After Piette[5.14]
a — Kinetic curve measured by ESR; b. — ESR spectrum of the radicals at steady state

nal sensitivity of the spectrometer. An example of the use of this technique will be given below.

5.3.8 INVESTIGATION OF ENERGY TRANSFER BY ESR

Energy transfer between triplet states can be investigated by measuring their concentration and lifetime using ESR. The kinetic equations of the energy transfer are the following

$$\frac{dN_d}{dt} = -k_d N_d + I_d$$

$$\frac{dN_a}{dt} = -k_a N_a + K N_a + I_a$$

5.39

where N_d, N_a are the concentration of the donor and acceptor triplet states, respectively, k_d, k_a are the rate constants, K is a factor representing the transitions to the ground state of the acceptor molecule and I_d, I_a are connected with the singlet state excitations. By appropriate simplifications the time dependence of the donor and acceptor triplet state concentrations can be calculated. By applying relatively short pulses of illumination and measuring the ESR signal amplitude of the donor and acceptor states one can follow the time dependence of the triplet state concentration. The solutions of the kinetic equation 5.39 in somewhat simplified form are

$$N_d(t) = \exp\{-k_d t\} + [N_d(0) - 1] \qquad 5.40$$

$$N_a(t) = \frac{KN_d(0)}{k_a - k_d}[\exp(-k_d t) - \exp(-k_a t)]$$

At equilibrium

$$\left(\frac{N_a}{N_d}\right)_{equ} \approx \frac{K}{k_a} \qquad 5.41$$

According to Equation 5.40 the concentration of the acceptor states should exhibit a maximum at

$$t_{max} = \frac{\ln(k_a/k_d)}{k_d - k_a} \qquad 5.42$$

By measuring the decay in the concentration of the donor and acceptor triplet states, and the concentrations at equilibrium, the rate constants k_a, k_d and K can be calculated.

A typical acceptor triplet state decay curve measured by Smaller, Avery and Remko[5.15] is shown in Fig. 5.17. The system consists of deuterated phenanthrene molecules as triplet donors and deuterated naphthalene molecules as acceptors. Triplet states are excited by filtered ultraviolet light at 87 °K. Since the decay of the triplet states is a fast process the spectrum accumulating technique described in Chapter 2 was used for recording the spectra. Experiments were repeated

50 times and the results were stored in memory units. The ESR signals are due to 'forbidden' transitions of $\Delta m = \pm 2$ (see the example in Chapter 3). In this case the large zero field splitting makes the 'forbidden' double quantum transitions more probable than the al-

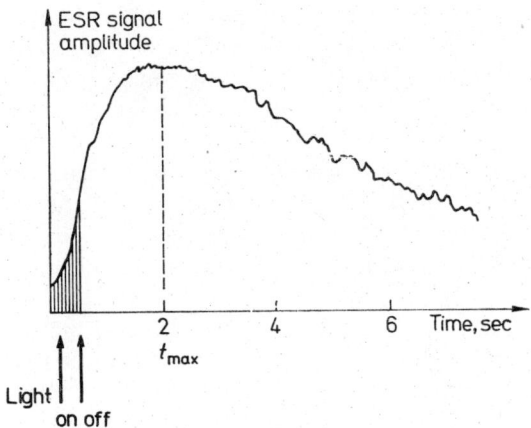

Fig. 5.17 The change in the concentration of excited triplet states during energy transfer. After Smaller, Avery and Remko[5.15]

lowed single quantum transitions, and thus the $\Delta m = \pm 2$ ESR line intensities are higher.

The curve of Fig. 5.17. corresponds to the ESR signal amplitude of the d-naphthalene triplet levels as a function of time. The maximum predicted by Equation 5.40 is clearly observed at about 2 sec. The driving light pulse is much shorter (0.6 sec).

5.4 POLYMERIZATION

The methods of radiofrequency and microwave spectroscopy have recently been applied to study polymerization and copolymerization reactions. ESR is widely used to detect radical products formed during the course of polymerization. In some cases NMR can also be applied to follow changes occurring in the chemical and physical structure during polymerization and copolymerization. Some information can also be derived from electric conductivity and dielectric loss measurements.

ESR has become the basic method for studying radical polymerization. As shown in Sections 5.1 and 5.3 above, the structure of intermediate radicals can readily be determined from the ESR spectra, provided that the line widths are small enough to resolve the hyperfine structure resulting from the interaction of the radical electron with the surrounding nuclei. It is also possible to follow the kinetics of formation and disappearance of the radicals provided that their concentrations are high enough to be detected by the given technique.

In radical polymerization the following radical types are of interest.

1. *Initiating radicals* $\dot{R}_i$. For initiating radical polymerization it is necessary to create certain types of radicals in the system containing the monomer. This can be done by chemical means, i.e. by introducing certain compounds, *initiators*, which by thermal decomposition or by other reactions produce initiating radicals.

These radicals react with the monomer molecules to form *monomer radicals* $\dot{R}_m$. Initiating radicals can also be formed by ultraviolet, X-ray and high-energy particle irradiations or by mechanical treatment. The concentration and structure of the initiating radicals can in most cases be investigated by ESR.

2. *Monomer radicals* $\dot{R}_m$. The initiating radicals may add to the monomer molecules or abstract atoms or groups from them to form monomer radicals. It is very important in the study of polymerization processes to know what monomer radicals are formed. This problem can also be investigated by ESR.

3. *Polymer radicals* $\dot{R}_p$. The reaction of the monomer radicals R_m with the monomer molecules results in new radicals, which consist of several monomer units. These radicals are connected with the growth of the polymer chains; as long as they live, the chain propagation goes on. The structure, concentration and lifetime of these radicals are therefore very important factors to be measured experimentally. In most cases the concentration of the polymer radicals during

the course of polymerization is too small to be detected by the usual ESR technique. It can only be done in highly viscous systems where the lifetime of the propagating radicals is high enough. In liquids, short-lived propagating radicals can be detected by the flow technique described in Chapter 2.

4. *Radicals formed by destruction of polymers* $\dot{R}_d$. Besides chain growth there is always a possibility of chain ruptures during the course of polymerization. The radicals present in the system may initiate reactions which tend to decrease the molecular weight of the product. The study of destruction of polymers is very important in polymer chemistry. Since it is a radical reaction, it can be investigated by ESR. The radical concentrations occurring during thermal destruction of polymers are usually too low to be detected. The destruction caused by ultraviolet light or by high-energy irradiation can, however, be investigated. Radicals formed by mechanical destruction of polymers can also be measured at low temperatures.

5.4.1 DETECTION OF INITIATING RADICALS

The method of investigating initiating radicals of polymerization depends on the way they are produced. A possible chemical way of producing initiating radicals is to perform such *redox reactions* as the following one

$$Ti^{3+} + H_2O_2 \rightarrow Ti^{4+} + OH^- + H\dot{O}$$

$$H\dot{O} + H_2O_2 \rightarrow H_2O + HO\dot{O} \qquad 5.43$$

The initiating radicals $H\dot{O}$ and $HO\dot{O}$ are short-lived; they can only be detected by the flow technique. In aqueous solutions containing some sulphuric acid with a flow rate of 4.5 ml/sec a double ESR line is observed[5.16] as shown in Fig. 5.18. The sharp line corresponds to $H\dot{O}$ radicals, the other to $HO\dot{O}$ radicals. By changing the concentration of the Ti^{3+} ions ($TiCl_3$) the [$H\dot{O}$] concentration changes in a way indicated in Fig. 5.18. At low Ti^{3+} concentrations only the sharp line corresponding to $H\dot{O}$ radicals is observed, the other line appears only at higher Ti^{3+} concentrations. From the experimental results the optimal concentration of the Ti^{3+} ions to produce initiating $H\dot{O}$ radicals can be determined.

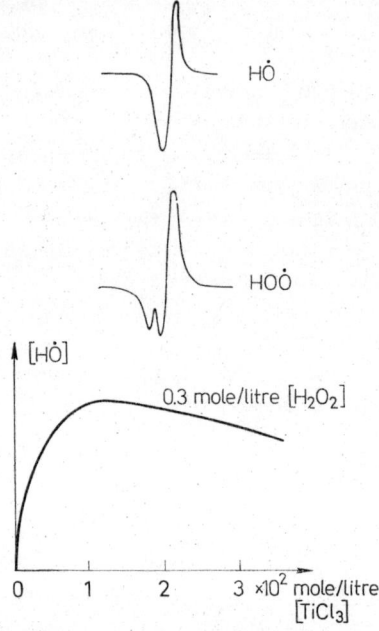

Fig. 5.18 ESR spectrum of initiating radicals $H\dot{O}$ and $HO\dot{O}$ formed during the reduction of Ti^{3+}. After Yoshida and Ranby[5.16]

Similar investigations were made by Fischer and Giacometti[5.17] in aqueous solutions of $TiCl_3$-t-butylhydroperoxide. The reaction leading to formation of initiating radicals is

$$Ti^{3+} + (CH_3)_3COOH \rightarrow Ti^{4+} + HO^- + (CH_3)_3C\dot{O}$$

$$(CH_3)_3C\dot{O} \rightarrow CH_3-CO-CH_3 + \dot{C}H_3 \qquad 5.44$$

The formation of $\dot{C}H_3$ radicals has in fact been observed in this system by the use of the flow technique. The spectrum obtained is similar to that shown in Fig. 5.10a with a slightly different splitting.

Initiating radicals formed by ultraviolet or high-energy radiation can be measured by the low temperature trapping technique described in the previous section. Polymerization is very often initiated by adding a small amount of benzoylperoxide to the monomer. Upon irradiation or heating benzoylperoxide will decompose to give initiating radicals:

$$\text{C}_6\text{H}_5-\underset{\underset{O}{\|}}{C}-O-O-\underset{\underset{O}{\|}}{C}-\text{C}_6\text{H}_5 \xrightarrow{\text{heat or u.v. irrad.}}$$

$$\longrightarrow \text{C}_6\text{H}_5-\underset{\underset{O}{\|}}{C}-\dot{O} \;+\; \dot{O}-\underset{\underset{O}{\|}}{C}-\text{C}_6\text{H}_5 \qquad 5.45$$

The benzoyloxy radicals formed are trapped in the rigid medium if the sample is irradiated at liquid nitrogen temperature. The trapped benzoyloxy radicals exhibit a not very well resolved 4-line spectrum.[5.18] The hyperfine interaction is due to the ring protons, probably to the *ortho-* and *para-*protons.

In many cases pure monomers polymerize upon irradiation. In such cases the initiating radicals are directly formed from the monomer molecules. The primary action of the radiation is in most cases abstraction of hydrogen atoms from the monomer molecule (see Section 5.3). In these cases monomer radicals R_m are formed in the first stage of the reaction. The hydrogen atoms abstracted can, however, also initiate polymerization by forming intermediate radicals with the monomer molecules. Examples of studying primary action of radiolysis have been given in the previous section. The same method can be applied to study initiating radicals in irradiated pure monomers.

5.4.2 INTERMEDIATE RADICALS TRAPPED IN VISCOUS MEDIA

A possible way of detecting monomer and polymer radicals R_m and $\dot{R}_p$ formed during the course of polymerization is to let them be trapped in crystalline or highly viscous media. The polymerization can be stopped at any moment by cooling down the system to such a low temperature that the mobility of the radicals is low. The viscosity of the polymer–monomer mixture becomes very high below a certain temperature known as the *glass-transition temperature* of the system. In the glassy state the mobility of the radicals is very low.

There are two common ways to investigate intermediate polymer or monomer radicals by the trapping technique. One is to create the initiating radicals at very low temperatures by ultraviolet or high-energy irradiation and measure the change in the ESR spectrum when temperature is increased gradually. The transformation of the initiating radicals into polymer or monomer radicals can, in principle, be observed in this way

$$\dot{R}_i \rightarrow \dot{R}_m \rightarrow \dot{R}_p$$

The other way is to let the system react at elevated temperatures for some time, then cool it down to liquid nitrogen temperature and take the ESR spectrum. By repeating this procedure at different reaction times the intermediate radicals can be detected.

Radicals trapped in viscous media have been investigated extensively by Hotta and Anderson[5,18] in various systems. Monomeric radicals have been observed by illuminating the monomer–initiator mixtures at liquid nitrogen temperature. In the case of styrene illuminated at 77 °K in the presence of benzoylperoxide initiator, a poorly resolved quartet is found with a *g*-value of 2.001 and with a splitting of 15 gauss. No ESR signal can be observed by illuminating pure styrene. The radicals observed are thus considered as being formed by addition of the benzoyl radicals to styrene molecules:

5.46

Monomeric radicals $\dot{R}_m$ can thus be measured by ESR at sufficiently low temperatures. The exact structure of the radicals can, however, in general not be determined with certainty.

The situation is more complicated when propagating polymeric radicals are also present in the system. At higher conversions, usually several radical species are present. This is illustrated by the following example.

Radiation polymerization of methyl methacrylate. Methyl methacrylate is the monomer most thoroughly investigated by the technique of ESR. Although the ESR measurements supply much useful information about the mechanism of polymerization, some problems have remained unresolved. So this example can also illustrate the limitations of the ESR trapping technique.

Upon illuminating a mixture of methyl methacrylate (MMA) and benzoylperoxide (BP) by ultraviolet light at liquid nitrogen temperature (77 °K) a rather complex ESR spectrum is obtained. According to the recent measurements of Komatsu and Sohma[5.19] a seven-line spectrum with a splitting of 23 gauss and a doublet with a spacing of 508 gauss appears (Fig. 5.19a). Illumination of MMA without BP gives only a very weak ESR signal.

Earlier measurements by Ingram[5.20] using OH· initiating radicals were interpreted differently. The spectrum in this case was considered to be a sextet. Hotta and Anderson[5.18] reported a quintet with a splitting of 20.5 gauss and an asymmetric doublet by using BP initiator. The difference in the interpretation is due to the small intensity of the side lines. The radicals are stable at 77 °K.

The intensity ratio of the spectra indicates that more than one radical species is present in the system. This can also be proved by heating the samples and observing the changes in the spectra. By keeping the sample at 115 °K for 3 minutes and re-cooling it to 77 °K the central peak of the spectrum is decreased but the total intensity is unchanged (Fig. 5.19b). This indicates that transformation of radicals occurs at 115 °K. At higher temperatures (151 °K) the spectra are changed more drastically. The total intensity begins to decrease and new lines appear in between the original ones. The final spectrum obtained by prolonged heat treatment is shown in Fig. 5.19c. This spectrum consists of a quintet and a superimposed weaker quartet. Such 5 + 4-line spectra have been observed by various authors in irradiated poly(methyl methacrylate)[5,21] or in partially thermally polymerized methyl methacrylate. Irradiation of pure poly(methyl meth-

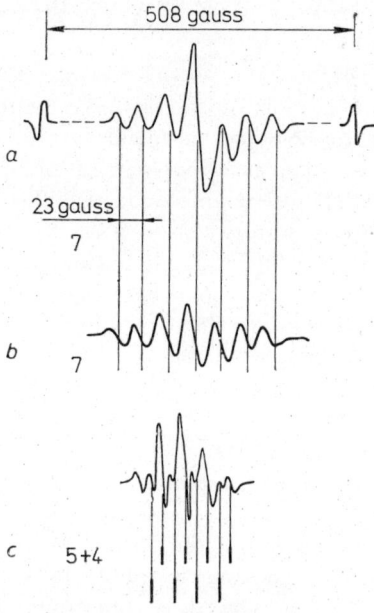

Fig. 5.19 ESR spectrum of methyl methacrylate containing ultraviolet irradiated benzoyl peroxide at 77 °K. After Komatsu and Sohma[5,19]

acrylate) from which the monomer was completely removed results in a quartet instead of the 5 + 4 spectrum. The temperature dependence of the 5 + 4 spectrum also shows that the quintet and the quartet corresponds to different radical species, or at least to different conformations of the same radical. The decay of the quartet is slower at elevated temperatures than that of the quintet.

Finally Fischer[5,22] has obtained very well resolved spectra during the course of redox polymerization of MMA in the liquid state, using the flow method. These results will be discussed later.

According to the experimental evidence obtained by the ESR trapping technique the following conclusions can be drawn.

1. In the first stage of the polymerization free H atoms are formed, which are stabilized at 77 °K (the doublet with a spacing of 508 gauss).

2. The central peak of the spectrum obtained at 77 °K is due to the benzoyloxy radicals 5.45. These radicals react with the monomer molecules to give monomeric radicals. This is why the total intensity of the spectrum is unchanged at a certain temperature interval, where polymerization is slow. The reaction in this stage can be the following

$$\underbrace{C_6H_5-\overset{\text{O}}{\underset{\|}{C}}-\dot{O}}_{\dot{R}_i} + \underset{H\ \ COOCH_3}{\overset{H\ \ CH_3}{C=C}} \longrightarrow$$

$$\underset{\dot{R}_m^{(1)}}{C_6H_5-\overset{\text{O}}{\underset{\|}{C}}-O-\underset{H}{\overset{H}{C}}-\underset{COOCH_3}{\overset{CH_3}{\dot{C}}}} \qquad 5.47$$

3. At higher temperatures polymer chains begin to grow. This would correspond to the decrease observed in the total radical concentration. The appearance of the quartet in between the lines can be attributed to the polymer radicals $\dot{R}_p$. In the case of radiation polymerization of MMA polymeric radicals can be formed either by propagation or by depropagation i.e. photodestruction. They also can be formed by simple hydrogen abstraction from the polymer chains. So far there is no direct observation of propagating polymer radicals in radiation polymerization of MMA.

4. The existence of free H atoms at 77 °K indicates that hydrogen abstraction also takes place. This can result in the following reactions.

$$\underset{\substack{H \quad CH_3 \\ | \quad\; | \\ C=C \\ | \quad\; | \\ H \quad COOCH_3}}{} \xrightarrow{u.v.} \begin{array}{c} \underset{\substack{H \quad \dot{C}H_2 \\ | \quad\; | \\ C=C \\ | \quad\; | \\ H \quad COOCH_3 \\ \dot{R}_m^{(2)}}}{} \;+\dot{H} \\ \\ \underset{\substack{H \quad CH_3 \\ | \quad\; | \\ C=C \\ | \quad\; | \\ H \quad COO\dot{C}H_2 \\ \dot{R}_m^{(3)}}}{} \;+\dot{H} \end{array} \qquad 5.48$$

These radicals would exhibit *triplet* ESR spectra.

The further reaction of these radicals and those formed by addition of the H atoms result in different monomer radical species. The addition reaction is the following

$$\underset{\substack{H \quad CH_3 \\ | \quad\; | \\ C=C \\ | \quad\; | \\ H \quad COOCH_3}}{} + \dot{H} \longrightarrow \underset{\substack{H \quad CH_3 \\ | \quad\; | \\ H-C-\dot{C} \\ | \quad\; | \\ H \quad COOCH_3 \\ \dot{R}_m^{(4)}}}{} \qquad 5.49$$

According to Komatsu and Sohma[5.19] the ESR spectrum can be considered as a superposition of a septet, quintet, triplet and singlet. According to the scheme given above these would correspond to the following radicals

singlet	$\dot{R}_i$ benzoyloxy radical
triplet	$\dot{R}_m^{(2)}$ or $\dot{R}_m^{(3)}$ radicals formed by $\dot{H}$ abstraction
quintet	$\dot{R}_m^{(1)}$ radicals formed by addition of $\dot{R}_i$ to the monomer

septet $\dot{R}_m^{(4)}$ radicals formed by addition of $\dot{H}$ to the monomer

The temperature dependence of the spectra seems to support this scheme. The decrease in the intensity of the singlet, for example, agrees quantitatively with the increase of the quintet indicating the existence of reaction 5.47.

The existence of $\dot{R}_m^{(4)}$ radicals have also been indicated by irradiating MMA with gamma rays.[5.23] In this case no initiator was present in the system. The spectrum obtained at 77 °K consisted of seven lines. In this case most probably only $\dot{R}_m^{(2)}$ and $\dot{R}_m^{(4)}$ radicals are present in the system.

The example of the polymerization of MMA shows that, although ESR can provide useful information about the radical species present, it is very difficult to identify them by using the trapping technique. Propagating polymeric radicals generally cannot be identified in this way. This can only be done by measurements *in situ* by applying the flow technique. Examples of this will be given below.

5.4.3 RADICAL INTERMEDIATES IN THE LIQUID STATE

The structure of intermediate radical products, formed during the course of polymerization in the liquid state, can be conveniently studied by the flow technique described in Chapter 2. This method was first applied to polymerization problems by Fischer[5.24] in 1963. A number of systems have been investigated since then with excellent results. The resolution of the spectra is high enough to distinguish between initiating, monomeric and polymeric radicals. The enormous increase in the resolution is indicated in Fig. 5.20 where the spectrum of polymeric radicals in methacrylic acid is shown in the liquid state (*a*) taken by the flow method and that trapped in the gelled state at 243 °K (*b*). The two spectra obviously have the same structure, but the one taken by the flow technique is further resolved. This experiment indicates that the 5 + 4 spectrum usually obtained in polymerization of methacrylates consists in reality of many more lines. The general type of these radicals is the following[5.24]

$$\dot{R}_p = R-CH_2-\underset{\underset{COOX}{|}}{\overset{\overset{CH_3}{|}}{C}}-CH_2-\underset{\underset{COOX}{|}}{\overset{\overset{CH_3}{|}}{C}}\cdot \qquad 5.50$$

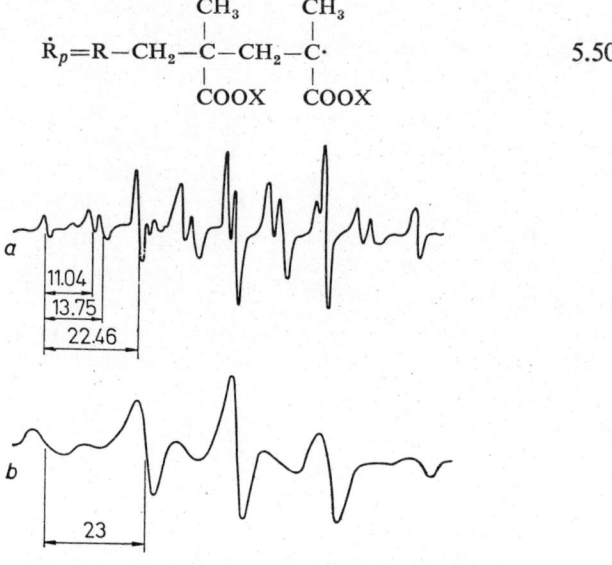

Fig. 5.20 ESR spectrum observed during polymerization of methyl methacrylate.
a — At high resolution measured by the flow technique. After Fischer[5.22] b — At low resolution measured in the glassy state. After Marx[5.23]

The hyperfine splitting is due to the coupling of the unpaired electron to the CH_2 and CH_3 groups. No splitting has been observed from the hydrogen atoms present in the COOX group.

Monomeric radicals. Monomeric radicals $\dot{R}_m$ are usually formed by addition of initiating radicals $\dot{R}_i$ to the monomer molecules M

$$\dot{R}_i + M = \dot{R}_m \qquad 5.51$$

According to Equation 5.43, for example, in the $Ti^{3+} - H_2O_2$ redox system OH radicals are formed. Addition of $\dot{O}H$ to monomers of type

$$M = \begin{array}{c} H\ X_1 \\ |\ \ | \\ C=C \\ |\ \ | \\ H\ X_2 \end{array} \qquad 5.52$$

results in formation of the following monomeric radicals

$$\dot{R}_m = HO-\begin{array}{c} H\ X_1 \\ |\ \ | \\ C-C \\ |\ \ | \\ H\ X_2 \end{array}\cdot \qquad 5.53$$

These radicals would give a characteristic ESR spectrum with a splitting due to the interaction of the unpaired electron with the groups CH_2, X_1 and X_2. The protons in the CH_2 group are equivalent because of rotation.

In the case of acrylic acid, for example, the following splitting constants have been observed [5.16]

$$HO-CH_2-\underset{COOH}{\overset{H}{\underset{|}{C}}}\cdot \quad \begin{array}{l} 20.45\ \text{gauss} \\ 27.58\ \text{gauss} \end{array} \qquad 5.54$$

By using the same monomer but different initiators the spectrum of the monomer radicals is different. Initiating $\dot{C}H_3$ radicals formed by reaction 5.44 would produce the following monomer radicals

$$CH_3-CH_2-\underset{COOH}{\overset{H}{\underset{|}{C}}}\cdot \quad \begin{array}{l} 20.17\ \text{gauss} \\ 23.78\ \text{gauss} \end{array}$$

Table 5.1

Splitting constants $a(CH_2)$ for $R-CH_2-\underset{\underset{X_2}{|}}{\overset{\overset{X_1}{|}}{C}}{}^{\cdot}$ type monomeric radicals

Group	R			
	HO–	CH_3–	NH_2–	HO–CH_2–
$R-CH_2-\underset{\underset{CN}{\|}}{\overset{\overset{H}{\|}}{C}}{}^{\cdot}$	28.15	25.19	23.75	22.89
$R-CH_2-\underset{\underset{COOH}{\|}}{\overset{\overset{H}{\|}}{C}}{}^{\cdot}$	27.58	23.78	25.03	22.81
$R-CH_2-\underset{\underset{CN}{\|}}{\overset{\overset{CH_3}{\|}}{C}}{}^{\cdot}$	23.92	21.83	19.24	17.82
$R-CH_2-\underset{\underset{COOH}{\|}}{\overset{\overset{CH_3}{\|}}{C}}{}^{\cdot}$	19.98	15.37	16.98	14.45
$R-CH_2-\underset{\underset{COOCH_3}{\|}}{\overset{\overset{CH_3}{\|}}{C}}{}^{\cdot}$	12.5			
$R-CH_2-\underset{\underset{COOCH_2CH_3}{\|}}{\overset{\overset{H}{\|}}{C}}{}^{\cdot}$	12.4			
$R-CH_2-\underset{\underset{CH=CH_2}{\|}}{\overset{\overset{H}{\|}}{C}}{}^{\cdot}$	12.6 13.8			

In Table 5.1 CH_2 splitting values are collected[5.16] for various monomeric radicals of the following type

$$R-CH_2-\underset{X_2}{\overset{X_1}{\underset{|}{\overset{|}{C}}}}.\qquad 5.55$$

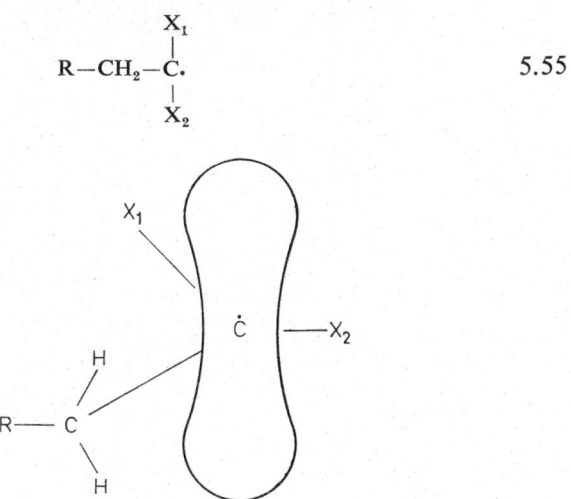

Fig. 5.21 Schematic representation of group $CH_2RCX_1X_2$

The horizontal rows show how the coupling between CH_2 protons and the unpaired electron depends on the group R. The vertical columns show the dependence of this coupling constant upon groups X_1 and X_2. The decrease observed in the coupling constants by making groups R, X_1 and X_2 larger is explained by geometrical arguments. Fig. 5.21 shows how the CH_2 group is oriented with respect to the radical. Substitution at X_1, X_2 or R makes the CH_2 plane twisted with respect to the $X_1-\dot{C}-X_2$ radical plane. The splitting caused by the CH_2 group is expressed as follows

$$A(CH_2) = A_0 \cos^2 \theta \qquad 5.56$$

where θ is the angle between planes $X_1-\dot{C}-X_2$ and that of $R-CH_2$. By introducing larger groups instead of X_1, X_2 or R the angle θ gets smaller and so does the coupling constant $a(CH_2)$.

As shown in Table 5.1 the relation 5.56 does not hold strongly. The coupling constant $a(CH_2)$ may be influenced by such other factors as intramolecular hydrogen bonding. This is most probably the reason for the high coupling constants observed in some radicals. The structure of the radicals can thus be described as follows:

$$
\begin{array}{cc}
\text{H} \quad\quad \text{H} & \text{H} \quad\quad \text{CH}_3 \\
| \quad\quad\quad | & | \quad\quad\quad | \\
\text{N}-\text{CH}_2-\text{C}\cdot & \text{N}-\text{CH}_2-\text{C}\cdot \\
| \quad\; 25.03 \quad | & | \quad\; 16.98 \quad | \\
\text{H}\ldots\text{HOOC} & \text{H}\ldots\text{HOOC}
\end{array}
$$

Propagating radicals. If the concentration of the initiating radicals $\dot{R}_i$ is high enough, as well as the monomeric radicals $\dot{R}_m$, the propagating polymeric radicals $\dot{R}_p$ can also be observed. The ESR spectra of $\dot{R}_m$ and $\dot{R}_p$ have different splittings and thus can easily be identified. The general structure of the polymeric radicals is the following

$$\dot{R}_p = R - \left(CH_2 - \underset{\underset{X_2}{|}}{\overset{\overset{X_1}{|}}{C}} \right)_n -- CH_2 - \underset{\underset{X_2}{|}}{\overset{\overset{X_1}{|}}{C}} \cdot \qquad 5.57$$

As can be expected the spectra of polymeric radicals are independent of the initiator used. It is not possible, however, to differentiate between propagating radicals having different chain lengths (different n-values).

The relative concentrations $[\dot{R}_m]$ and $[\dot{R}_p]$ can be changed by changing the monomer concentration [M]. At low monomer concentrations only the monomer radicals can be observed. Increasing the monomer concentration the ESR signal corresponding to radicals appears and becomes more and more intense. A typical curve first measured by Fischer and Giacometti[5.17] is shown in Fig. 5.22. The intensity of the $\dot{R}_m$ and $\dot{R}_p$ spectra is plotted against the monomer concentration. The sharp increase in $[\dot{R}_m]$ at low monomer concentrations corresponds to reaction 5.51. At higher monomer concentrations the polymer chains begin to grow and $[\dot{R}_p]$ is increased. The curves correspond to polymerization of acrylic acid in the presence of $\dot{C}H_3$-initiating radicals.

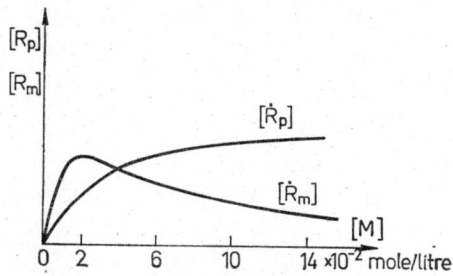

Fig. 5.22 The change of the monomer and polymer radical concentration during polymerization of acrylic acid. After Yoshida and Ranby[5.16]

5.4.4 KINETIC STUDIES

Besides the study of free radicals formed during the course of polymerization, it is possible to investigate their kinetics of transformation as well. In favourable cases it is possible to follow how the initiating radicals are transformed to propagating ones and how propagating radicals are trapped or transformed to non-radical products. The kinetic curves obtained for radicals can be compared with the polymerization curves measured by conventional chemical methods. It is also possible to follow the changes in the physical structure of the material during the course of the polymerization. This is very important in highly viscous systems (gels, glassy systems) in the solid state, where the physical structure plays a decisive role in the kinetics of polymerization. Nuclear magnetic resonance can also be used for analyzing the products formed.

ESR measurements of radical-transformation kinetics can be performed rather easily in the solid state or in highly viscous liquid states where the probability of chain-termination is low enough to assure the required radical concentrations. As shown earlier, the resolution of ESR in such systems is rather poor as a result of the poor averaging of the local dipole fields. Correspondingly in such systems, where transformations of radicals are easy to investigate, radical structures are usually less well determined. Results have been obtained in the study of kinetics of radiation (u.v., X or gamma)-initiated polymerization in the solid phase and in highly viscous media. It is also possible to measure the formation and decay of intermediate (propagating) radicals in solution in the presence of inhibitors. Some examples will be given below.

Inhibited polymerization of styrene. In the presence of molecular inhibitors, in addition to propagating polymer radicals $\dot{R}_p$, transient radicals are formed by reaction of the polymer radicals with the monomer molecules. The concentration of these radicals [$\dot{R}_t$] may be high enough to be detected by ESR. The polymerization scheme is roughly the following:

$$\dot{R}_i + M \to \dot{R}_p$$
$$\dot{R}_p + I \to \dot{R}_t \qquad 5.58$$
$$\dot{R}_t + \dot{R}_p \to \text{polymer}$$

where $\dot{R}_i$, $\dot{R}_p$, $\dot{R}_t$ are the initiating radicals, polymer radicals and transient radicals, respectively, formed by reaction of the propagating polymer radicals with the inhibitor molecules I. M is the monomer molecule.

Tüdős et al.[5.25] have studied the polymerization of styrene in the presence of such molecular inhibitors as *p*-nitrosophenol. Using the technique of ESR they could identify $\dot{R}_p$ and $\dot{R}_t$ radicals in the liquid state without using the continuous flow technique. The transient radical is

where $(R)_n$ is the growing polystyrene chain. Since the ESR spectra are measured in the liquid state the resolution is good. The splitting constants are the following:

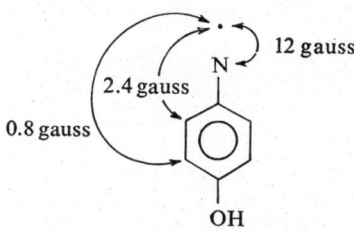

Coupling with the nitrogen produces a triplet with 1 : 1 : 1 intensity ratio split by 12 gauss. Interaction with the 2 *ortho*- and 2 *meta*-ring protons results in 1 : 2 : 1 splittings by 2.4 and 0.8 gauss, respectively.

The kinetics of the change in radical concentration $[\dot{R}_t]$ are simply measured by keeping the sample at the temperature of the reaction and recording the ESR spectrum continually. Plotting the integrated line intensities directly against time gives the required curves. The kinetic equations are the following

$$\frac{d[\dot{R}_t]}{dt} = ak[\dot{R}_p][I] - k'[\dot{R}_p][\dot{R}_t]$$

$$\frac{d[I]}{dt} = -k[\dot{R}_p][I]$$

5.59

where $[\dot{R}_t]$ and $[\dot{R}_p]$ are the concentration of the transient radicals and polymer radicals, respectively; [I] is the inhibitor concentration; k and k' are the rate constants of the reaction; a is a probability factor determined by the vibrational state of the transient radicals $\dot{R}_t$.

In the quasi-stationary state

$$\frac{d[\dot{R}_t]}{dt} = \text{const.}$$

the corresponding transient radical concentration is the following

$$[\dot{R}_t] = a \frac{k}{k'} [I]_0 \left(1 - \frac{t}{t_0}\right) \qquad 5.60$$

where $[I]_0$ is the initial inhibitor concentration, t_0 is the length of the induction period, t is the reaction time.

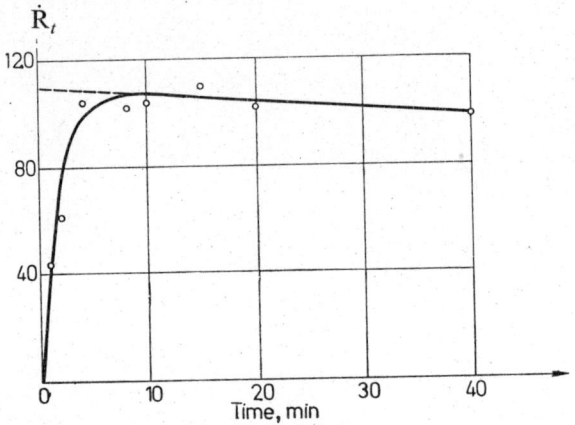

Fig. 5.23 The formation of transient radicals during inhibited polymerization of styrene. After Tüdős et al.[5.25]

An experimental plot of $[\dot{R}_t]$ against reaction time is shown in Fig. 5.23. Upon mixing the monomer styrene, initiator (azobisisobutyronitrile), and inhibitor (p-nitrosophenol) the sample is placed in the ESR cavity heated to the temperature of the reaction. The magnetic field is swept through the spectral range repeatedly. The derivative spectra are twice integrated and the integral values are plotted against reaction time. In the quasi-stationary portion, Equation 5.60 holds fairly well. By measuring the change of the monomer concentration $[I]$ independently the reaction rate constants k and k' can be determined. Repeating the measurement at different temperatures the activation energies can also be calculated.

NMR measurements during polymerization. As shown in Chapter 1 the nuclear magnetic relaxation times depend on the motional degree of freedom of the molecules, where the magnetic nuclei are located. During the course of polymerization these nuclei change their environment gradually as they pass from the mobile monomer to the less mobile polymer molecules. di Stefano, Bonera and Rigamonti[5,26] succeeded in detecting changes in the proton spin lattice relaxation time T_1 during the course of polymerization. The relaxation time was measured by the adiabatic fast passage method, described in Section 1.4. Two distinct relaxation processes could be distinguished, corresponding to protons in monomer molecules and in polymer chains.

From the temperature dependence of the motional correlation times (see Section 3.4)

$$\tau_c = \tau_0 \exp(E_M/kT)$$

the activation energy E_M of the motion could be determined. τ_c can be calculated from the measured value of T_1 according to Equation 3.23 in Chapter 3. In the case of styrene and methyl methacrylate the motional activation energies were found to be increased gradually with increasing conversion.

Generally it is advisable to measure both relaxation times T_1 and T_2. As will be discussed later (Section 5.5) the spin–lattice relaxation time T_1 is chiefly sensitive to translational degrees of freedom, while T_2 is sensitive to rotational ones.

Polymerization in the solid state. In the solid state the mobility of the polymer radicals is low enough to be observed directly by ESR. In solid state polymerization the following main problems can be investigated by radio frequency spectroscopy

1. Formation and decay of initiating radicals.
2. Kinetics of transformation of the initiating into propagating polymer radicals.
3. Changes in the physical structure of the material during the course of the polymerization.

As an example of the transformation of initiating radicals into propagating ones the radiation polymerization of acrylonitrile at low temperatures is discussed by Bensasson, Bodard and Marx.[5.27]

Irradiated acrylonitrile at $-190\ °C$ exhibits a poorly resolved ESR spectrum. It is attributed to the following monomer radical

$$H_3C-\overset{\overset{H}{|}}{\underset{\underset{CN}{|}}{C}}\cdot$$

Upon heating the sample to $-130\ °C$ the spectrum is gradually transformed into 3 lines attributed to the polymer radicals

$$\sim \overset{\overset{H}{|}}{\underset{\underset{H}{|}}{C}}-\overset{\overset{H}{|}}{\underset{\underset{CN}{|}}{C}}-\overset{\overset{H}{|}}{\underset{\underset{H}{|}}{C}}-\overset{\overset{H}{|}}{\underset{\underset{CN}{|}}{C}}\cdot$$

The ratio of these two radical species is measured by comparing the amplitude of the central line with that of the side lines. The result is shown in Fig. 5.24. The relative amplitudes, thought to be proportional to $[\dot{R}_p]/[\dot{R}_m]$, are plotted against temperature. The polymer conversion curve is also shown. At about $-130\ °C$ the increase of the conversion is accompanied by the transformation of the 5-line spectrum to the 3-line one, indicating the transformation of the monomer radicals $\dot{R}_m$ into polymer radicals $\dot{R}_p$.

Graft copolymerization. ESR can be applied to study the kinetics of radiation-induced graft copolymerization. In these experiments polymers irradiated with X-rays or ^{60}Co gamma rays are treated with liquid or gaseous monomers. The radicals trapped in the solid polymer act as initiators for the polymerization of the monomer. Since the initiating radicals are in these cases fixed in the solid matrix the monomer must diffuse to them. The growing polymer radicals are again more or less fixed. It is also possible to irradiate the polymer swollen in the liquid monomer to induce graft copolymerization.

Since the initiating radicals are trapped, the most straightforward way for investigating reaction kinetics is to measure the change of this trapped radical concentration as a function of reaction time at fixed temperatures.[5.28] A characteristic curve is shown in Fig. 5.25. The

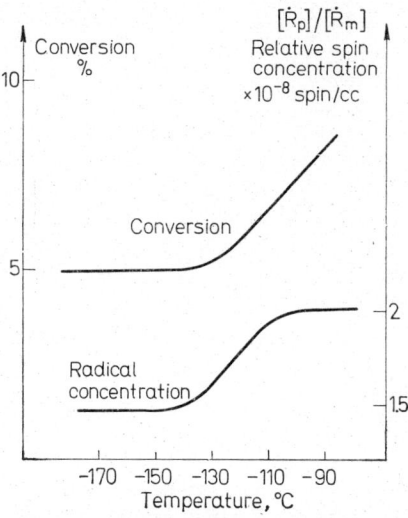

Fig. 5.24 The change of the radical concentration in irradiated acrylonitrile during solid state polymerization. After Bensasson, Bodard and Marx[5.27]

change in the peroxy radical concentration in polytetrafluoroethylene is shown as a function of the time of reaction with liquid styrene monomer. The decrease in the peroxy radical concentration is found to be strictly linear up to a point, where the reaction slows down sharply. The penetration of the grafting front into the polymer can be observed under a microscope with dyed micro-cuts taken at different degrees of grafting. These measurements have shown that at the point when the radical decay is slowed down, the grafting front has penetrated the polymer. At this point the grafting as measured by dilatometric methods or by weight gain after extraction is practically stopped. The grafting curve is also shown in Fig. 5.25.

The linearity of the grafting curve can be interpreted by considering the quasi-stationary state resulting from the diffusion of the monomer

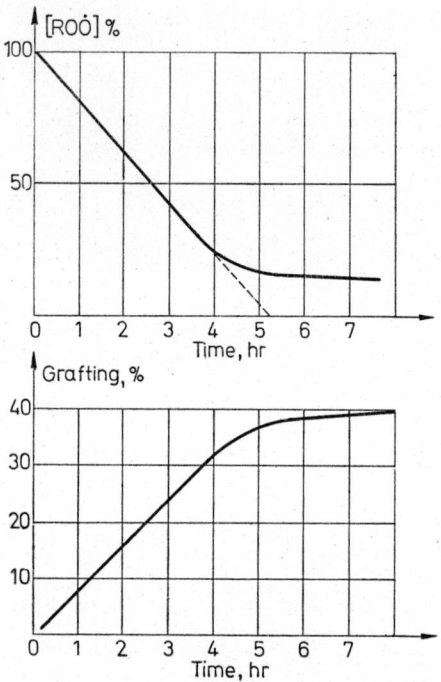

Fig. 5.25 The change in the peroxy radical concentration in irradiated polytetrafluoroethylene during the course of grafting[5.28]

through the polymer. As the grafting front proceeds, the monomer comes in contact with new initiating radicals. The grafting reaction is found to proceed further at a strongly reduced rate, after the polymer has been completely grafted through. During the course of this reaction an appreciable change in the crystallinity of the polymer is observed.

The relatively high amount of residual peroxy radicals remaining in the system indicates that the monomer cannot get into the crystalline phase, where the rest of the peroxy radicals are located. By partial destruction of the crystallinity, however, these radicals are progressively released and therefore the reaction goes on slowly.

Repeating the measurement at different temperatures the activation energy of the initiating radical decay can be determined. For irradiation with a total dose of 30 Mrad the activation energy is found to be 10.5 ± 0.7 kcal/mole.

From the conventional kinetic curves only the overall activation energy of the grafting can be determined. ESR provides an independent method for measuring the activation energy of the initiation, which is of much value in the study of the reaction mechanism.

5.4.5 CHANGE OF PHYSICAL STRUCTURE DURING SOLID STATE POLYMERIZATION

The main difficulty in studying the kinetics of solid state polymerization is that the reaction is highly influenced by the physical structure of the material and this structure is continuously changing during the course of the polymerization.[5.29]

Such effects have been studied extensively by Hardy et al.[5.30] in solid solution and in eutectic mixtures. The changes in the physical state have been measured by the X-ray diffraction method.

Some information can, however, be derived about the change of the physical structure by radio frequency spectroscopic methods as well. These methods seem to have advantages over the usual ones, because changes in the physical structure can be followed during the course of the reaction. Only a few preliminary measurements have been made in this field as yet. The main possibilities are the following.

1. As shown in Section 3.3, ESR spectra are strongly dependent on the crystallinity of the medium. In single crystals or in oriented structures the g-values and the hyperfine splittings are anisotropic. It is possible to follow the destruction of crystallinity during the course of the reaction by observing the change in the anisotropy of the ESR signals.

2. The widths of NMR lines in the crystalline state are much greater than those in the amorphous state (dipole–dipole interaction). A solid material containing both amorphous and crystalline phases usually exhibits a broad NMR line with a superimposed narrow one. The ratio of the areas under these lines is proportional to the amorphous/crystalline ratio.

3. The change in the physical structure of the material during solid state polymerization should be reflected in the dielectric spectra or in the d.c. conductivities or photo-conductivities of the samples.[5.31]

As an example of how the anisotropy of ESR signals is changed during the course of solid state polymerization, ESR spectra of N-vinylsuccinimide single crystals are shown in Fig. 5.26.[5.32]

In spectrum a the highly anisotropic hyperfine splitting of the monomer radicals is shown. The monomer has been irradiated at -78 °C with a ^{60}Co source with a total dose of 10 Mrad. The spectrum has been measured at -130 °C. Upon rotation of the crystal the spectrum is appreciably changed.

Spectrum (b) corresponds to the same sample measured at -5 °C. This sample has already polymerized to a conversion of about 5%. The signal is not sensitive to the rotation of the crystal any more, showing that the crystalline structure has already been destroyed by the growing polymer chains. At intermediate temperatures the gradual change of spectrum a into b can be followed together with the corresponding polymer conversion. The isotropic 3-line spectrum of Fig. 5.26 is attributed to the growing polymeric radicals.

The isotropy of the splitting indicated that the environment of these radicals is completely amorphous. At lower conversion a mixed signal of polymeric and monomeric radicals is obtained. The anisotropy is most likely due to monomer radicals located in the crystalline phase. The system has not yet been completely analyzed.

Another preliminary experiment[5.33] has shown that the electrical conductivity changes occurring by irradiating the sample with X-rays, referred to as radiation-induced conductivities, show a definite decrease with increasing polymerization. The sample is again N-vinylsuccinimide. The solid monomer is subjected to X-ray pulses and the current response under a constant voltage is recorded.

The amplitude of the response pulse is found to be only slightly dependent on the temperature. Since the charge carriers in the solid are

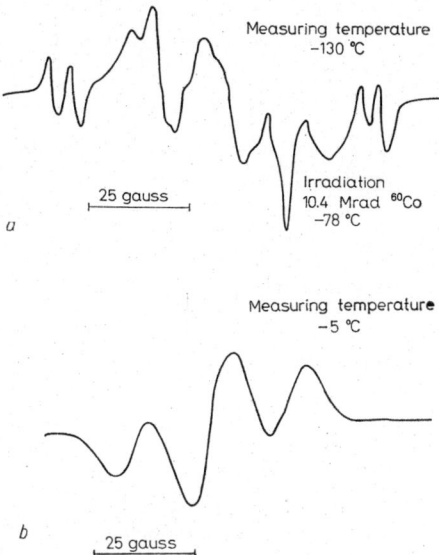

Fig. 5.26 The change in the anisotropy of the ESR hyperfine splitting during solid state polymerization of N-vinylsuccinimide. Courtesy of Hardy, Nagy and Gécs[5.32]

generated by the X-rays, the current amplitudes are mainly correlated with the mobility of the carriers, which is strongly dependent on the physical structure.

In Fig. 5.27 the amplitudes of the current response pulses are shown as a function of the polymerization time. This experiment is done in the following way: the X-rays are switched on and the change in the current is recorded. X-irradiation is kept on for a few minutes, and then it is switched off. The change in the current is observed again. This procedure is repeated at different total doses. The corresponding conversion curve is measured on identical separate samples which have received the same irradiation doses.

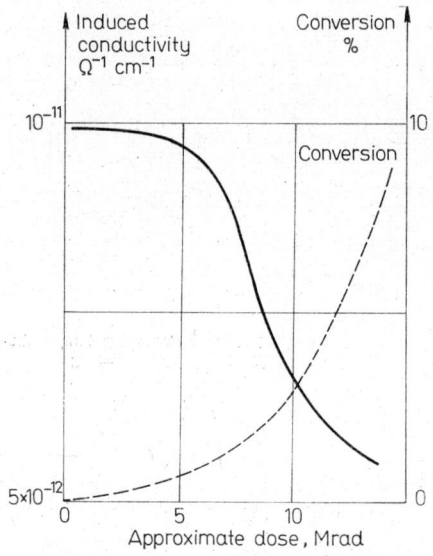

Fig. 5.27 The change of the radiation induced electrical conductivity of N-vinylsuccinimide during polymerization in the solid state[5.33]

More recently similar investigations have been made by using X-rays chopped by a rotating disc.[5.34] The current-response pulses are simply amplified by a broad-band amplifier and displayed on a cathode ray oscilloscope. These fast-response pulses exhibit a similar change during the course of polymerization to that shown in Fig. 5.27.

5.4.6 DESTRUCTION OF POLYMERS

In most polymers reactions are initiated upon prolonged standing in natural light in air or upon processing. These reactions usually result in a serious reduction of the average molecular weight of the polymer. The concentration of the radicals in the natural process is too low to be observed by ESR. However, in extreme conditions the natural degradation process can be accelerated and the radicals detected. The following main possibilities are considered:

1. Illumination of the polymers with high intensity ultraviolet light and studying the reactions of the radicals formed. It is also possible to study the effects of chemical protectors on the radical formation.
2. Study of the reaction of the radicals formed by ultraviolet illumination with gases or vapours.
3. Study the radical formation by heat-treatment of the polymers.
4. Study of radicals formed by mechanical destruction of the polymers at low temperatures.

Examples of this are given in Section 3.3. It is also possible to heat the polymers gradually and study the reaction of the trapped radicals. In some cases reactions can be observed at very low temperatures. Although these experiments differ from the actual conditions where degradation reactions are interesting, the conclusions drawn from them are very valuable. Some examples are given below.

ESR during heat treatment of polymers. Storing polymers at elevated temperatures results in a gradual transformation of their structure. During the course of this process ESR signals are observed. Upon heating polyvinyl chloride samples for example to 200 °C a sharp ESR signal appears[5.35] with increasing amplitude until a saturation value is reached. An illustrative curve is shown in Fig. 5.28. If the sample is kept *in vacuo* at elevated temperature it becomes deep yellow and loses weight gradually (HCl is evolved). The paramagnetic centre concentration is closely related to the loss of weight due to dehydrochlorination. In the presence of air the radical concentration is much higher and the time for reaching saturation is longer (dotted line in Fig. 5.28).

The nature of the paramagnetic centres is not known at present, although their connection with the degradation reaction is evident. The lines are sharp (4.2 gauss) with g-factors very close to the free spin value. The concentrations are rather high (10^{17} to 10^{18} spins/ml). HCl evolution ceases as the spin concentration (*in vacuo*) reaches saturation.

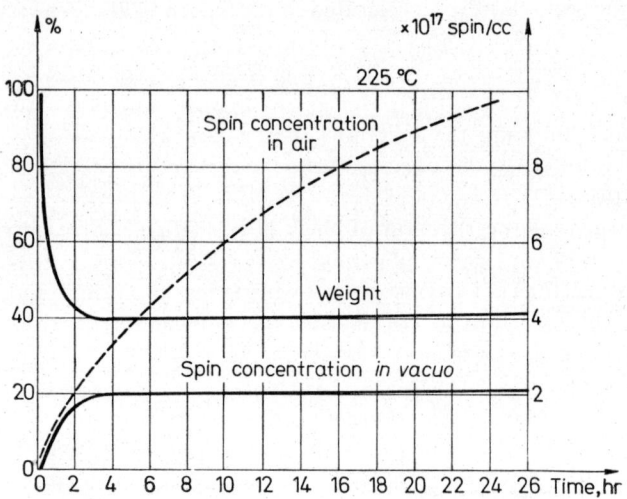

Fig. 5.28 The appearance of radicals during thermal treatment of polyvinyl chloride. After Ohnishi et al.[5.36]

Similar ESR signals have been observed in other pyrolyzed polymers as in the example given in Section 5.5; pyrolyzed polyacrylonitrile.

Radiation oxidation of polymers. Another important problem in studying degradation processes is how free radicals trapped in polymers react with atmospheric oxygen. An example of this has already been given in Section 5.1, where oxidation of fluorocarbon radicals in polytetrafluoroethylene is discussed.

Oxidation of irradiated polymers can be measured by observing the transformation of type $\dot{R}$ radicals obtained by irradiation *in vacuo* to peroxy radicals $R\dot{O}\dot{O}$. In some polymers this reaction can be observed at room temperature or higher, as done in polytetrafluoroethylene. In other cases peroxy-type radicals can only be trapped at low temperatures. In polyethylene irradiated at −185 °C *in vacuo*, radicals are gradually transformed into ROȮ type radicals upon admission of air at slightly elevated temperatures (−113 °C). An example of this transformation is shown in Fig. 5.29 after Ohnishi, Sugimoto and

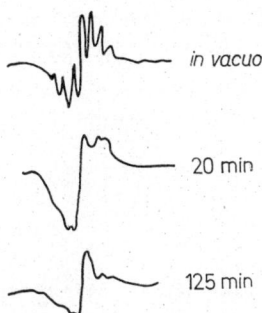

Fig. 5.29 The change in the ESR spectrum of irradiated polyethylene during the course of oxidation. After Ohnishi et al.[5.36]

Nitta.[5.36] The initial radical obtained by irradiation *in vacuo* is the following

$$\dot{R} = \sim \underset{H}{\overset{H}{C}} - \underset{H}{\overset{\cdot}{C}} - \underset{H}{\overset{}{C}} = \underset{H}{\overset{}{C}} - \underset{H}{\overset{H}{C}} \sim$$

This radical exhibits a fairly well resolved structure as shown in Fig. 5.29. The hyperfine splitting is due to the interaction of the radical electron with the 3 allylic and 4 methylene protons.

The gradual change of the spectrum into a single asymmetric line is due to the formation of peroxy radicals. Two types are probably formed

$$\sim \underset{H}{\overset{H}{C}} - \underset{OH}{\overset{H}{C}} - \underset{O}{\overset{}{C}} - \underset{O-\dot{O}}{\overset{H}{C}} - \underset{H}{\overset{H}{C}} \sim$$

and

$$\begin{array}{ccc} \text{H} & \text{H} & \text{H} \\ | & | & | \\ \text{C} - \text{C} - \!\!\!\!\!- \text{C}\sim \\ \| & | & | \\ \text{O} & \text{O}-\dot{\text{O}} & \text{H} \end{array}$$

Such radicals exhibit asymmetric ESR spectra with g-factors different from the free electron value, because these types of radicals are rather strongly localized on the oxygen atom. The probable scheme of the oxidation is the following[5,36]

$$\dot{\text{R}} \xrightarrow{\text{O}_2} \underbrace{\sim\overset{\text{H}}{\underset{\text{H}}{\text{C}}}-\overset{\text{O}-\dot{\text{O}}}{\underset{\text{H}}{\text{C}}}-\text{C}=\text{C}-\overset{\text{H}}{\underset{\text{H}}{\text{C}}}\sim}_{(\text{RO}\dot{\text{O}})_1}$$

$$(\text{RO}\dot{\text{O}})_1 \longrightarrow \underbrace{\sim\overset{\text{H}}{\underset{\text{H}}{\text{C}}}-\overset{\text{OOH}}{\underset{\text{H}}{\text{C}}}-\dot{\text{C}}=\text{C}-\overset{\text{H}}{\underset{\text{H}}{\text{C}}}\sim}_{(\dot{\text{R}}\text{OH})}$$

$$\dot{\text{R}}\text{OH} \longrightarrow \underbrace{\sim\overset{\text{H}}{\underset{\text{H}}{\text{C}}}-\overset{\text{H}}{\underset{\text{OH}}{\text{C}}}-\overset{}{\underset{\text{O}}{\text{C}}}-\dot{\text{C}}-\overset{\text{H}}{\underset{\text{H}}{\text{C}}}\sim}_{(\dot{\text{R}}\text{OOH})}$$

$$\dot{\text{R}}\text{OOH} \xrightarrow{\text{O}_2} \text{RO}\dot{\text{O}}$$

$$\text{RO}\dot{\text{O}} \longrightarrow \text{stable peroxides}$$

At intermediate cases when $\dot{\text{R}}$ and $\text{RO}\dot{\text{O}}$ radicals are present in the system the saturation method can be used for separating them. The $\text{RO}\dot{\text{O}}$ radicals have much shorter spin–lattice relaxation times than

R radicals. Increasing the microwave power of the ESR set-up would thus saturate signals of Ṙ radicals at much lower power level than signals of ROȮ radicals. A typical saturation curve is shown in Fig. 5.30 for irradiated polypropylene. The signal amplitudes are plotted

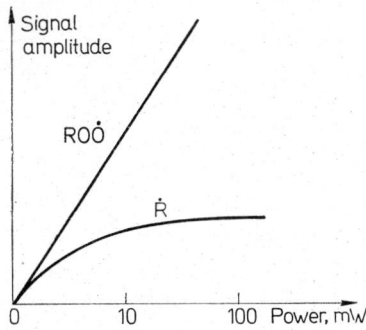

Fig. 5.30 Separation of peroxy radicals from hydrocarbon radicals by saturation. Polypropylene

against microwave power for Ṙ radicals (*in vacuo*) and for ROȮ radicals obtained after admission of air (Ṙ and ROȮ-type spectra of polypropylene are shown in Fig. 3.25). According to this plot the decay of radical concentration [Ṙ] is to be measured at the lowest possible microwave power level, while formation of ROȮ-type radicals is to be measured above 100 mW where the Ṙ spectrum is saturated.

Using this technique it is possible to measure the kinetics of transformation of Ṙ radicals into ROȮ. Radical decay curves of initial Ṙ-type radicals are shown in Fig. 5.31 for irradiated polyethylene. Radical concentrations [Ṙ] and [ROȮ] are plotted against reaction time. The spontaneous recombination of Ṙ radicals kept *in vacuo* at the same temperature is also shown.

Similar measurements have been made on other polymers such as polyvinyl chloride, polyvinyl alcohol, cellulose, poly(methyl methacrylate), polyethylene glycol, polyamides and on polyethylene terephthalate.

Using similar techniques the reaction of trapped radicals in irradiated polymers with other gases or vapours can also be studied.

Mechanical destruction of polymers. Examples of radical formation during mechanical destruction of polymers at low temperatures have

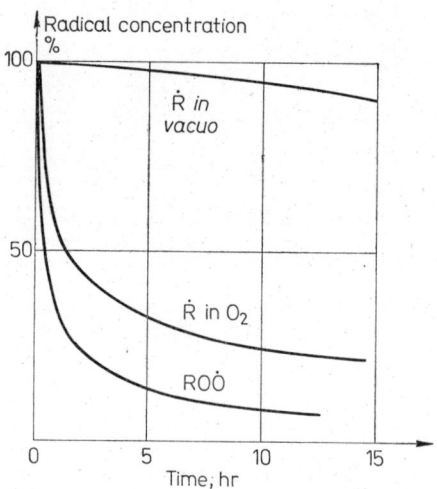

Fig. 5.31 The kinetics of radical oxidation in irradiated polyethylene. After Ohnishi et al.[5,36]

been given in Section 3.3. Although the resolution is not very good, in some cases the radical structure can be identified. Since these radicals are formed on the surface of the polymer fragments they are highly reactive. Their kinetics of recombination, reactions with gases and with solid or liquid substances can be studied rather easily. As an example of this the reaction of polystyrene radicals formed by mechanical destruction at low temperatures with methyl methacrylate monomer is discussed.[5.37] Mechanical destruction of polystyrene is made at liquid nitrogen temperature with the instrument shown in Fig. 3.30. The polystyrene fragments exhibit an ESR spectrum (shown in

Fig. 5.32a) attributed to polystyrene radicals (ṖS)

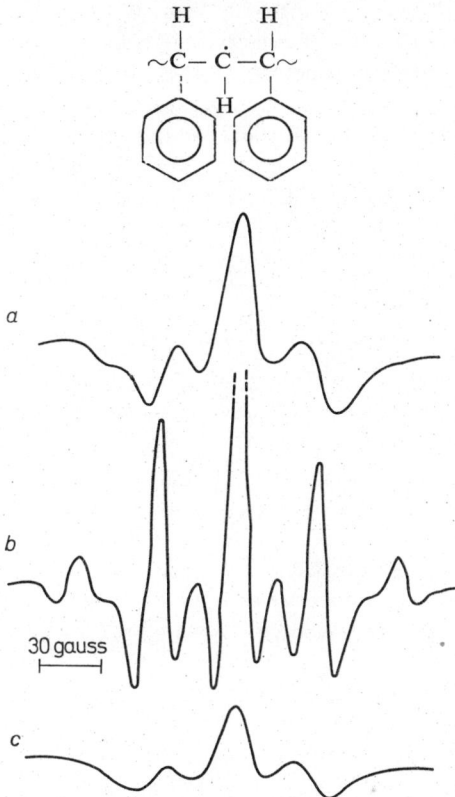

Fig. 5.32 The reaction of polystyrene radicals formed by mechanical destruction at 77 °K with methyl methacrylate.
a — Initial polystyrene radicals; b — Radicals obtained after admission of methyl methacrylate; c — Sample heated to +65 °C, F. Szőcs, private communication

The spectrum of Fig. 5.32 is the second derivative of the absorption lines. If deoxygenated methyl methacrylate monomer (MMA) is admitted to the sample the spectrum is transformed to that shown in Fig. 5.32b. This is the same spectrum as obtained by mechanical destruction of poly(methyl methacrylate). Similar spectra are observed in irradiated poly(methyl methacrylate). The next step is to heat the sample up to +65 °C. At this temperature the PMMA radical spectrum is gradually transformed back to the ṖS radical spectrum (Fig. 5.32c).

The reaction scheme is the following

$$PS \xrightarrow{\text{mechanical destruction}} \dot{P}S$$

$$\dot{P}S + MMA \rightarrow \dot{M}MA + PS$$

$$\dot{M}MA + PS \xrightarrow{\text{heating}} MMA + \dot{P}S$$

The observed migration of the radical from polystyrene to methyl methacrylate can be explained as follows.

1. MMA monomer reacts with the ṖS radicals present at the surface of the PS fragments to form a PS–MMA copolymer at the surface (grafting). The growing ṀMA radicals are trapped at the low temperature.

2. Upon heating the sample up to +65 °C ṀMA radicals are released and abstract hydrogen atoms from the PS molecules. In this way ṀMA radicals vanish and ṖS radicals appear again.

As shown by this example radicals formed by mechanical treatment of polymers may exhibit peculiar behaviour because they are located at the surface of the polymer in a relatively high concentration.

Chemomechanical formation of radicals. According to the experimental evidence obtained recently,[5.28] in some systems the change in the physical structure during polymerization may lead to formation of radicals by chain rupture. By grafting styrene on to irradiated polytetrafluoroethylene at high conversion the appearance of new fluorocarbon radicals has been observed. As discussed earlier in this section, grafting proceeds very slowly in the crystalline part of the polymer accompanied by a gradual destruction of the crystalline structure. At very high conversions the pressure of the growing polymer chains is high enough to result in ruptures of the fluorocarbon chains. The

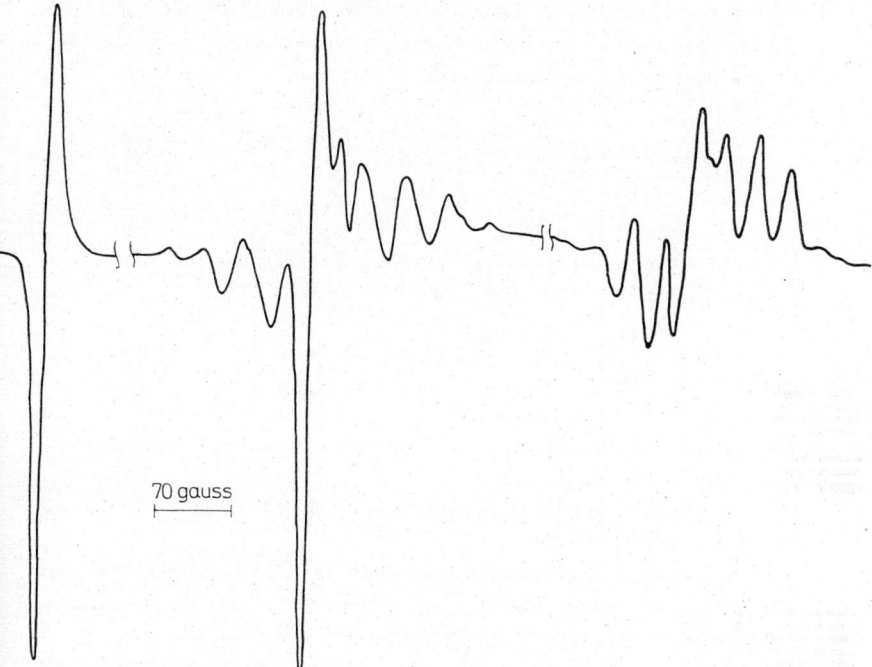

Fig. 5.33 Chemo-mechanical formation of radicals during the course of radiation grafting of styrene onto polytetrafluoroethylene

corresponding radical is

$$\sim\underset{F}{\overset{F}{\underset{|}{C}}}-\underset{F}{\overset{F}{\underset{|}{C}}}-\underset{F}{\overset{\cdot}{\underset{|}{C}}}-\underset{F}{\overset{F}{\underset{|}{C}}}-\underset{F}{\overset{F}{\underset{|}{C}}}\sim$$

A typical set of ESR spectra is shown in Fig 5.33. The first single line corresponds to ROO type radicals formed in polytetrafluoroethylene during irradiation in air. The second spectrum is measured at 80 °C after the polymer has completely grafted through (see Fig. 5.25 for the grafting curves). The third spectrum corresponds to the same sample after prolonged grafting. In this case the mechanical effects can be seen directly; the surface of the sample becomes uneven — it has swelled enormously and partially disintegrated.

5.5 CATALYSIS

The methods of electron spin resonance and nuclear magnetic resonance have recently been applied to study catalyst systems. The following main problems can be investigated by these methods.

1. *Adsorption.* Elementary processes on the surface of solids can be studied by measuring dielectric and nuclear magnetic relaxation times. The most straightforward method is the spin–echo technique described in Section 2.3.

2. *Nature of active centres of catalyst.* Most of the catalytic active centres exhibit electron paramagnetic and/or semiconductive properties. Recently attempts have been made to correlate ESR and electric conductivity data of catalysts with their activities. The electron paramagnetism can be studied by electron spin resonance. Results have been obtained for complex catalyst systems containing such paramagnetic ions as Ti^{+++}, Cu^{++}, Cr^{+++} and also for organic semiconductor catalysts.

3. *Study of the reactions leading to formation of active catalysts.* The most successful technique for doing this is ESR. Using the continuous flow technique described in Section 2.2 the fate of paramagnetic centres (ions, radicals, trapped electrons) can be measured during the course of reactions. Results have been obtained for various Ziegler–Natta catalyst systems and some ion-radical catalysts.

4. *Study of the catalytic action itself.* Although there is no direct way of investigating the mechanism of catalysis, information derived from ESR and electric conductivity measurements is of much help. It is possible to study such problems as energy and charge transfer, effect of additives (catalyst poisons). Only a few preliminary results have been obtained in this field as yet.

The capability of radio frequency spectroscopic methods is illustrated by the examples given below.

5.5.1 ADSORPTION

The possibility of distinguishing between molecules in solid, liquid or gaseous and sorbed phase lies in their different motional freedom. Atoms, ions or molecules in the solid phase may have in most cases only rotational degrees of freedom although rotation is hindered. In the liquid state translation is possible, rotation is relatively free. In the sorbed phase the molecular layer at the surface of the solid adsorbent is rather strongly bound. In the subsequent layers the molecules are more and more mobile. The situation is illustrated schematically in Fig. 5.34, the case of adsorption of liquids at a solid surface. The molecules of the liquid are shown as ellipsoids in order to indicate their orientation.

The difference in motional freedom of the sorbed molecules from those located in the solid or liquid phase is reflected in their molecular electric dipole relaxation time and also in their nuclear magnetic dipole–dipole relaxation time. The effect on the dielectric (dipole) relaxation is illustrated in Fig. 5.35 for water sorbed on silica gel. The dielectric loss (imaginary part ε'' of the complex permittivity) is plotted against temperature at microwave frequencies.[5.38] As shown, the dielectric loss of ice is very small. At the melting temperature, 0 °C, it is increased sharply as a result of the increase of the mobility of the water molecules. In the sorbed phase a different curve is obtained as a result of the difference in dielectric relaxation times. Deviation from the $\varepsilon''(T)$ plot of the pure liquid indicates the presence of the sorbed phase. The relative amount of molecules sorbed can be estimated.

The method of measuring nuclear magnetic relaxation times is much more sensitive. The spin–spin relaxation times are very short in the solid state and long in the liquid state. In the sorbed phase intermediate values are found. It is possible to distinguish between the different

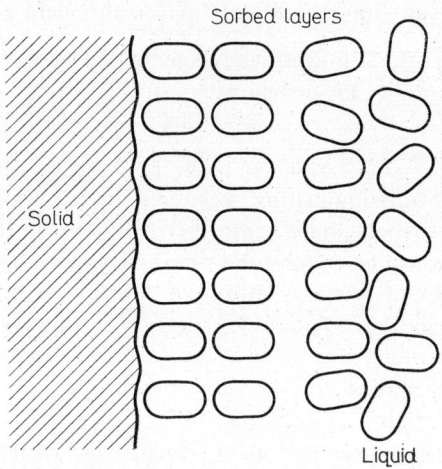

Fig. 5.34 Schematic representation of sorbed monomolecular layers on a solid surface

molecular layers sorbed on the solid surface by measuring the relaxation times as a function of the liquid content. An example is shown in Fig. 5.36 for water sorbed in silica gel. The spin–lattice (T_1) and spin–spin (T_2) relaxation times, measured by the spin–echo method, are plotted against relative water content.[5.39]

The spin–spin relaxation time, T_2, is short at low water content, because all the molecules are in the sorbed phase and thus their mobility is reduced. Water content is expressed as a number of monomolecular layers at the surface of the silica gel. The sharp rise of T_2 between layers 2 and 3 shows that the sorbed layers are saturated with water and the excess is becoming more and more mobile. T_2 is mostly sensitive to rotational degrees of freedom, since the dipole–dipole interaction depends on the relative orientations of the molecules (see the discussion in Chapter 2, and Equation 2.9). Thus the molecules sorbed in the second and third monomolecular layers are fixed strongly, they cannot even rotate; for those in the layers further away rotational freedom is permitted.

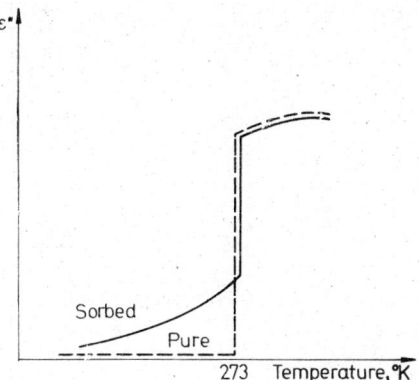

Fig. 5.35 The dielectric loss of water sorbed on silica gel as a function of temperature

The spin–lattice relaxation time T_1 behaves in the opposite way. It is high at small water content, i.e. in the strongly sorbed phase, and is decreased very sharply before the first monomolecular layer is filled up. As shown in Chapter 2 spin–lattice relaxation time is connected with the lifetime of the nuclear spins. The longer the lifetime in a given nuclear energy state is, the longer the spin–lattice relaxation time T_1 will be. The sharp decrease of the spin–lattice relaxation time with increasing water content can be interpreted by considering the exchange between the sorbed and unsorbed molecules. At very low water content this exchange is small. As the first layer gets more and more populated the probability of exchange gets higher and higher, resulting in the observed decrease of the spin–lattice relaxation time. Thus T_1 in sorbed systems is connected with the translational motion of the molecules, and T_2 with their rotational motions. The fact that T_1 is decreased sharply before the first monomolecular layer is filled up indicates that translational motion of the molecules is much less affected by adsorption than is rotation.

As shown by this example the spin–echo method is a very powerful tool for investigating adsorption. The method is very sensitive, since

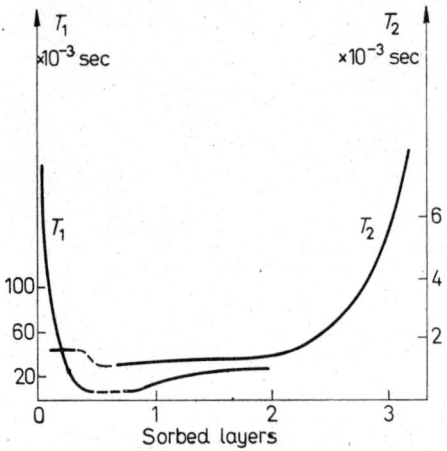

Fig. 5.36 Proton relaxation times of water sorbed on silica gel measured by the spin–echo technique as a function of the water content

differences in the different sorbed monomolecular layers can be detected. It is also sensitive to the surface properties of the sorbent.

On the basis of the difference in motional freedom of the sorbed molecules from those in the liquid or gaseous phase, it is possible to study adsorption of radicals on solid surfaces. The spin–spin interaction among unpaired electrons is just as sensitive to rotational freedom as the nuclear spin–spin interaction. Since the spin–echo technique is rather difficult in ESR (although it is possible) the variation of the relaxation times is in most cases followed by measuring the variation of the line widths. As a result of the dipole–dipole broadening ESR lines are in most cases broader in the sorbed than in the liquid phase.

Thus in heterogeneous catalyst systems it is possible to decide whether the radical products measured by ESR are in the liquid, solid or sorbed phase.

In solid catalysts containing paramagnetic ions there is an indirect way of investigating adsorption phenomena. The ESR spectrum, of the paramagnetic ions may be sensitive to the local magnetic fields of the sorbed molecule. Therefore adsorption can be indicated by measuring the variation of the line widths. Kazanskii[5.40] has shown, for example, that in chromium alum catalysts the Cr^{5+} ions at the surface react with oxygen to produce Cr^{6+} ions:

$$Cr^{5+} \underset{\diagdown O_2 \diagup}{----} Cr^{5+} \longrightarrow Cr^{6+} \underset{\diagdown O-O \diagup}{} Cr^{6+}$$

This process can be followed by measuring the variation of the ESR line widths upon admission of oxygen.

5.5.2 INVESTIGATION OF CHROMIUM OXIDE CATALYSTS BY ESR

As an example of ESR investigation of heterogeneous catalyst systems $Al_2O_3 - CrO_3$ and $SiO_3 - CrO_3$ are discussed in some detail. Chromium trioxide is not paramagnetic. Other chromium compounds such as $\alpha\text{-}Cr_2O_3$ and $Cr(OH)_3$ are paramagnetic, exhibiting anisotropic ESR lines about 500 gauss and 130 gauss wide, respectively. During the course of the formation of active catalysts the following main types of paramagnetic centres are formed[5.41]

1. δ-*centres*. Isolated Cr^{3+} ions exhibiting ESR spectra similar to that of ruby. The line is located at 1500 gauss in the X-band.

2. β_N-*centres*. Solid solution of Cr_2O_3 in Al_2O_3. The line widths are between 200 and 750 gauss, the g-value is 1.975.

3. β_W-*centres*. Amorphous or polycrystalline Cr_2O_3 exhibiting a very wide (1300 gauss) line at $g = 1.98$.

4. γ-*centres*. Cr^{5+} ions at the surface of the catalysts. The corresponding line widths are 50 to 140 gauss depending on the concentration of the Cr^{5+} ions.

5. *Trapped electrons* in Cr_2O_3 exhibiting a sharp line (5 gauss) at $g = 2.001$.

Although the interpretation of these lines is subject to discussion, the fact is that such types of lines are observed in the active catalyst systems.

The next step is to find if there is any correlation between the paramagnetic centres observed and the activity of the catalysts. Chromium oxide catalysts are used in polymerization of ethylene and for dehydrogenation of cyclohexane. The catalysts are prepared by saturating the base materials Al_2O_3 and SiO_3 with chromic acid and are activated by thermal treatment. The ESR signals are changed during the course of the activation.

The comparison between ethylene polymerization and ESR signals of the catalysts leads to the following conclusions.[5,41]

Catalysts exhibiting γ-type spectra are found to be not necessarily active in ethylene polymerization. Depending on the CrO_3 content two types of γ-signals can be observed

1. γ_D-signals (anisotropic): $(\Delta H)_0 = 127-137$ gauss

$$g_{max} = 1.987-1.991$$
$$g_\perp = 1.996-1.999$$
$$g_{||} = 1.918-1.932$$

These signals are attributed to Cr^{5+} ions dispersed homogeneously at the surface of the base material. Catalysts exhibiting γ_D-type ESR signals are found to be inactive in ethylene polymerization.

2. γ_M-signals: $(\Delta H)_0 = 55$ gauss
$$g_{max} = 1.970-1.974$$

These catalysts contain more than 3% CrO_3. The signals are attributed to mono-layers of Cr^{5+} ions formed on the surface of the base material. Catalysts exhibiting γ_M-type signals are usually found to be active, but this is only a necessary condition for the activity. More detailed analysis has shown that catalysts containing Cr^{6+} ions are most active. During the course of polymerization the Cr^{6+} ions at the surface of the catalyst are being reduced to Cr^{5+} with a corresponding increase of the γ_M-type ESR signal amplitude. Experimentally an increase of a factor 3 has been found.

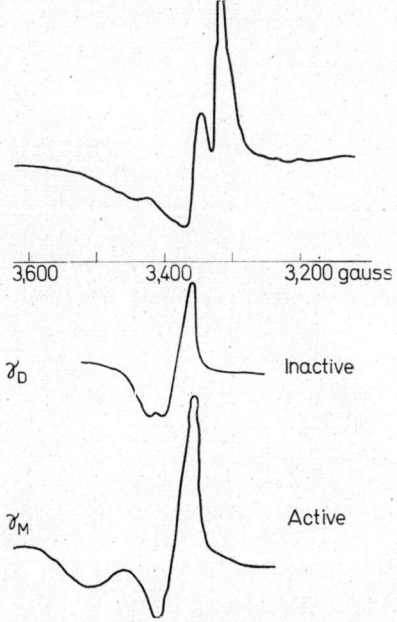

Fig. 5.37 Main types of ESR lines observed in chromium aluminium catalysts. After Benbenek[5.41]

γ_D- and γ_M-type centres are usually both present in the sample at the same time. The relative intensities can only be determined by analyzing the observed line shapes. A typical spectral line is shown in Fig. 5.37.[5.41] The first spectrum is a mixed $\gamma_M + \gamma_D$-type signal. The second one is a γ_D-type corresponding to an inactive catalyst. The third spectrum is a γ_M-signal constructed by subtraction of γ_D from $\gamma_M + \gamma_D$. This signal is not observed separately.

The conclusion is that the activity is connected with Cr^{6+} ions at the surface which cannot be observed directly by ESR. During the course of polymerization they are reduced to Cr^{5+} ions which can be observed as sharp (50 gauss) γ_M signals. It seems to be important that the chromium ions should form a mono-layer on the surface of the catalyst; if their concentration is too low for that, the system will be inactive.

5.5.3 ESR STUDY OF ZIEGLER–NATTA CATALYST SYSTEMS

Ziegler–Natta catalysts used in polymerization of olefines at atmospheric pressure are formed by reaction of titanium salts with aluminium alkyls. The exact reaction scheme is not known.

$TiCl_4$ is not paramagnetic while $TiCl_3$ is. However, no ESR signals have been obtained from $TiCl_3$ at room temperature probably because of dimer formation. On the other hand, it has been shown that the reduction of titanium ions does not involve radical formation. Despite this, formation of paramagnetic centres in the catalysts is observed by ESR. By using the continuous flow technique Adema[5.42] succeeded in detecting intermediate paramagnetic centres during the course of the following reactions

$$TiCl_4 + Al(C_2H_5)_2Cl$$
$$TiCl_4 + [Al(C_2H_5)Cl_2]_2$$
$$(C_5H_5)_2TiCl_2 + Al(CH_3)_3$$
$$(C_5H_5)_2TiCl_2 + Al(CH_3)_2Cl$$
$$(C_2H_5)_2TiCH_3Cl + AlCH_3Cl_2$$
$$(C_5H_5)_2TiCl_2 + AlCH_3Cl_2$$
$$(C_5H_5)_2TiCH_3Cl + AlCl_3$$
$$TiCl_3CH_3 + [Al(C_2H_5)Cl_2]_2$$
$$TiCl_3 + [Al(C_2H_5)_2Cl_2]_2$$

$$C_5H_5 = \text{cyclopentadienyl}$$

The ESR spectra are poorly resolved in the heterogeneous systems and fairly well resolved in the homogeneous systems. The paramagnetic centres are connected with the Ti–Al complexes formed during the course of reduction, such as for example

(C$_5$H$_5$)$_2$TiCl$_2$Al(CH$_3$)$_2$ broad singlet
(C$_5$H$_5$)$_2$TiCl$_2$AlCH$_3$Cl   6 lines, poorly resolved
(C$_5$H$_5$)TiCl$_2$AlCl$_2$ 6 lines, well resolved

The structures of the complexes are not known for sure. From the ESR data the following general structure is probable for the complexes from Ti

```
    H₅C₅         Cl         R₁
        \       /  \       /
         Ti       Al
        /       \  /       \
    H₅C₅         Cl         R₂
```

The unpaired electron is located near the Ti and Al atoms, partially delocalized. Upon increasing the concentration an exchange narrowing of the ESR line is observed indicating delocalization. The estimated delocalization area is 50 Å^2. Substitution of C$_5$H$_5$ or R$_1$, R$_2$ groups does not affect the spectra appreciably.

In the last reaction (with TiCl$_3$) the probable structure of the complex is

```
            Cl    Cl
             \   /
              Al
             /
      Cl  Cl      Cl  Cl
       \  |        |  /
        Ti--------Ti
       /  \        /  \
      Cl   Cl    Cl
```

Although ESR data do not reveal the exact structure of the Ti–Al complexes, important information can be derived by measuring the kinetics of formation of these paramagnetic centres and the dependence of their concentrations on the initial [Ti] and [Al] concentrations and on other factors. The concentrations can be compared with the polarization rates obtained by using the corresponding catalyst. Typical curves are shown in Fig. 5.38 for catalysts containing various [Ti] [Al] concentration ratios of 0.5 and 1.5. The first diagram corresponds to the change in the concentration of the paramagnetic centres as a function of the reaction rate, the second to the ethylene polymerization rates.

As an example the kinetics of the reduction of $TiCl_3CH_3$ and $TiCl_4$ with $[Al(C_2H_5)Cl_2]_2$ are discussed in some detail on the basis of the investigations of Adema (1965).[5,43]

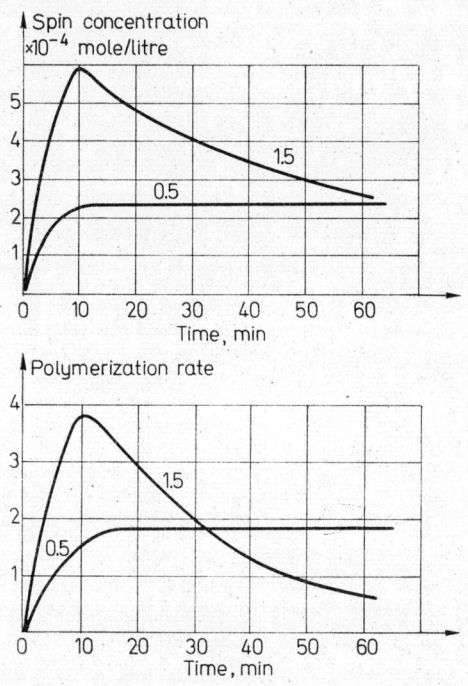

Fig. 5.38 Radical formation and polymerization rate for Ziegler–Natta catalysts. Figures in the curves indicate [Ti]/[Al] concentration ratios. After Adema[5,43]

Titanium compounds and aluminium alkyls are brought together in heptane solution in nitrogen atmosphere and the mixture is pumped through the cavity resonator of the ESR spectrometer. Water and oxygen traces have to be eliminated carefully. A symmetrical ESR signal is found with a g-value of 1.925. The line width is 21 gauss.

The concentration of the paramagnetic centres is increased with reaction time linearly until a saturation value is reached. A typical curve is shown in Fig. 5.39. The rate of formation of the centres, a function of the initial [Ti] and [Al] concentrations can be expressed

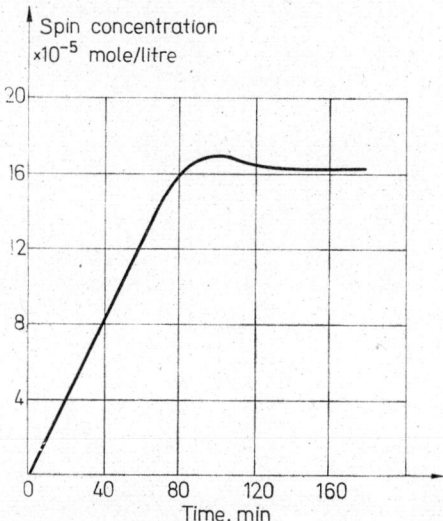

Fig. 5.39 The build-up of radicals in a Ziegler-Natta catalyst system. After Adema[5,43]

as follows

$$\frac{d[\dot{R}]}{dt} = k[Ti]_0^{3/2}[Al_2]_0^{1/2} \qquad \text{(for TiCl}_3\text{CH}_3\text{)}$$

where $[Ti]_0$ and $[Al_2]_0$ are the initial concentrations of $TiCl_3CH_3$ and $[Al(C_2H_5)Cl_2]_2$, respectively; $[\dot{R}]$ is the concentration of the paramagnetic centres; k is the rate constant of the reaction.

The saturation value of $[\dot{R}]$ depends linearly on the $[Al_2]$ concentration, where $[\dot{R}] = 8 \times 10^{-3}[Al_2]$.

In the case of the $TiCl_4-[Al(C_2H_5)Cl_2]_2$ system the formation of paramagnetic centres is described by the following equation

$$\frac{d[R]}{dt} = k[Ti]_0^{1/2}[Al_2]_0^{3/2} \qquad \text{(for } TiCl_4\text{)}$$

the rate constant is $k = 1.4 \times 10^{-2}$ litre mole $^{-1}$ min^{-1} at 20 °C.

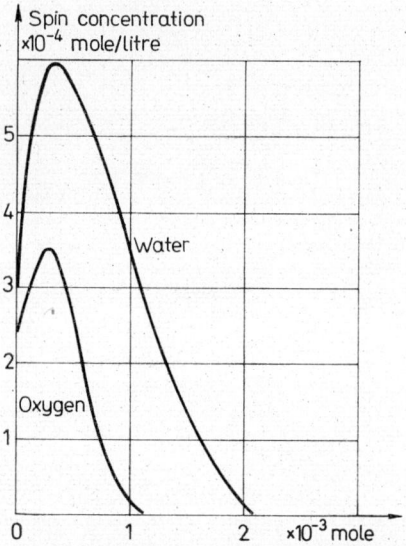

Fig. 5.40 The effect of water and oxygen on the radical concentration in Ziegler–Natta catalysts. After Adema[5,43]

The activation energy of the reaction can also be determined by repeating the kinetic measurements at various temperatures. For the $TiCl_4$ system it is found to be 12.8 ± 0.8 kcal/mole.

The catalytic action and also the formation of paramagnetic centres are strongly affected by small amounts of water or oxygen. The polymerization rates are appreciably increased by adding 5 to 10 vol.% water or oxygen to the system. The effect on the concentration of paramagnetic centres is shown in Fig. 5.40. The curves correspond to the system $TiCl_4-[Al(C_2H_5)_2Cl]_2$ with 100 mmole/litre $[TiCl_4]$ and 25 mmole/litre $[Al(C_2H_5)_2Cl]_2$ concentrations measured at 20 °C by the continuous flow technique. The paramagnetic centre concentration against water or oxygen content curves exhibit a maximum between 5 and 10 vol.% to [Al] and so do the polymerization rate curves. This effect has been extensively studied by Adema[5.43] by performing continuous polymerization and ESR measurements in two reactors: one with water or oxygen, the other without them. Samples were taken and pumped through the ESR system from both reactors at definite times. Thus the variation of paramagnetic centres in the doped and undoped systems could easily be followed.

5.5.4 ANIONIC CATALYSTS. DIMERIZATION OF 1,1-DIPHENYLETHYLENE

ESR technique has been used successfully to study reactions catalyzed by radical anions. A series of such investigations has been made by A. G. Evans and J. C. Evans.[5.44] An example will be discussed here: the dimerization of the 1,1-diphenylethylene radical ion.

1,1-Diphenylethylene is dissolved in cyclohexane in the presence of metallic sodium or potassium and is left standing for several hours, then filtered. The following radical ions are formed

Solutions containing 10^{-3} mole/litre diphenylethylene and potassium have a blue color which is not removed by filtering. The solution exhibits a rather well resolved ESR spectrum decreasing in time according to second-order kinetics. From the hyperfine splitting the following interactions are revealed

$$\text{Ph}_2\text{C}^{(-)}\text{—CH}_2 \quad \text{K}^{(+)}$$
(0.9 gauss, 8.13 gauss)

The second-order decay of these radicals is due to the following dimerization reaction

$$\text{Ph}_2\overset{(-)}{\text{C}}\text{—}\overset{\cdot}{\text{C}}\text{H}_2 \;\; \text{K}^{(+)} \;\; + \;\; \text{Ph}_2\overset{(-)}{\text{C}}\text{—}\overset{\cdot}{\text{C}}\text{H}_2 \;\; \text{K}^{(+)}$$

$$\longrightarrow \;\; \text{Ph}_2\overset{(-)}{\text{C}}\text{—CH}_2\text{—CH}_2\text{—}\overset{(-)}{\text{C}}\text{Ph}_2 \quad \text{K}^+ \;\; \text{K}^+$$

The rate constant calculated from the radical decay curve is 1.67×10^{-2} litre mole^{-1} sec^{-1} at 31 °C. Repeating the measurements at different temperatures the activation energy of the reaction could be obtained (6.5 ± 0.3 kcal/mole).

5.5.5 CATALYSIS BY ORGANIC SEMICONDUCTORS

Organic semiconductors have long been known as catalysts of biochemical reactions. The photo-oxidation of organic compounds irradiated by ultraviolet or visible light, for example, can be accelerated by addition of organic dyes. The semiconductive properties of dyes have been extensively studied by Terenin[5.45] and Meier.[5.46] By the method of pulsed photoconductivity similar to that described in Section 1.5 the sign of the charge carriers of most organic dyes has been determined. Semiconductors are usually classified into two groups according to the sign of the carriers. If the concentration of positive charge carriers is considerably higher than that of negative carriers, the material is

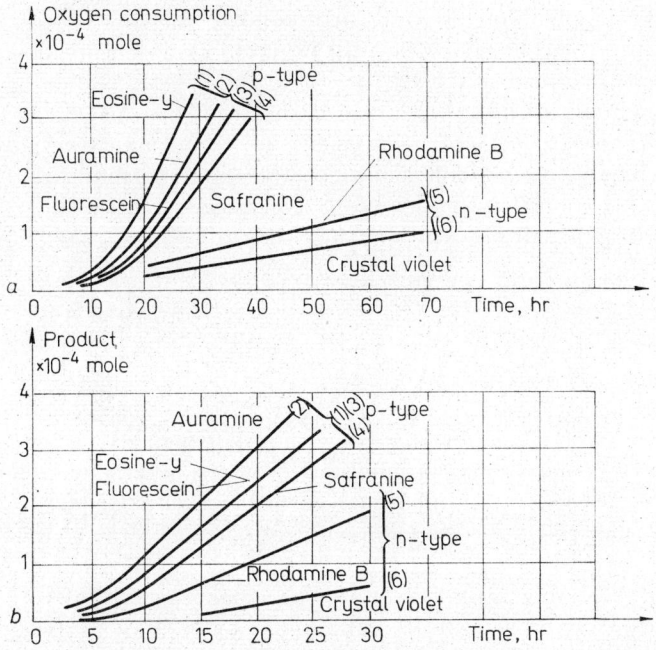

Fig. 5.41 The catalytic action of organic semiconductors. After Inoue, Hayashi and Imoto[5.47]

called a p-type semiconductor. If negative charge carriers are in excess it is called an n-type semiconductor.

Dye-catalyzed photo-oxidation of isopropyl alcohol. Recently Inoue, Hayashi and Imoto[5.47] have investigated dye-catalyzed photo-oxidation of isopropyl alcohol in the liquid phase. They found a correlation between the catalytic activity and the type of semiconductivity of the dyes. Some of their results are shown in Fig. 5.41. The oxygen consumption of the system is plotted against reaction time for different dye catalysts. The first group of curves, numbered 1 to 4, shows high rates of oxygen consumption. All dyes corresponding to these curves are identified as p-type conductors. The final product of the photo-oxidation in these cases is mainly acetone.

Curves 5 and 6 of low rate correspond to the catalysts of n-type conductors. The final product of the photo-oxidation in these cases is found to be pinacol.

The amount of product formed is shown in Fig. 5.41b.

Catalysis by polymeric semiconductors. Semiconductive polymers having conjugated double bonds exhibit catalytic action. The most thoroughly studied system is pyrolyzed polyacrylonitrile.

The structure of polyacrylonitrile is the following

$$\sim CH_2-\underset{\underset{C\equiv N}{|}}{\overset{\overset{H}{|}}{C}}-CH_2-\underset{\underset{C\equiv N}{|}}{\overset{\overset{H}{|}}{C}}-CH_2 \sim$$

The electrical conductivity of this polymer is rather low, 10^{-14} ohm^{-1} cm^{-1}. On keeping the polymer at elevated temperatures of 200 to 700 °C in argon or nitrogen atmosphere a cyclization is observed corresponding to an increase in the conductivity. Samples pyrolyzed above 300 °C exhibit electron spin resonance spectra, indicating the formation of paramagnetic centres.

The process of cyclization is the following

$$\sim CH_2-\underset{\underset{C\equiv N}{|}}{\overset{\overset{H}{|}}{C}}-CH_2-\underset{\underset{C\equiv N}{|}}{\overset{\overset{H}{|}}{C}} \sim \rightarrow$$

Further dehydrogenation results in the following cyclic structure:

The electrical conductivity is changed by eight orders of magnitude between pyrolysis temperatures of 400 °C and 700 °C. The activation energies calculated from the temperature dependence of the samples obtained at the given pyrolysis temperature are also changed.

The pyrolyzed samples are dark brown, at higher temperatures deep black. They exhibit fairly intense ESR signals corresponding to spin concentrations of 10^{19}–10^{20} spins/g, depending on the pyrolysis temperature.

The catalytic activity of the semiconductive polymers has been tested by Dawans et al.[5,48] for decomposition of N_2O. For n-type semiconductive catalysts the following scheme is proposed

$$N_2O + e^- \rightarrow N_2 + O^-$$

$$2O^- \rightleftarrows O_2 + 2e^-$$

$$N_2O + O^- \rightarrow N_2 + O_2 + e^-$$

where e^- is the electron taken up from the semiconductive material.

The catalytic activity of pyrolyzed polyacrylonitrile has also been tested by observing the decomposition of cyclohexanol. The activities are found to be correlated with the spin concentration, as shown in Fig. 5.42. The spin concentrations and the activities per unit surface of the catalysts are plotted against the pyrolysis temperature. The spin concentration exhibits a maximum at about 420 °C, the activity at about 380 °C.

Similar correlation has been found by polyisoprene containing benzoquinone. The initial structure is the following:

Upon pyrolysis, highly conjugated systems such as the following are formed

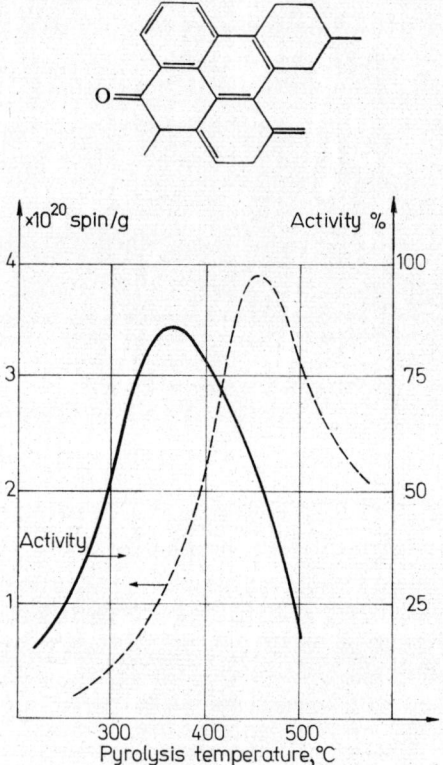

Fig. 5.42 Radical concentration and catalytic activity in pyrolyzed polyacrylonitrile. After Dawans[5,48]

As in the case of polyacrylonitrile there is a strong increase in the conductivity as the pyrolysis temperature is increased with a simultaneous appearance and increase in the spin concentration. The corresponding curves are shown in Fig. 5.43. Curve *a* corresponds to the variation of the electrical d.c. conductivity of the polymer as a func-

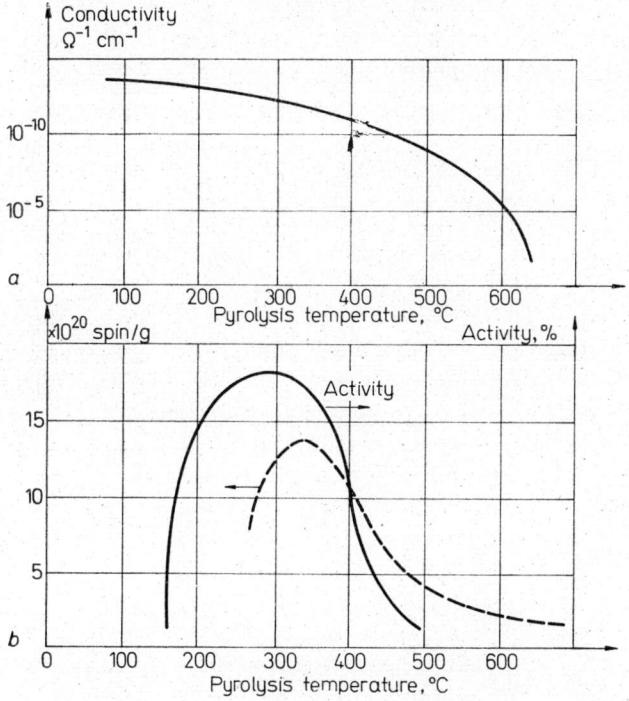

Fig. 5.43 The correlation of the electrical conductivity, radical concentration and catalytic activity of pyrolyzed polyisoprene. After Dawans[5,48]

tion of the pyrolysis temperature. Fig. 5.43b shows the correlation found between the concentration of paramagnetic centres formed during the course of pyrolysis and the catalytic activity of the sample.

5.6 BIOCHEMICAL REACTIONS

Radio frequency spectroscopy, especially ESR, has gained a considerable interest in biochemistry. Free radicals are known to play a very important role in photosynthesis, in enzyme reactions and in effects of high-energy irradiation on biological substances. In addition, NMR is used for analyzing biochemical reaction products. The investigation of electrical properties of biological systems has also gained much interest in connection with charge transfer and energy transfer phenomena.

5.6.1 PHOTOSYNTHESIS

Photosynthesis is the building of carbohydrates from carbon dioxide under the action of visible light. The whole process is not yet understood. ESR techniques have been applied to detect paramagnetic centres formed during the course of photosynthesis since 1954 when Commoner, Townsend and Pake[5.49] succeeded in detecting ESR signals in lyophilized leaves illuminated by visible light.

The ESR signals appearing during the course of photosynthesis usually have a poorly resolved structure and therefore it is rather difficult to identify the paramagnetic centres formed. The following main possibilities are usually considered.

   1. Electrons released by the illumination and trapped in the system.
   2. Photodissociation leading to formation of free radicals which recombine when illumination is off.
   3. Formation of paramagnetic excited triplet states by illumination.
   4. Secondary radicals formed during the course of the reaction initiated by the illumination.

According to the experimental facts it is improbable that the observed signals correspond to triplet states because they should be much broader than the observed ones. The observed line widths are in the order of 10 gauss with g-factors close to the free spin value. Direct photodisintegration is also rather improbable, because of the small energy of the light quanta. The most reasonable assumption about the origin of the paramagnetic centres is that electrons are liberated during the course of photosynthesis and trapped in the solid or highly viscous media. This assumption is supported by the fact that the build-up of paramagnetic centres is very rapid even at low temperatures.

An illustrative example is shown in Fig. 5.44. The ESR signal of spinach chloroplasts[5.50] is shown in Fig. 5.44a. The dotted line cor-

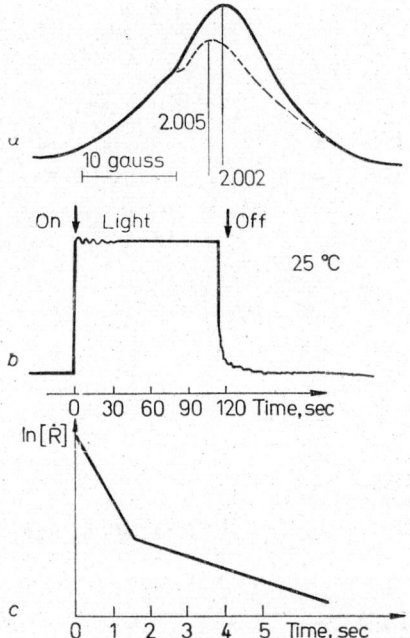

Fig. 5.44 Radical formation in illuminated spinach chloroplasts. After Tollin, Sogo and Calvin[5,50]

responds to the integrated line observed in the dark, the full line to that under illumination. As shown, the g-factor is shifted to a lower value upon illumination. The response of the paramagnetic centre formation to the light is shown in Fig. 5.44b. The magnetic field of the spectrometer is set to the maximum of the dark line and the response to illumination is recorded directly.

As shown, the response is very fast and reversible. The decay of the paramagnetic centre concentration is shown in Fig. 5.44c in semilog scale. Two distinct exponential portions are observed with decay rates depending on the temperature.

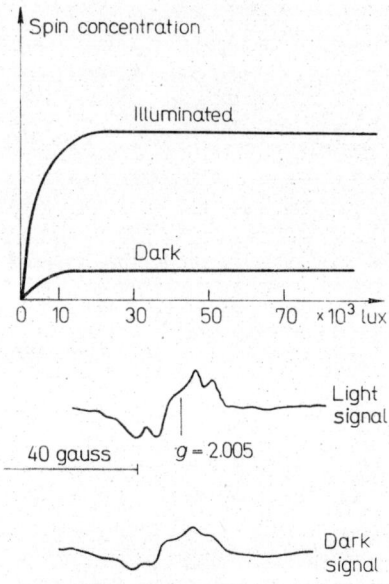

Fig. 5.45 Radical formation in illuminated living *Chlorella*. After Commoner et al.[5,51]

The ESR signals observed in chloroplasts depend on their preparation. The highest differences between the light and dark signals are observed in the unwashed undialysed samples. The signals exhibit poorly resolved structures. In carbon dioxide the rate of decay of the concentration of paramagnetic centres is considerably increased.

ESR in living chlorella. The improvement in the sensitivity of the ESR spectrometers has made it possible to observe signals during the course of photosynthesis in living cells. Several algae have been investigated. The technical difficulty is that the living species should be kept wet during the ESR measurement and thus only small amounts can be introduced into the cavity resonator without reducing its quality factor seriously.

The ESR signal intensities in illuminated living *Chlorella* show saturation on increasing the light intensity.[5.51] A characteristic curve is shown in Fig. 5.45. The signal intensities are plotted against light intensity. The corresponding dark signal intensities obtained after switching the light off are also shown. Upon decreasing the temperature the light signal is saturated at lower intensities and the dark signals obtained after the light has been switched off become dependent on the intensity of the illumination, i.e. the centres created by illumination get partially frozen. Both the light and dark signals of chlorella are decreased upon introducing carbon dioxide.

5.6.2 ENZYME REACTIONS

Enzyme-catalyzed reactions are extremely important in biology and in general organic chemistry as well. Although much effort is being made to get a fairly uniform picture of the mechanism of enzyme reactions, the problem remains unsolved for the time being.

Enzyme reactions can be studied by ESR either by observing the formation, decay and transformation of the radicals formed during the course of the reaction or by following the fate of the paramagnetic ions which the enzymes may contain. Since enzyme reactions always involve charge transfer, the study of the electrical conductivity of these systems is also of interest.

An enzyme-catalyzed reduction can be written schematically as follows

$$AH_2 \xrightarrow{\text{dehydrogenase}} \dot{A}H + \dot{H} \rightarrow A + 2\dot{H}$$

The semiquinone-like radical $\dot{A}H$ can usually be observed by ESR. Commoner, Townsend and Pake[5.49] succeeded in detecting ESR signals of 6 to 8 gauss width during the course of such reactions. The radical concentrations are in the order of 10^{-6} to 10^{-8} mole/litre. The g-factors are close to the free electron value indicating that the $\dot{A}H$-type radicals are strongly delocalized.

In favourable cases, well resolved hyperfine splitting is found in enzyme systems, making it possible to analyze the structure of the intermediate radicals. The semiquinone-type radicals of coenzyme Q_{10}, for example, exhibit a well resolved hyperfine structure in solution in ethanol.[5.52] The radical structure is the following

$$\underset{\substack{CH_3O \\ CH_3O}}{\overset{\overset{\cdot}{O}}{\bigcirc}}\underset{O^-}{\overset{CH_3}{\underset{CH_2-CH=\overset{|}{C}-CH_2-(CH_2CH=\overset{\overset{CH_3}{|}}{C}-CH_2)_9H}{}}}$$

The splitting is due to the CH_3 and CH_2 groups with isotropic splitting constants of 2.2 gauss and 1.1 gauss, respectively. The g-value is 2.00467 ± 0.00002.

Charge transfer during enzyme reactions. In oxidation–reduction processes catalyzed by enzyme systems, transfer of charge from the substrate to coenzymes is found to be of basic interest. The process is schematically the following

$$SH_2 + C \rightleftharpoons S + C-H + H^+ + e^-$$

where S is the substrate, C is the coenzyme, e^- is the electron liberated during the process.[5.53] The most important coenzymes of such reactions are diphospho-pyridine-nucleotide (DPN) and triphospho-pyridine nucleotide (TPN). During the course of the reaction ESR signals can be observed with g-values close to 2 and line widths in the order of 10 gauss. The classical example is

$$CH_3CH_2OH + DPN^+ \xrightarrow{enzyme} CH_3CH_2O + DPNH + H^+ + e^-$$

A few minutes after mixing the components an ESR signal appears with $g = 2.006$, having a line width of 13 gauss. The signal is increased to reach a steady state level for 10 minutes, then decays down to zero with a half-life of about 6 minutes. No signals can be observed in the separate components. The ESR signal is thus connected with the enzyme reaction. Similar ESR spectra have been observed in other systems such as DPNH-glyceraldehyde phosphate dehydrogenase, TPNH-glucose-6-phosphate dehydrogenase.[5.54] The origin of the ESR spectra in these systems is not known. Since the lines show no hyperfine splitting and the g-factors are close to the free spin value, it is possible that the signals are due to the liberated electrons which are present in the system in concentrations depending on the reaction scheme.

5.6.3 CHARGE TRANSFER COMPLEXES

As shown in Section 5.2 charge transfer reactions can be investigated by measuring the electrical conductivity of the system and also by ESR. According to Isenberg and Szent-Györgyi[5.55] in the system irboflavin (R)–tryptophan (T) the following charge transfer occurs

$$R + T \rightleftarrows R^- + T^+$$
$$R^- + T^+ \rightleftarrows [R^-T^+]$$
$$[R^-T^+] \rightleftarrows R + T$$

where $[R^-T^+]$ is the complex formed. Formation of paramagnetic centres in this system has been observed by M. Sidran.[5.56] The following results have been obtained.

1. Riboflavin-5-phosphate itself exhibits a single 23.4 gauss wide ESR line with $g = 2.0040$. No signal is obtained in tryptophan. The highest spin concentration observed was 10^{16} spin/gauss.

2. Mixtures of riboflavin–tryptophan show an ESR signal with an intensity depending on the preparation.

3. Illumination of R and R–T mixtures results in an increase in the concentration of the radicals. In samples stored in darkness after illumination the radical concentration is decreased as shown in Fig. 5.46.

In pure riboflavin, radicals generated by ultraviolet or visible light illumination decay with a half-life of about 5 days. In the $[R^-T^+]$ complex decay is accelerated. The complex is less sensitive to illumination with visible light.

4. The lifetime of the radicals produced by illumination is considerably decreased in the presence of water. Samples treated

with water and dried before, after, or during illumination exhibit different spin concentrations.

Since the ESR spectrum consists of a single line the structure of the radical cannot be determined. The line is Gaussian showing dipole-

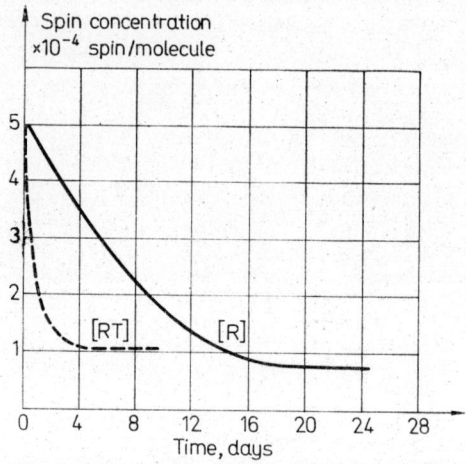

Fig. 5.46 The decay of radicals in illuminated riboflavin [R] and riboflavin-triptophan molecular complex [RT]. After Sidran[5,56]

dipole broadening characteristics. It is possible that it corresponds to riboflavin radicals formed during the course of the reaction and not to triplet states and to the $[R^-T^+]$ complex.

In solutions it is possible to resolve the hyperfine splitting of riboflavin. The spectrum is shown in Fig. 5.47. The probable structure of the radical is

The hyperfine structure is due to the interaction of the radical electron with the nitrogen nuclei. Similar spectra have been observed[5,57] in solutions of reduced flavo-mononucleotides, flavin-adenine-dinucleotides and lumiflavin (for the spectrum see Fig. 5.47b). The

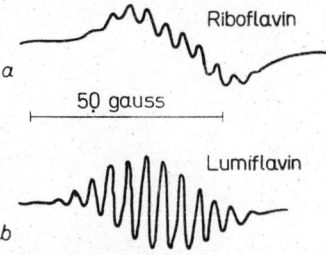

Fig. 5.47 The ESR spectrum of illuminated riboflavin and lumiflavin. After Ehrenberg[5,57]

structure of the lumiflavin radical is the following

Charge transfer complexes of serotonin-flavo-mononucleotide have been studied by Isenberg and Szent-Györgyi.[5,55] Besides the hyperfine splitting of flavo-mononucleotide, spectra attributed to serotonin radicals have been observed in acidic solutions.

The experiments were carried out as follows: a solution of 10^{-2} mole flavo-mononucleotide (FMN) was placed in the cavity resonator of the ESR set-up in a quartz capillary. No signal was observed. On adding a small amount of dry serotonin and stirring the sample with a thin platinum wire a rather well resolved signal was observed. Increasing the amount of dry serotonin gradually the ESR signal was first increased, then a saturation value was reached. The ESR spectrum in second derivative representation is shown in Fig. 5.48. It is thought to

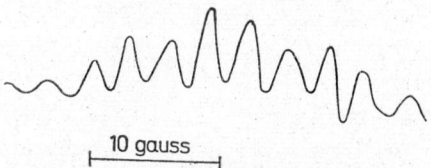

|—— 10 gauss ——|

Fig. 5.48 The second derivative ESR spectrum of the serotonin-flavonucleotide molecular complex. After Isenberg and Szent-Györgyi[5.55]

consist of the spectrum of riboflavin–semiquinone radicals and another radical attributed to the serotonin. The proposed reaction scheme is the following

$$S + R \rightleftarrows S^+ R^- \rightleftarrows \dot{S} + \dot{R}$$
$$\qquad\qquad \updownarrow \quad \updownarrow$$
$$\qquad P_2 \rightleftarrows P_1 \quad P_3$$

Thus the observed spectrum should be the superposition of $\dot{R}$ and $\dot{S}$ spectra. From the radical concentration and the amount of dry serotonin needed to reach maximum concentration it was concluded that one free radical corresponds to 250 FMN molecules.

Charge transfer complexes of polynuclear carcinogenic aromatic hydrocarbons also exhibit ESR spectra. Szent-Györgyi[5.58] has shown that ability to form charge transfer complexes is connected with carcinogenicity. The spin concentrations measured in some complexes with I_2 are given in Table 5.2. The components alone do not exhibit ESR signals; these only appear on mixing with I_2. As shown in the table, fairly high spin concentrations are formed.

Table 5.2
Electron spin concentration in some carcinogenic charge transfer complexes

Complexes with I_2	Colour in trinitrobenzene	Carcinogenicity	Spin concentration spin/gauss
Indole	yellow	0	1×10^{19}
2-aminofluorene	dark purple	++	1×10^{18}
p-dimethylazobenzene	purple	++	4×10^{18}
o-aminazotoluene	red-brown	++	1×10^{19}
9,10-dimethylbenzanthracene	red	+++	1×10^{19}
3,4-benzpyrene	red	+++	1×10^{18}
Methylcholanthrene	dark purple	+++	3×10^{18}

5.6.4 RADIOLYSIS IN BIOLOGICAL SYSTEMS

Radiolysis is far from being thoroughly understood even in simple organic materials. In complicated biological systems it is much more difficult to investigate. In this field the use of ESR is absolutely necessary. As shown in Section 5.4 ESR is a unique method of studying radicals formed during the course of radiolysis. In biological systems, unfortunately, hyperfine structures are in most cases not resolved. Thus there is no exact way to determine radical structures.

The extent of delocalization of the radical electron is usually determined by accurate measurement of the g-values. As shown in Chapter 1 the deviation from the free spin value is due to spin–orbit coupling. The extent of delocalization can thus be characterized by the deviation from the free spin value. Additional information can be obtained by analyzing line shapes.

Many materials of biological interest have been irradiated and the radicals detected by ESR. The method for investigating radicals trapped in living tissues, compounds of biological interest, is essentially the same as for those in ordinary simple organic compounds discussed in Section 5.3. It is possible to detect radicals *in vivo*, although the sensitivity is seriously reduced by the water content of the samples.

The most straightforward problem in investigating radicals trapped in irradiated biological materials is to study the saturation of the spin concentration [Ṙ] as a function of the irradiated dose. Typical saturation curves for two amino acids, glycine and alanine, are shown in Fig. 5.49. The corresponding spectra are also shown.[5.59] As can be seen there is a considerable difference in the saturation values. Radicals in

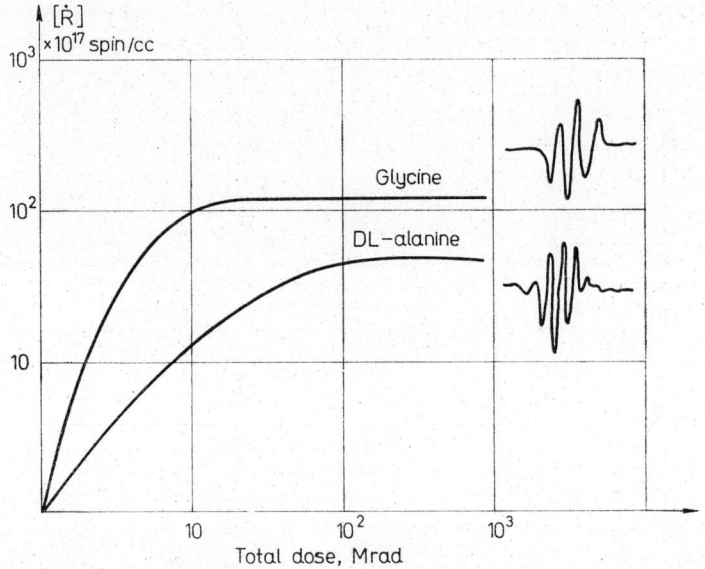

Fig. 5.49 The dependence of the radical concentration in glycine and DL-alanine on the irradiation dose[5,59]

glycine are saturated at about 10 Mrad while those in alanine at 200 Mrad.

Similarly it is interesting to study the dependence of the trapped radical concentration on the dose rate. An example is shown in Fig. 5.50 for irradiated deoxyribonucleic acid (DNA), after the measurements of Gordy et al.[5.60] As shown in Fig. 5.50 radicals in DNA exhibit a definite dose rate dependence. At 22 °C the samples irradiated at higher dose rate (5.6×10^6 rad/hr) exhibit larger radical concentrations, at the same total dose, than those irradiated at lower rates (0.027 Mrad/hr). By increasing the total dose, the difference gets larger. Such behaviour is generally not observed in ordinary organic compounds.

It is also possible to study the process of recombination of the trapped radicals by heating the sample gradually and recording the decay of the ESR signals. In DNA the decay of the trapped radicals also exhibits a dose rate dependence as shown in Fig. 5.50b. The

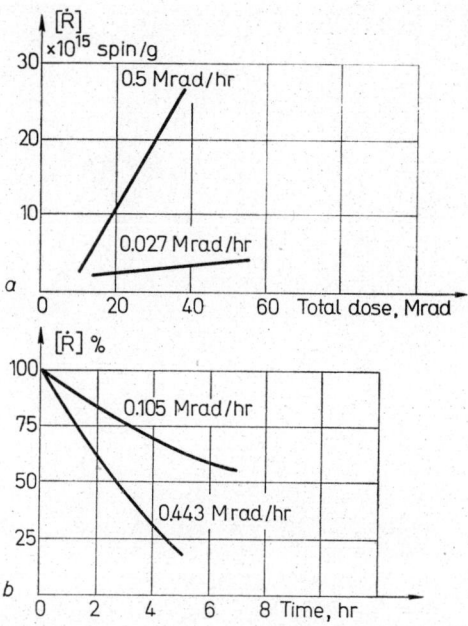

Fig. 5.50 Radical formation in irradiated desoxyribonucleic acid.
a — Spin concentration against irradiation dose; b — The decay of the radicals. After Gordy[5.60]

sample irradiated at higher dose rates (0.443 Mrad/hr), but receiving the same total dose, decays faster than that irradiated at the lower dose rate of 0.105 Mrad/hr.

For a detailed analysis of ESR in biological systems see the book by G. Schoffa (Bibliography).

REFERENCES

5.1 Schimmel, F. M. and Heineken, F. M., *Physica* **23**, 781 (1957).
5.2 Yamazaki, I. and Piette, L. H., *J. Amer. Chem. Soc.* **87**, 986 (1965).
5.3 Bamford, C. H. and Ward, J. C., *Trans. Faraday Soc.* **58**, 97 (1962).
5.4 Tsvetkov, Yu. D., Lebedev, Ya. S. and Voevodsky, V. V., *Vysokomol. Soed.* **1**, 1635 (1959).
5.5 Rexroad, H. N. and Gordy, W., *J. Chem. Phys.* **30**, 399 (1959).
5.6 Eastman, I. W., Engelsma, G. and Calvin, M., *J. Amer Chem. Soc.* **84**, 1339 (1962).
5.7 Atherton, N. M. and Goggins, A. E., *Trans. Faraday Soc.* **61**, 1399 (1965).
5.8 Bloom, M., Reeves, L. W. and Wells, E. J., *J. Chem. Phys.* **42**, 1615 (1965).
5.9 Carr, H. Y. and Purcell, E. M., *Phys. Rev.* **94**, 630 (1954).
5.10 Voevodsky, V. V., *Proc. Symp. on Radiation Chemistry, Tihany, Hungary 1962*, p. 110 (1964).
5.11 Fessenden, R. W. and Schuler, R. H., *J. Chem. Phys.* **39**, 2147 (1963).
5.12 Chachaty, C. and Schmidt, M. C., *J. Chim. phys.* 527 (1965).
5.13 Marx, R., *J. Chim. Phys.* 767 (1965).
5.14 Piette, L. H., NMR and EPR spectroscopy, *Varian Ass.*, 3rd Annual Workshop, Palo Alto, California, Pergamon Press, Oxford (1960) 241.
5.15 Smaller, B., Avery, E. C. and Remko, J. R., *J. Chem. Phys.* **43**, 922 (1965).
5.16 Yoshida, H. and Ranby, B., *Paper presented at the IUPAC Conference, Prague*, preprint No P71 (1965).
5.17 Fischer, H. and Giacometti, G., *Paper presented at the IUPAC Conference, Prague*, preprint No 329 (1965).
5.18 Hotta, K. and Anderson, R. S., 4th International Symposium on Free Radical Stabilization, *Nat. Bur. Stand.* Washington (1959).
5.19 Komatsu, T. and Sohma, J., *Paper presented at the IUPAC Conference, Prague*, preprint No 82 (1965).
5.20 Ingram, D. J. E., Symons M. C. R. and Townsend, M. G., *Trans. Faraday Soc.* **54**, 409 (1959).
5.21 Kourim, P. and Vacek, K., *Paper presented at the IUPAC Conference Prague*, preprint No 74 (1965) with the references cited therein.
5.22 Fischer, H., *Z. Naturforsch.* **18a**, 1142 (1963); *Polymer Letters* **2**, 529 (1964).
5.23 Marx, R., *J. Chim. phys.* 767 (1965).
5.24 Fischer, H., *Z. Naturforsch.* **18a**, 1142 (1963); *ibid* **19a**, 267 (1964).
5.25 Tüdős, F., Kende, I., Bereznich, T., Solodovnikov, S. P. and Voevodsky, V. V., *Magyar Kémiai Folyóirat* **69**, 371 (1963); Tüdős, F. and Smirnov, *Acta Chim. Hung.* **15**, 401 (1958).
5.26 di Stefano, P., Bonera, G. and Rigamonti, A., *Compt. Rend. Colloque Ampère* **10**, 375 (1961).

5.27 Bensasson, R. V., Bodard, M. and Marx, R., *Symposium on Radiation Chemistry, Tihany, Hungary 1962*, p. 82, (1964).
5.28 Dobó, J. and Hedvig, P., *Makromol. Chem.* **82,** 289 (1965).
5.29 Dobó, J. and Hedvig, P., *Paper presented at the IUPAC Conference,* Prague (1965). *J. Polymer Sci.* Part C, No 16, 2577 (1967).
Charlesby, A., *Rep. Progr. Phys.* **28,** 463 (1965).
5.30 Hardy, G., Varga, J., Nagy, G., Cser, F. and Erő, M., *Paper presented at the IUPAC Conference, Prague,* preprint No P304, (1965).
5.31 Sazhin, B. I., *Elektroprovodnost polymerov.* Izd. Khimia, Leningrad (1964).
5.32 Hardy, G., Nagy, G. and Gécs, M., unpublished results.
5.33 Hedvig, P., *Paper presented at the 20th IUPAC Conference Moscow* (1965).
5.34 Hedvig, P., *Proc. 2nd Tihany Symposium on Radiation Chemistry,* Akadémiai Kiadó 599 (1967).
5.35 Ouchi, I., *J. Polymer Sci.* A3, 2685 (1965).
5.36 Ohnishi, S., Sugimoto S. and Nitta, I., *J. Polymer Sci.* A1, 605 (1963).
5.37 Butyagin, P. Yu., *Doklady Akad. Nauk. S. S. S. R.* **140,** 145 (1961).
Butyagin, P. Yu., *Paper presented at the IUPAC Conference, Prague,* preprint No P236 (1965).
Szőcs, F., (private communication).
5.38 Kamiyoshi, J., *J. Chem. Phys.,* **21,** 1295 (1953).
5.39 Winkler, H., *Compt. Rend. Colloque Ampère* **10,** 219 (1961).
5.40 Kazanskii, V. B. and Pecherskaya, Yu. I., *Kinetika i Kataliz* **2,** 454 (1961); ibid **3,** 111 (1962).
5.41 Benbenek, S., Malinowski, S., Kosek, S. and Iwicka, D., *Przemysl Chemisztry,* **44,** 385 (1965), and the reference cited therein.
5.42 Adema, E. H., Bartelink, H. J. M. and Smidt, J., *Rec. Trav. chim.* **80,** 173 (1961), ibid **81,** 73 (1962); **81,** 226 (1962).
5.43 Adema, E. H., *Paper presented at the IUPAC Conference, Prague,* preprint No P310 (1965).
5.44 Evans, A. G. and Evans, J. C., *Trans. Faraday Soc.* **61,** 1202 (1965).
5.45 Terenin, A., *Proc. Chem. Soc.* 321 (1961).
5.46 Meier, H., *Z. wiss. Phot.* **53,** 1 (1958).
5.47 Inoue, H., Hayashi, S. and Imoto, E., *Bull. Chem. Soc. Japan* **37,** 326 (1964).
5.48 Dawans, F., Gallard, J., Teyssie, Ph. and Traynard, Ph., *Paper presented at the IUPAC Conference, Paris* (1964).
5.49 Commoner, B., Townsend, J. and Pake, G. E., *Nature* **174,** 689 (1954).
5.50 Tollin, G., Sogo, P. B. and Calvin, M., *Ann. N. Y. Acad. Sci.* **74,** 310 (1958).
5.51 Commoner, B. et al., *Proc. Nat. Acad. Sci.* **42,** 710 (1956); *Science* **126,** 57 (1957).
5.52 Blois, M. S. and Maling, I. E., *Biochem. Biophys. Research Comm.* **3,** 132 (1960).
5.53 Mahler, H. R. and Brand, L. *Free Radicals in Biological Systems* (ed. M. S. Blois), Academic Press, New York, 157 (1691).

5.54 Commoner, B., Lippincott, B. B. and Passonneau, J. V., *Proc. Nat. Acad. Sci.* **44**, 1099 (1958).
5.55 Isenberg, I. and Szent-Györgyi, A., *Proc. Nat. Acad. Sci. (Wash.)* **44**, 857 (1958); ibid. **46**, 1307 (1960).
5.56 Sidran, M., *Bull. Colloque Ampère* **9**, 283 (1960).
5.57 Ehrenberg, A., *Archiv. Chem.* **19**, 97 (1962).
5.58 Szent-Györgyi, A., Isenberg, I. and Baird, S. L., *Proc. Nat. Acad. Sci. (Wash.)* **46**, 1444 (1960); ibid. **46**, 1334 (1960).
5.59 Ebert, M. and Howard, A. (Eds.), *Radiation Effects in Chemistry and Biology*, 194 (1963). Uebersfeld, J., *Theses Fac. Sci.* Paris (1956); Uebersfeld, J. and Erb, E., *Compt. Rend. Acad. Sci.* Paris **242**, 478 (1956).
5.60 Gordy, W., *Radiation Research Suppl.* **1**, 491 (1959).

Appendix

Table 6.1
MAGNETIC RESONANCE PROPERTIES OF NUCLEI

Nucleus	Abundance %	Spin I $h/2$	Magnetic moment in nuclear magnetons	Relative sensitivity at constant field	Resonance frequency at 15 kgauss Mc/s
Hydrogen ^1_1H	99.9844	1/2	2.79270	1.000	63.8655
^2_1H	1.56×10^{-2}	1	0.85738	9.64×10^{-3}	9.8040
Lithium ^6_3Li	7.43	1	0.82191	8.51×10^{-3}	9.3975
^7_3Li	92.57	3/2	3.2560	0.294	24.8205
Beryllium ^9_4Be	100	3/2	−1.1774	1.39×10^{-2}	8.9745
Boron $^{10}_5B$	18.83	3	1.8006	1.99×10^{-2}	6.8625
$^{11}_5B$	81.17	3/2	2.6880	0.165	20.490
Carbon $^{13}_6C$	1.108	1/2	0.70216	1.59×10^{-2}	16.0575
Nitrogen $^{14}_7N$	99.635	1	0.40357	1.01×10^{-3}	4.6140
$^{15}_7N$	0.365	1/2	−0.28304	1.04×10^{-3}	6.4725
Oxygen $^{17}_8O$	3.7×10^{-2}	5/2	−1.8930	2.91×10^{-2}	8.6580
Fluorine $^{19}_9F$	100	1/2	2.6273	0.834	60.0825
Sodium $^{23}_{11}Na$	100	3/2	2.2161	9.27×10^{-2}	16.8930
Magnesium $^{25}_{12}Mg$	10.05	5/2	−0.85471	2.68×10^{-2}	3.909
Aluminium $^{27}_{13}Al$	100	5/2	3.6385	0.207	16.6410
Silicon $^{29}_{14}Si$	4.70	1/2	−0.55477	7.85×10^{-2}	12.690
Phosphorus $^{31}_{15}P$	100	1/2	1.1305	6.64×10^{-2}	25.8525
Sulphur $^{33}_{16}S$	0.74	3/2	0.64274	2.26×10^{-3}	4.899
Chlorine $^{35}_{17}Cl$	75.4	3/2	0.82089	4.71×10^{-3}	6.258
$^{37}_{17}Cl$	24.6	3/2	0.68329	2.72×10^{-3}	5.208
Potassium $^{39}_{19}K$	93.08	3/2	0.39094	5.08×10^{-4}	2.9805
Calcium $^{43}_{20}Ca$	0.13	7/2	−1.3153	6.39×10^{-2}	4.2975
Scandium $^{45}_{21}Sc$	100	7/2	4.7491	0.301	15.5145
Titanium $^{47}_{22}Ti$	7.75	5/2	−0.78712	2.10×10^{-3}	3.600
$^{49}_{22}Ti$	5.51	7/2	−1.1023	3.76×10^{-3}	3.6015

Continuation of Table 6.1

Nucleus	Abundance %	Spin I h/2	Magnetic moment in nuclear magnetons	Relative sensitivity at constant field	Resonance frequency at 15 kgauss Mc/s
Vanadium $^{51}_{23}$V	~100	7/2	5.1392	0.383	16.7895
Chromium $^{53}_{24}$Cr	9.54	3/2	—0.4735	1.00×10^{-4}	3.609
Manganese $^{55}_{25}$Mn	100	5/2	3.4610	0.178	15.8295
Cobalt $^{59}_{27}$Co	100	7/2	4.6388	0.281	15.1545
Copper $^{63}_{29}$Cu	69.09	3/2	2.2206	9.38×10^{-2}	16.9275
$^{65}_{29}$Cu	30.91	3/2	2.3790	0.116	18.135
Zinc $^{67}_{30}$Zn	4.12	5/2	0.8735	2.86×10^{-3}	3.9525
Gallium $^{69}_{31}$Ga	60.2	3/2	2.0108	6.93×10^{-2}	15.327
$^{71}_{31}$Ga	39.8	3/2	2.5549	0.142	19.476
Germanium $^{73}_{32}$Ge	7.61	9/2	—0.8768	1.40×10^{-3}	2.2275
Arsenic $^{75}_{33}$As	100	3/2	1.4349	2.51×10^{-2}	10.938
Selenium $^{77}_{34}$Se	7.50	1/2	0.5333	6.97×10^{-3}	12.1965
Bromine $^{79}_{35}$Br	50.57	3/2	2.0990	7.86×10^{-2}	16.0005
$^{81}_{35}$Br	49.43	3/2	2.2626	9.84×10^{-2}	17.2470
Rubidium $^{85}_{37}$Rb	72.8	5/2	1.3483	1.05×10^{-2}	6.1665
$^{87}_{37}$Rb	27.2	3/2	2.7415	0.177	20.898
Strontium $^{87}_{38}$Sr	7.02	9/2	—1.0893	2.69×10^{-3}	2.7675
Zirconium $^{91}_{40}$Zr	11.23	5/2	—1.3	9.40×10^{-3}	6.0
Niobium $^{93}_{41}$Nb	100	9/2	6.1435	0.482	15.6105
Cadmium $^{111}_{48}$Cd	12.86	1/2	—0.5922	9.54×10^{-3}	13.542
$^{113}_{48}$Cd	12.34	1/2	—0.6195	1.09×10^{-2}	14.166
Indium $^{113}_{49}$In	4.16	9/2	5.4960	0.345	13.965
$^{115}_{49}$In	95.84	9/2	5.5072	0.348	13.9935
Tin $^{115}_{50}$Sn	0.35	1/2	—0.9132	3.50×10^{-2}	19.830
$^{117}_{50}$Sn	7.67	1/2	—0.9949	4.53×10^{-2}	23.655
$^{119}_{50}$Sn	8.68	1/2	—1.0409	5.18×10^{-2}	23.805
Antimony $^{121}_{51}$Sb	57.25	5/2	3.3417	0.160	15.285
$^{123}_{51}$Sb	42.75	7/2	2.5334	4.57×10^{-2}	8.277
Tellurium $^{123}_{52}$Te	0.89	1/2	—0.7319	1.80×10^{-2}	17.385
$^{125}_{52}$Te	7.03	1/2	—0.8824	3.16×10^{-2}	20.175
Iodine $^{127}_{53}$I	100	5/2	2.7939	9.35×10^{-2}	12.7785
Xenon $^{129}_{54}$Xe	26.24	1/2	—0.7726	2.12×10^{-2}	17.670
$^{131}_{54}$Xe	21.24	3/2	0.6868	2.77×10^{-3}	5.235
Caesium $^{133}_{55}$Cs	100	7/2	2.5642	4.74×10^{-2}	8.3775

Continuation of Table 6.1

Nucleus	Abundance %	Spin I h/2	Magnetic moment in nuclear magnetons	Relative sensitivity at constant filde	Resonance frequency at 15 kgauss Mc/s
Barium $^{135}_{56}$Ba	6.59	3/2	0.837	4.99×10^{-3}	6.375
$^{137}_{56}$Ba	11.32	3/2	0.936	6.97×10^{-3}	7.140
Lanthanum $^{139}_{57}$La	99.911	7/2	2.7615	5.92×10^{-2}	9.0210
Praseodymium $^{141}_{59}$Pr	100	5/2	3.8	0.234	16.95
Neodymium $^{143}_{60}$Nd	12.20	7/2	-1.1	2.81×10^{-3}	3.3
Europium $^{151}_{63}$Eu	47.77	5/2	3.4	0.168	15.0
$^{153}_{63}$Eu	52.23	5/2	1.5	1.45×10^{-2}	6.9
Ytterbium $^{171}_{70}$Yb	14.27	1/2	0.45	4.19×10^{-3}	10.35
$^{173}_{70}$Yb	16.08	5/2	-0.65	1.18×10^{-3}	2.970
Lutetium $^{175}_{71}$Lu	97.40	7/2	2.6	4.94×10^{-2}	8.55
Tantalum $^{181}_{73}$Ta	100	7/2	2.1	2.60×10^{-2}	6.9
Rhenium $^{187}_{75}$Re	37.07	5/2	3.1437	0.133	14.379
$^{189}_{75}$Re	62.93	5/2	3.1760	0.137	14.5260
Osmium $^{189}_{76}$Os	16.1	3/2	0.6507	2.24×10^{-3}	4.9605
Platinum $^{195}_{78}$Pt	33.7	1/2	0.6004	9.94×10^{-3}	13.7295
Mercury $^{199}_{80}$Hg	16.86	1/2	0.4993	5.72×10^{-3}	11.4180
$^{201}_{80}$Hg	13.24	3/2	-0.607	1.90×15^{-3}	4.620
Thallium $^{203}_{81}$Tl	29.52	1/2	1.5960	0.187	36.495
$^{205}_{81}$Tl	70.48	1/2	1.6114	0.192	36.855
Lead $^{207}_{82}$Pb	21.11	1/2	0.5837	9.13×10^{-3}	13.3485
Bismuth $^{209}_{83}$Bi	100	9/2	4.0389	0.137	10.263

(Courtesy of Varian Associates)

Table 6.2

INTENSITY RATIOS OF ESR HYPERFINE LINES AND NMR SPIN-SPIN COUPLINGS CAUSED BY EQUIVALENT PROTONS

No. of protons	Intensity ratio
1	1 : 1
2	1 : 2 : 1
3	1 : 3 : 3 : 1
4	1 : 4 : 6 : 4 : 1
5	1 : 5 : 10 : 10 : 5 : 1
6	1 : 6 : 15 : 20 : 15 : 6 : 1
7	1 : 7 : 21 : 35 : 35 : 21 : 7 : 1
8	1 : 8 : 28 : 56 : 70 : 56 : 28 : 8 : 1
9	1 : 9 : 36 : 84 : 126 : 126 : 84 : 36 : 9 : 1
10	1 : 10 : 45 : 120 : 210 : 252 : 210 : 120 : 45 : 10 : 1

Bibliography

Microwave Molecular Spectroscopy

Gordy, W. Smith, W. V. and Trambarulo, R., *Microwave Spectroscopy*, J. Wiley, New York (1953).
Ingram, D. J. E., *Spectroscopy at Radio and Microwave Frequencies*, Butterworths, London (1955).
Townes, C. H. and Schawlow, A. L., *Microwave Spectroscopy*, McGraw-Hill, New York (1955).
Freymann, R. and Soutif, M., *La Spectroscopie Hertzienne Appliquée à la Chimie*, Dunod, Paris (1960).
Maier, W., Microwave Spectroscopy, *Pure and Applied Chemistry*, **4**, 157 (1962).
Sheridan, J., *Progr. Dielectrics* **4**, 2 Heywood, London (1962).

Electron Spin Resonance

Bleaney, B. and Stevens, K. W. H., *Rep. Progr. Phys.* **16**, 108 (1953); Bowers, K. D. and Owen, J., ibid., **18**, 304 (1955).
Uebersfeld, J. and Combrisson, J., *J. Phys. Rad.* **14**, 104 (1953).
Wertz, J. E., *Chem. Rev.* **55**, 829 (1955).
Ingram, D. J. E., *Free Radicals as Studied by Electron Spin Resonance*, Butterworths, London (1958).
Low, W., *Solid State Physics*, Suppl. 2, Academic Press, New York (1960).
NMR and EPR Spectroscopy, *Varian Ass. 3rd Annual Workshop, Palo Alto, California*, Pergamon Press, Oxford (1960).
Altschuler, S. A. and Kozyrev, B. M., *Paramagnetic Resonance* (in Russian), Gos. Izd. Fiz. Mat. Moscow (1961).
Breinert, B. and Sands, D., *Free Radicals in Biological Systems*, Academic Press, New York (1961).
Pake, G. E., *Paramagnetic Resonance*, W. A. Benjamin Inc., New York (1962)
Blyumenfeld, L. A., Voevodsky, V. V. and Semenev, A. G., *Application of Electron Paramagnetic Resonance* (in Russian), Izd. Sibirsk. Otd. Akad. Nauk. S.S.S.R., Novosibirsk (1962).
Schoffa, G., *Elektronenspinresonanz in der Biologie*, Verlag G. Braun, Karlsruhe (1964).

Nuclear Magnetic Resonance

Bloembergen, N., *Thesis*, Utrecht (1948).
Gutowsky, H. S., *Disc. Faraday Soc.* **19**, 187 (1955).
Andrew, E. R., *Nuclear Magnetic Resonance* Cambridge University Press, Cambridge (1955).
Gutowsky, H. S., *Analytical Applications of NMR in Physical Methods in Chemical Analysis*, vol. 3 (ed. W. G. Berl.) Academic Press Inc., New York (1956).
Jackman, L. M., *Applications of Nuclear Magnetic Resonance Spectroscopy in Organic Chemistry*, Pergamon Press, New York, London (1959).
Pople, J. A., Schneider, W. G. and Bernstein, H. J., *High Resolution Nuclear Magnetic Resonance*, McGraw-Hill, New York (1959).
Lösche, A., *Kerninduktion*, Leipzig (1959).
Abragam, A., *The Principle of Nuclear Magnetism*, Oxford Press (1961).

Nuclear Quadrupole Resonance

Buyle-Bodin, M., *Resonance Nucléaire Quadripolaire*, Thesis, Masson, Paris (1955).
Das, T. P. and Hahn, E. L., *Nuclear Quadrupole Resonance Spectroscopy*, Academic Press, New York (1958).
Scrocco, E., *Ricerca Sci.* **5**, Suppl. **30**, 49 (1960).
O'Konski, C. T., *Determination of Organic Structures by Physical Methods* **2**, 661 (1961).
Boudouris, G., *J. Phys. Rad.* **23**, 43 (1962).

Dielectric Spectroscopy

Freymann, R., *Les Ondes Hertziennes et la Structure Moleculaire*, Hermann, Paris (1936).
Fröhlich, H., *Theory of Dielectrics*, Oxford Press (1949).
Yager, W. A. and Baker, W. O., *J. Amer. Chem. Soc.* **64**, 2164 (1942).
Bauer, M., *Cahier de Physique* **4**, 1 (1944); **5**, 27 (1945).
Smith, J. W., *Electric dipole moments*. Butterworths, London (1955).
Freymann, R. and Soutif, M., *La Spectroscopie Hertzienne Appliquée à la Chimie*, Dunod, Paris (1960).

Electrical Conductivity in Organic Solids

Herwing, H. U., *Z. Electrochem.* **63**, 360 (1959).
Ikonuchi, H. and Akamatu, K., *Solid State Physics* **12**, 93 (1961).

Kallmann, H. and Silver, M. (eds), *Symposium on Electrical Conductivity in Organic Solids*, Interscience (1961).
Brilluin, L., *Cahiers de Physique* 413 (1961); Becher, M. and Mark H. F., *Angew. Chem.*, **73**, 641 (1961).
Brophy, I. I. and Buttrey, I. W. (eds), *Organic Semiconductors*, Macmillan, New York (1962).
Kallweit, I. H., *Kolloid Z.* **188**, 97 (1963).
Topchiev, A. V. (ed.), *Organic Semiconductors* (in Russian), Izd. Akad. Nauk. S. S. S. R., Moscow, (1963).
Hamann, C., Organische Halbleiter, *Phys. Status Solidi* **12**, 483 (1965).

Index

Abragam–Pryce equation, 158
Absorption cell for microwave spectrometers, 40
Absorption coefficient of gases, 41
Absorption NMR spectrometer, 83
Accumulation of spectra, 60
Acetylenic group, determination by NMR, 257
Acids, NMR shifts in, 264
Acrylic acid, polymerization, ESR 371
Acrylonitrile, polymerization, ESR 381
Activation energies for conduction, 135—6
Addition of ESR spectra, 70
Adiabatic fast passage, 100—103
Adipic acid, irradiated, ENDOR, 196
Admittance comparators, 125—7
Adsorption, 397—401
Alcohols, NMR shifts in, 264
Aliphatic alcohols, NMR, 251
Aliphatic radicals 214—17
n-Alkenes ^{13}C resonance, 258
Alkyl amides, NMR shifts in, 264
o-Aminoazotoluene, ESR, 425
N-Aminocarbazyl, ESR, 166—7
2-Aminofluorene, ESR, 425
Ammonia, 14
Ammonium tartrate, irradiated, ESR, 178—9
Androstane, NMR, 303—5
Anionic catalysts, 409
Anisotropic hyperfine interaction in ESR, 160

Anisotropy
 of magnetic shielding, 262
 of the splitting factor, 67, 188
Anthracene radical ion ESR, 209
Aromatic groups, determination by NMR, 245
Aromatic radical ions, ESR, 208—14
Aromatic stereoisomers, NMR, 292
dl-Aspartic acid, irradiated, ESR, 181
Atomic gases, 219—24
Automatic frequency control for ESR, 54—5

Bands of microwave molecular spectroscopy, 36
Bauer–Fröhlich equation, 122
Benzene
 derivatives, NMR, 246—9
 radical ion, ESR, 208
 radicals, ESR, 352
 spin polarization, 101
 spin–spin coupling, NMR, 253
1,4-Benzo-semiquinone ion radical, ESR, 161—3, 171—3
Benzoyloxy radicals, ESR, 363
Benzoylperoxide, decomposition, ESR, 363
3,4-Benzpyrene, ESR, 425
Biochemical reactions, 415—28
Bohr magneton, 15
Boron, nuclear resonances of, 276—7
t-Butyl-benzosemiquinone, ESR, 325
s-Butylhydroperoxide, photolysis, ESR, 356—7
t-Butylhydroquinone, ESR, 324

Caprolactam, irradiated, ESR, 179—81
Carbonyl groups, determination by NMR, 257
Carboxylic acid ethers, ESR, 214—6
Carcinogenic charge transfer complex, 425
Catalysis, 396—410
 by organic semiconductors, 410
Cation exchange, intermolecular, 338
Cavity resonators for ESR, 52—4
Cellulose acetate, dielectric, 313
CF_2BrCBr_2CN, NMR, 290—1
Charge carrier mobilities, 137—8
Charge transfer complexes, 421
 in enzyme reactions, 420
Chemical exchange, 93, 331—8
Chemical shift of NMR lines, 23, 87—9
$CHFCl_2$, nuclear–nuclear double resonance of, 114
 Overhauser effect, 115
Chloranil molecular complex, ESR, 333
Chlorella, living illuminated, ESR, 418
Chlorine, atomic, ESR, 220
Chloroform, hydrogen bonding in different solvents, NMR, 268
2-Chloroprene microwave molecular double resonance of, 117
Cholesterol, NMR, 301—2
Chromium alum catalysts, ESR, 401
Chromium oxide catalysts, 401
Chromium oxide catalysts, ESR, 401—4
Cis-trans isomerization, NMR, 283—9
Cole–Cole diagram, 124
Collision broadening of spectrum lines, 42
Colour centres, 16, 142, 341
Complex admittance, 121
Complex permittivity, 120
Configurational interaction, in radicals, 159

Contact hyperfine interaction, 158
Coproporphyrin-I methyl ester, NMR, 245
Correlation time for radicals in solution, 201
Cupric sulphate pentahydrate, ESR, 150
Cuprous chloride, ESR, 68
Cycloalkenes ^{13}C resonances, NMR, 258
Cyclohexane
 ^{13}C resonances, NMR, 258
 stereoisomers, 295—300
 type stereoisomers, 295—301

Debye equations, 122
Delocalization of electrons, 19
Destruction of polymers, 386
Dewar systems for ESR, 72—4
Dextrose, charred, ESR, 149
Dezoxyribonucleic acid, irradiated, ESR, 426—7
Diamagnetic anisotropy in NMR 235—7
Diamagnetic shielding, 226—30
 in NMR, 228
2,3-Dibromoprene, NMR, 254
2,4-Dichloropentane, NMR, 312
Dielectric spectroscopy, 28—31
 at intermediate frequencies, 125
 at microwaves, 128
 at very low frequencies, 131
 technique of, 119—31
Diffusion of radicals in solution, 201
Dimethyl acetylene dicarboxalate, NMR, 258—9
p-Dimethylazobenzene, ESR, 425
9,10-Dimethylbenzanthracene, 425
p-Dioxane ^{13}C satellites in NMR, 260
Diphenyl derivatives, NMR, 292—4
1,1-Diphenylethylene dimerization, ESR, 409
Diphenyl nitrosyl radicals, ESR, 204—5

α,α-Diphenyl-β-picrylhydrazyl
  (DPPH), 146—8, 156, 166
Dipole absorption, 29—30
  bands, 123
Dipole–dipole, broadening, 62
  coupling, 62
  hyperfine interaction, 160
  interaction, 61—3
Dipole orientation, 29
Dipole relaxation, 29—30
  time, 122
p,p'-Disulpho-α-diphenyl-β-picrylhydrazyl, SDPPH, 148—9
Di-t-butyl-1,4-benzo-semiquinone radical ion, 169
DL-alanine, irradiated, ESR, 426
Doppler broadening, 42
Double channel ESR cavity, 154
Double modulation in ESR, 60
Double resonance methods, 108—13
  in ESR, 195
  in microwave molecular spectroscopy, 116
  in NMR, 113
Dye catalysts, 411

Effective shielding constants for methylene groups, 241—2
Electrical conduction, 31—3
Electrical conductivity, 31—4
Electric dipole ESR, 218
  scheme of, 221
Electrolytic cell for ESR, 77
Electron
  exchange reactions, 331
  transfer, 202, 332
Electron–nuclear resonance, 111, 195—7
  apparatus, 112
  results, 191—7
Electron spin resonance, 15—22
  spectrometer, 49
  technique, 48—78
Emission spectra of molecules, 47

ENDOR, see electron–nuclear double resonance
Energy transfer, 357
  ESR, 357—9
Enzyme reactions, 419—21
Erroneous operation of ESR spectrometer, 78
Ethane radiolysis, ESR, 350, 354
Ethanol
  hydrogen bonding in, NMR, 267
  proton exchange in, NMR, 93, 336—8
Ethylbenzene, NMR, 98
Ethylene, NMR spin–spin coupling, 253
Exchange interaction, 63
Exciton waves, 342

F-centres 16, 142, 341
Flavo-mononucleotide-serotonin complex, ESR 423—4
Flow technique in ESR, 75—6
Fluorine resonance spectra, 270—4
Fluorobenzenes, NMR, 271—4
Formamide, NMR, 286
Fraenkel equation in ESR, 172—3
Free radical
  concentrations, determination of, 142—52
  intermediates, 363—75
  structures, determination of, 155—61
Free radicals, 141—224
  formed by mechanical treatment, 197
  in solutions, 208—14
  in the gaseous state, 217—24
  in the liquid state, 199—214
  in the solid state, 174—97
  trapping of, 175—7, 363—9
Free spin value of the g-factor, 17
Frequency sweep method for spin–spin decoupling, 96
Furan, NMR, 265—6

g-anisotropy
  in crystalline and in amorphous substances, 188
  in polymers, 190—2
Gaussian spectrum line, 63—4
Glycine, irradiated, ESR, 426
Graft-copolymerization, 380—3
Gyromagnetic ratio, 81

Haemin, ESR, 68
Haemoglobin, ESR, 68
Heterocyclic compounds, NMR, 249—50
Heterogeneous catalysts, ESR, 401—4
Hexachlorocyclohexane stereoisomers, NMR, 295—300
HOÓ radicals, ESR, 362
HÓ radicals, ESR, 362, 370
Hydrogen
  atomic, ESR, 219
  bonding, NMR, 266—70
  bromide, NMR, 264
  chloride, NMR shift in, 264
  frozen atoms, ESR, 344—5
  iodide, NMR, 264
  molecular, NMR shift in, 264
Hydroquinone, oxidation, ESR, 322
Hyperconjugation in radicals, 167
Hyperfine interaction
  in ESR, 156—70
  second order, 206—7
  with many nuclei, 161
Hyperfine levels, 18
Hyperfine splitting, 17—8

Indole, ESR, 425
Induction method in NMR, 82
Induction NMR spectrometer, 85
Inhomogeneous broadening, 65—6, 189
Integration of spectra, 97—8
Intensity of ESR lines, 21
Internuclear double resonance, INDOR, 260

Inversion of molecules, 14—5
Isopropyl alcohol, photo oxidation of, 411—2

Karplus–Fraenkel equations in ESR, 214
Kinetic measurements by ESR, 353—7

Lead nuclear resonances, 280
Lithium, metallic, Overhauser effect of 108
Lock-in detector, 45—7
  scheme of, 46
Lorentzian spectrum line, 63—4
Loss angle tangent, 120
Lumiflavin, illuminated, ESR, 423

Magnetic energy splitting, 17
Maser action, principle of, 47
McConnel equation, 159
Methacrylic acid
  irradiated, ESR, 346—7
  polymerization, ESR, 369—70
Methane, NMR shift in, 264
Methine groups, determination by NMR, 243—5
Methylcholantrene, ESR, 425
Methylene groups, determination by NMR, 241
Methyl groups, determination by NMR, 237—41
Methyl methacrylate
  polymerization, ESR, 365—70
  radicals, ESR, 351
Methylpentanediol, dielectric, 124
Methyl radicals, ESR, 345—7, 371
Microwave
  components, 38—40
  generators, 38
  inversion, 14
  molecular spectrometers, 35—45
  rotation, 10
  spectrometer scheme, 37
  spectroscopy, 10—5

INDEX

MnCl$_2$, ESR, 174
Mn^{++}-doped magnesium oxide, 150—1, 174
MnSO$_4$, ESR, 174
Modulation broadening, 69
Molecular relaxation times, 122—5
Monomeric radicals, ESR, 372
Myosmine, NMR, 256—7

Naphthalene radical ion, 331
Negative spin densities, 170—1
Nitrobenzene radical ions, ESR, 209—14
Nitrogen
 nuclear resonances of, 280
 trapped, atomic, ESR, 345
Noise level, 41
Nuclear magnetic resonance, 22—6
 spectrometer, 98
 technique, 80—99
Nuclear magneton, 16
Nuclear–nuclear, double resonances, 113—5
 multiple resonances, 118
Nuclear Overhauser effect, 115
Nuclear quadrupole, coupling in NMR, 92
 detection, 85
 energy states, 27
 resonance, 26—8
 spectrometer, 86, 99

OD radicals, electric dipole, ESR, 223
OH radicals, electric dipole, ESR, 222
Oriented polymers, ESR, 181—4
Oriented radicals, ESR, 190—2
Overhauser effect, 108—10
Oxidation, 320—31
Oxygen effect in ESR, 205—6

Peroxylamine disulphonate, ESR 147—8

Peroxy radicals
 ESR, 320—1
 in polyethylene, 389
 in poly(tetrafluoroethylene), 328
 in sodium methacrylate, 327
Phase transitions, 30
Phenantrene–naphthalene triplet energy transfer, ESR, 357—9
Phenol–formaldehyde resins, NMR, 307—10
Phenol, irradiated, ESR, 175—6
Phenols, NMR shifts in, 264
Phosphorus
 nuclear resonances, 274—6
 trapped atoms, ESR, 344—5
Photolysis, 340
 at liquid helium temperature, 343
Photosynthesis, 416—9
Phosphine
 NMR shift in, 264
 photolysis, ESR, 344
Polarization of nuclei, 100
Polyacrylonitrile
 catalyst, 413—5
 pyrolyzed, 412—4
Poly(butyl methacrylate), dielectric, 314
Polychlorotrifluoro ethylene, irradiated, ESR, 175—7
Poly(ethylene glycol) NMR, dielectric, 316—7
Polyethylene
 irradiated, ESR, 347
 mechanical destruction, ESR, 199
 oxidation, ESR, 389—92
Poly(isobutylene) mechanical destruction, ESR, 199
Polyisoprene–benzoquinone catalyst, 413
Polymers
 dielectric spectroscopy of, 312—7
 NMR of, 306—17

Poly(methylacrylate), mechanical destruction, ESR, 199
Poly(methyl methacrylate), dielectric polarization of, 133
Poly(methyl methacrylate), mechanical destruction, ESR, 198—9
Poly(methylstyrene), mechanical destruction, ESR, 199
Polyphoshates, NMR, 274—5
Polypropylene
  irradiated, ESR, 181—4
  oxidation, ESR, 391
Polystyrene
  mechanical destruction of, ESR, 199, 392
  reaction with methyl methacrylate, 393
Poly(tetrafluoroethylene),
  chemomechanical radical formation 395
  grafting, ESR, 381—3
  irradiated, ESR, 327—8
  oxidation, ESR, 327—8
Poly(vinyl acetate), mechanical destruction of, ESR, 199
Poly(vinyl alcohol) mechanical destruction of, ESR, 199
Polyvinyl chloride
  chlorinated, dielectric, 315
  heat treatment of, ESR, 387
  NMR, 311
Propagating radicals during polymerization, ESR, 374—5
Propenylbenzene, NMR, 285
Propylene, NMR, 283—4
Proton
  exchange, 334
  NMR, 334—40
  field meter scheme, 84
  resonance shifts, NMR, 229
Pyrazine radical ion, ESR, 338—40
Pyridine, NMR, 249
Pyrophosphate, NMR, 275
Pyrrole, NMR, 263—6

Q-factor of ESR cavities, 53
Quadrupole coupling constant, 28
Quadrupole hyperfine splitting, 28
Quadrupole transitions, 28
Quantitative analysis by NMR, 97—8
Quinoline, NMR, 250

Radiation-induced conductivity during solid state polymerization 385—6
Radicals, see free radicals
Radiolysis, 340—57
Redox reactions, ESR, 361
Resolution
  of ESR, 61—71
  of microwave spectra, 42—3
  of NMR, 89—92
Resonance methods in dielectric spectroscopy, 129
Riboflavin-5-phosphate, ESR, 421
Riboflavin-tryptophan complex, 421—3
Rotational energies of molecules, 11
Rotational energy levels, 11—2
Rotational isomers, 283—95
Rotational quantum number, 11
Ruby, ESR 151

Saturation
  broadening of, 103—14
  method of measuring T_1, 103
  microwave spectrum lines, 42—3
Second order hyperfine splitting in ESR, 206—7
Semiquinones, ESR, 321
Sensitivity
  diagram for ESR, 57
  of ESR, 55—61
  of microwave spectrometers, 40—2
  of NMR, 96—8
Shape of spectrum lines, 63
Shoolery rule in NMR, 241—3
SH radicals, electric dipole, ESR, 222—3

Sideband method for spin–spin decoupling, 95
Sigma-pi parameters in ESR, 173
SiH_4, NMR shift, 264
Silicon, nuclear resonances, 280
Sodium
 citrate pentahydrate, irradiated, ESR, 192—5
 methacrylate, ESR, 327
Solid state polymerization, 379—81, 383—6
Solvent effects in ESR, 200—5
Spinach chloroplasts, illuminated, ESR, 417
Spin
 densities, calculation of, 159
 lattice relaxation in ESR, 23—4, 201
 polarization, 100
 temperature, 21, 101
Spin–echo
 method, 104—8
 sequences in water, 107
 spectrometer, 106
Spin–spin
 coupling in hydrocarbons, 250—4
 coupling in NMR, 230—4
 decoupling in NMR, 94—6
 relaxation, 24—5
Splitting factor, g, 17
Standard materials
 for ESR, 144—55
 positioning of, 152—5
Stark
 modulation
  spectrometer, 44
  systems, 43—5
 splitting, 13, 44
Stepwise radical recombination, 175—7
Stereochemical analysis by NMR, 281—303
Stereoid structures, 301—6
Stereoisomers, NMR analysis, 282—301

Stretched polymers, ESR, 190—1
Styrene
 inhibited polymerization, ESR, 376—8
 NMR, 255—6
 polymerization, ESR, 364
Superheterodyne detection, 59

Tacticity of polymers,
 dielectric, 314
 NMR, 310
Tetracene radical ion, ESR 164—5
Tetramethyl-1,4-benzo-semiquinone radical ion, 168—9
Tetramethylsilane, NMR, 229
Theoretical splitting in ESR, 185—8
Thio-indigo radicals, ESR, 203—4
Thiophenols, NMR shifts, 264
Tin, nuclear resonances of, 280
Titanium containing catalysts, ESR, 361
Toluene, NMR, 249
Transmission line method in dielectric spectroscopy, 130
Triplet excitations, ESR, 357—9

N-Vinyl succinimide
 induced conductivity of, 386
 polymerization, ESR, 385

Water,
 dielectric absorption of, 130
 NMR shift in, 264
 sorbed on silica gel, 399—400
 spin–echo sequences, 107
Wide line NMR
 application of, 316
 spectrometers, 93—4

Xylene, NMR, 247—9

Zeeman splitting, 13
 of a microwave spectrum line, 217

mm
6/16/70

OHIO UNIVERSITY